Implementing Data Analytics and Architectures for Next Generation Wireless Communications

Chintan Bhatt
Charotar University of Science and Technology, India

Neeraj Kumar
Thapar University, India

Ali Kashif Bashir
Manchester Metropolitan University, UK

Mamoun Alazab
Charles Darwin University, Australia

IGI Global
PUBLISHER of TIMELY KNOWLEDGE

A volume in the Advances in Wireless Technologies and Telecommunication (AWTT) Book Series

Published in the United States of America by
IGI Global
Information Science Reference (an imprint of IGI Global)
701 E. Chocolate Avenue
Hershey PA, USA 17033
Tel: 717-533-8845
Fax: 717-533-8661
E-mail: cust@igi-global.com
Web site: http://www.igi-global.com

Library of Congress Cataloging-in-Publication Data

Names: Bhatt, Chintan M., 1988- editor. | Kumar, Neeraj (Computer scientist) editor. | Bashir, Ali Kashif, 1982- editor. | Alazab, Mamoun, 1980- editor.
Title: Implementing data analytics and architectures for next generation wireless communications / Chintan Bhatt, Neeraj Kumar, Ali Kashif Bashir, and Mamoun Alazab, editors.
Description: Hershey, PA : Information Science Reference (an imprint of IGI Global), [2021] | Includes bibliographical references and index. | Summary: "This book covers the existing and emerging theoretical and practical challenges in the design, development and implementation of big data algorithms, protocols, architectures, and applications for next-generation wireless communications and their applications in smart cities"-- Provided by publisher.
Identifiers: LCCN 2021000192 (print) | LCCN 2021000193 (ebook) | ISBN 9781799869887 (hardcover) | ISBN 9781799869894 (paperback) | ISBN 9781799869900 (ebook)
Subjects: LCSH: Computer network architectures. | Big data. | Wireless communication systems. | Internet of things. | Artificial intelligence.
Classification: LCC TK5105.52 .I47 2021 (print) | LCC TK5105.52 (ebook) | DDC 621.39/81--dc23
LC record available at https://lccn.loc.gov/2021000192
LC ebook record available at https://lccn.loc.gov/2021000193

This book is published in the IGI Global book series Advances in Wireless Technologies and Telecommunication (AWTT) (ISSN: 2327-3305; eISSN: 2327-3313)

British Cataloguing in Publication Data
A Cataloguing in Publication record for this book is available from the British Library.

Advances in Wireless Technologies and Telecommunication (AWTT) Book Series

Xiaoge Xu
University of Nottingham Ningbo China, China

ISSN:2327-3305
EISSN:2327-3313

Mission

The wireless computing industry is constantly evolving, redesigning the ways in which individuals share information. Wireless technology and telecommunication remain one of the most important technologies in business organizations. The utilization of these technologies has enhanced business efficiency by enabling dynamic resources in all aspects of society.

The **Advances in Wireless Technologies and Telecommunication Book Series** aims to provide researchers and academic communities with quality research on the concepts and developments in the wireless technology fields. Developers, engineers, students, research strategists, and IT managers will find this series useful to gain insight into next generation wireless technologies and telecommunication.

Coverage

- Radio Communication
- Digital Communication
- Broadcasting
- Grid Communications
- Telecommunications
- Wireless Technologies
- Virtual Network Operations
- Mobile Technology
- Mobile Web Services
- Network Management

IGI Global is currently accepting manuscripts for publication within this series. To submit a proposal for a volume in this series, please contact our Acquisition Editors at Acquisitions@igi-global.com or visit: http://www.igi-global.com/publish/.

The Advances in Wireless Technologies and Telecommunication (AWTT) Book Series (ISSN 2327-3305) is published by IGI Global, 701 E. Chocolate Avenue, Hershey, PA 17033-1240, USA, www.igi-global.com. This series is composed of titles available for purchase individually; each title is edited to be contextually exclusive from any other title within the series. For pricing and ordering information please visit http://www.igi-global.com/book-series/advances-wireless-technologies-telecommunication/73684. Postmaster: Send all address changes to above address.

Titles in this Series

For a list of additional titles in this series, please visit:
http://www.igi-global.com/book-series/advances-wireless-technologies-telecommunication/73684

Handbook of Research on Advances in Data Analytics and Complex Communication Networks
P. Venkata Krishna (Sri Padmavati Mahila Universit, India)
Engineering Science Reference • © 2022 • 420pp • H/C (ISBN: 9781799876854) • US $345.00

Emerging Materials and Advanced Designs for Wearable Antennas
Vinod Kumar Singh (SR Group of Institutions, India) Vikas Dubey (Bhilai Institute of Technology, India) Anurag Saxena (SR Group of Institutions, India) Ratnesh Tiwari (Bhilai Institute of Technology, India) and Himani Goyal Sharma (Department of Electrical Engineering, Poornima College of Engineering, Jaipur, India)
Engineering Science Reference • © 2021 • 210pp • H/C (ISBN: 9781799876113) • US $225.00

Principles and Applications of Narrowband Internet of Things (NBIoT)
Sudhir K. Routray (Bule Hora University, Ethiopia) and Sasmita Mohanty (University of Aveiro, Portugal)
Information Science Reference • © 2021 • 436pp • H/C (ISBN: 9781799847755) • US $195.00

Design Methodologies and Tools for 5G Network Development and Application
P. Suresh (Vel Tech Rangarajan Dr. Sagunthala R&D Institute of Science and Technology, India) G. Vairavel (Vel Tech Rangarajan Dr. Sagunthala R&D Institute of Science and Technology, India) and U. Saravanakumar (Muthayammal Engineering College, India)
Information Science Reference • © 2021 • 291pp • H/C (ISBN: 9781799846109) • US $195.00

Harnessing the Potential of AI for Modeling 6G Networks
Moses John Strydom (University of South Africa, South Africa)
Engineering Science Reference • © 2021 • 300pp • H/C (ISBN: 9781799849698) • US $215.00

Evolution of Software-Defined Networking Foundations for IoT and 5G Mobile Networks
Sunil Kumar (Amity University, Noida, India) Munesh Chandra Trivedi (National Institute of Technology, Agartala, India) Priya Ranjan (Amity University, Noida, India) and Akash Punhani (Amity University, Noida, India)
Information Science Reference • © 2021 • 235pp • H/C (ISBN: 9781799846857) • US $195.00

Managing Resources for Futuristic Wireless Networks
Mamata Rath (Birla School of Management, Birla Global University, India)
Information Science Reference • © 2021 • 338pp • H/C (ISBN: 9781522594932) • US $195.00

IGI Global
PUBLISHER of TIMELY KNOWLEDGE

701 East Chocolate Avenue, Hershey, PA 17033, USA
Tel: 717-533-8845 x100 • Fax: 717-533-8661
E-Mail: cust@igi-global.com • www.igi-global.com

Table of Contents

Detailed Table of Contents

The objective of green networking is to minimize greenhouse gas emissions while maintaining the same level of performance. Green networking refers to all processes used to optimize networking and internetworking functions to make it more energy efficient. Green networking concepts can be extended to cover any method that reduces latency, save bandwidth, and/or decrease computation time, as a reduction in these factors invariably leads to power savings. These savings can directly translate into lowering greenhouse gas emissions and reduce computing's carbon footprint and its impact on the environment. Energy-awareness is critical in the networking infrastructure, especially in wireless 5G networks and beyond. Research on blockchain for 5G wireless networks is still in its infancy. But it is obvious that blockchain will significantly uplift the shape and experience of future mobile applications and services. Identifying the green networking analytics will lead to sustainable energy policy planning for the future.

The food and beverage industry is one of the most water-intensive industries, with water required for various processes (e.g., washing, cooking, cleaning) at almost every stage of the production, as well as being a key constituent in many food and drink products. Therefore, a real-time efficient water management strategy is imperative, and the novel internet of things (IoT)-based technologies can be of significant help in developing it. This chapter presents the architecture of an IoT-based water-monitoring system followed by the demonstration of a case study of a beverage factory wherein the monitoring system helped understand the detailed water usage as well as finding solutions and addressing overconsumption of water during the manufacturing processes. The successful deployment of IoT helped reduce the annual water consumption by 6.7%, monitor water usage in real-time, and improve it.

Network attacks are continuously surging, and attackers keep on changing their ways in penetrating a system. A network intrusion detection system is created to monitor traffic in the network and to warn regarding the breach in security by invading foreign entities in the network. Specific experiments have been performed on the NSL-KDD dataset instead of the KDD dataset because it does not have redundant data so the output produced from classifiers will not be biased. The main types of attacks are divided into four categories: denial of service (DoS), probe attack, user to root attack (U2R), remote to local attack (R2L). Overall, this chapter proposes an intense study on linear and ensemble models such as logistic regression, stochastic gradient descent (SGD), naïve bayes, light GBM (LGBM), and XGBoost. Lastly, a stacked model is developed that is trained on the above-mentioned classifiers, and it is applied to detect intrusion in networks. From the plethora of approaches taken into consideration, the authors have found maximum accuracy (98.6%) from stacked model and XGBoost.

5G is a new universal wireless standard, a new form of mobile network engineered to bring everyone and everything virtually together. 5G is not only for mobile phones, but it is also the foundation for virtual reality (VR), the internet of things (IoT), and autonomous driving, connecting many electronic devices to the internet. Having good healthcare is very important as it affects all parts of human life and social well-being. Moreover, it is crucial to have a great healthcare system if we want economic growth, workforce productivity, and society to advance. Despite all the hard work done by scientists and medical professionals, today's healthcare is mainly inefficient, and a significant overhaul is required. This chapter discusses the primary advantages, including the 5G's main features in healthcare and their limitations and probable solutions and applications to the latest scenario.

Since the early 1990s, there has been a lot of enthusiasm for using high-speed connectivity to develop local community links through education, employment possibilities, fostering community events, and enhancing overall sociability within a local region. 5G is the 5th iteration of a broadband network operating on cellular systems. 5G is not only for mobile phones, but it is also the foundation for virtual reality (VR); the internet of things (IoT); and autonomous transport, immersive services, and public infrastructure; and connecting many electronic devices to the internet. In this chapter, first, the authors have discussed the evolution of 1G network to 6G networks by focussing on its potential impact on the quality of life. Further, 5G applications in IoT, autonomous transport, immersive services, and public infrastructure have been discussed. Then the chapter discusses popular advantages, limitations in the current technologies, implementations, and future perspective.

Handling unpredictable attack vulnerabilities in self-proclaiming secure algorithms in WSNs is an issue. Vulnerabilities provide loop holes for adversary to barge in the privacy of the network. Attacks performed by the attacker can be active or passive. Adversary may listen to the sensitive information and exploit its confidentiality which is passive, or adversary may modify sensitive information being transferred over a WSN in case of active attacks. As Internet of things has basically three layers, middle-ware layer, Application layer, perceptron layer, most of the attacks are observed to happen at the perceptron layer in case of both wireless sensor network and RFID Tag implication Layer. Both are a major part of the perceptron layer that consist a small part of the IoT. Some of the major attack vulnerabilities are exploited by executing the attacks through certain flaws in the protocol that are difficult to identify and almost complex to identify in complicated bigger protocols. As most of the sensors are resource constrained in terms of memory, battery power, processing power, bandwidth and due to which implementation of complex cryptosystem to keep the data being transferred secure is a challenging phase. The three main objectives studied in this scenario are setting up the system, registering user and the sensors via multiple gateways. Generating a common key which can be used for a particular interaction session among user, gateway and the sensor network. In this paper, we address one or more of these objectives for some of the fundamental problems in authentication and mutual authentication phase of the WSN in IoT deployment. We prevent the leakage of sensitive information using the rabin cryptosystem to avoid attacks like Man-in-the-middle attack, sensor session key leakage, all session hi-jacking attack and sniffing attacks in which data is analyzed maliciously by the adversary. We also compare and prove the security of our protocol using proverif protocol verifier tool.

E-health is a cloud-based system to store and share medical data with the stakeholders. From a security perspective, the stored data are in encrypted form that could further be searched by the stakeholders through searchable encryption (SE). Practically, an e-health system with support of multiple stakeholders (that may work as either data owner [writer] or user [reader]) along with the provision of multi-keyword search is desirable. However, the existing SE schemes either support multi-keyword search in multi-reader setting or offer multi-writer, multi-reader mechanism along with single-keyword search only. This chapter proposes a multi-keyword SE for an e-health system in multi-writer multi-reader setting. With this scheme, any registered writer could share data with any registered reader with optimal storage-computational overhead on writer. The proposed scheme offers conjunctive search with optimal search complexity at server. It also ensures security to medical records and privacy of keywords. The theoretical and empirical analysis demonstrates the effectiveness of the proposed work.

With the rise in use of internet in various fields like education, military, government sector, banking, the security and privacy of the info has been the foremost concern. As in today's era, most of the handling of data and transactions are done online. When the data is transferred from the one end of sender to the other end of receiver online, it's eavesdropped by an intruder and thus could be a threat to the secrecy or confidentiality of the info. The hottest technique that protects the confidentiality of the data is cryptography which converts the plain text into scrambled form which is unreadable. Then the receiver applies a reverse mechanism to decrypt the unreadable data to readable form. This mechanism is known as encryption-decryption process or cryptography. Cryptography can be both symmetric and asymmetric. Here the authors discuss symmetric and asymmetric algorithms.

In this chapter, the authors explore a cost model and the come about cost-minimization client booking issue in multi-level mist figuring organizations. For an average multi-level haze figuring network comprising of one haze control hub (FCN), different fog access nodes (FANs), and user equipment (UE), how to model the cost paid to FANs for propelling assets sharing and how to adequately plan UEs to limit the cost for FCN are still issues to be settled. To unravel these issues, multi-level cost model, including the administration delay and a straight backwards request dynamic installment conspire, is proposed, and a cost-minimization client planning issue is defined. Further, the client planning issue is reformulated as an expected game and demonstrated to have a Nash equilibrium (NE) arrangement.

IoT includes many sensors that have to collect the data and send it to the superior nodes; for such interaction between the IoT devices, various wireless technologies are available, like infrared, Li-Fi, WI-Fi, Zigbee, Bluetooth, etc. Among all the available, Bluetooth proved the most promising short-range wireless communication technology due to various factors. To fulfil the increasing demand for wireless connectivity, the Bluetooth SIG must continuously perform up-gradation. Here, analysis of

Preface

Developments in the fields of Internet technologies, mobile telecommunications, and ubiquitous computing have created new frontiers for tomorrow's data-centric society. These futuristic applications would require fast, efficient, and reliable internet connectivity. Such extremely diversified Internet traffic and applications will need dynamic and highly adaptive network environments with high reliability and ultra-low latency.

Innovations in artificial intelligence, machine learning, and network data analytics paves way for novel analytical applications in mobility management, resource allocation, network control and application management. Novel analytical techniques and services will help revolutionize the existing networks to be future-ready. The intelligent networks of tomorrow will reach tremendous levels of automation and optimization through the gathering, processing, learning, and controlling of data in a rational manner.

The scope of this book includes all aspects of artificial intelligence, machine learning, and data analytics that would enable and enhance the next generation networks. This book addresses conventional measurements such as traffic management, Quality of Experience, service quality, etc. as well as futuristic network behavior management such as, energy aware routing, predictive maintenance through intelligent services.

This book involves 11 chapters. The work *Blockchain and green networking analytics in 5G networks and beyond by Janet Light* is related to Blockchain and green networking analytics in 5G networks and beyond. The work *Improving Water Efficiency in the Beverage Industry with the Internet of Things by Sandeep Jagtap and team* does brief review of the architecture of IoT-based water-monitoring system followed by the demonstration of a case study of a beverage factory wherein the monitoring system helped understand the detailed water usage as well as finding solutions and addressing overconsumption of water during the manufacturing processes. The work *Network Intrusion Detection using Linear and Ensemble ML Modeling by Shilpi Parikh and team* proposes an intense study on linear and ensemble models such as Logistic Regression, Stochastic Gradient Descent (SGD), Naïve Bayes, Light GBM(LGBM) and XGBoost. A stacked model is developed in same chapter, that is trained on the above-mentioned classifiers, and it is applied to detect intrusion in networks. The focus of the work *5G in HealthCare: Features, Advantages, Limitations and Applications by Vijay Prakash and team* is to discusses the primary advantages, including the 5G's main features in healthcare and their limitations and probable solutions and applications to the latest scenario. The work *High-Speed Connectivity: Potential Impact on the Quality of Life by Vijay Prakash and team* discussed the evolution of 1G network to 6G networks by focusing on its potential impact on the quality of life. Further, 5G applications in IoT, autonomous transport, immersive services and public infrastructure have been discussed. The work *A Rabin Cryptosystem based Lightweight authentication protocol and session key generation scheme for IoT deployment:*

authentication in IoT by PRIYANKA AHLAWAT and team address some of the fundamental problems in authentication and mutual authentication phase of the WSN in IoT deployment. The focus of the work *Multi-Keyword Searchable Encryption for E-Health System with multiple data writers and readers by Dhruti P Sharma and team* is to propose a multi-keyword SE for an E-Health system in multi-writer multi-reader setting. With this scheme, any registered writer could share data with any registered reader with optimal storage-computational overhead on writer. The work *A Comparative study on Symmetric & Asymmetric Key Encryption Techniques: Symmetric & Asymmetric Key Encryption Techniques by Sneha Padhiar and team* is to discuss about symmetric & asymmetric algorithms. The focus of the work *Demystifying Multi-tier Cost Model for Scheduling in Fog Communication Networks by JAGADESH T and team* is to explore a cost model and the come about cost-minimization client booking issue in multi-level mist figuring organizations. The work *Analysis of Bluetooth Versions (4.0, 4.2, 5, 5.1 and 5.2) for IoT Applications by S. D. Padiya and team* does analysis of Bluetooth versions discussed based on the characteristics such as speed, bandwidth, range, power, message capacity, beacon provision, compatibility, reliability, errors detection, correction capability, advertisement packets, duty cycle, slot availability masks and many more. The focus of the work *Evaluation of Turbo Decoder Performance Through Software Reference Model by Manjunatha K N and team* is to model a software reference model for turbo encoder and decoder using MATLAB Simulink. Performance of the proposed model is estimated and analyzed on various parameters like frame length, number of iterations and channel noise.

We hope this book introduces capable concepts and outstanding research results to support further development of AI techniques and services in Next Generation Wireless Networks.

Chintan Bhatt
Charotar University of Science and Technology, India

Neeraj Kumar
Thapar University, India

Ali Kashif Bashir
Manchester Metropolitan University, UK

Mamoun Alazab
Charles Darwin University, Australia

Chapter 1
Blockchain and Green Networking Analytics in 5G Networks and Beyond

Janet Light
University of New Brunswick Saint John, Canada

ABSTRACT

The objective of green networking is to minimize greenhouse gas emissions while maintaining the same level of performance. Green networking refers to all processes used to optimize networking and internetworking functions to make it more energy efficient. Green networking concepts can be extended to cover any method that reduces latency, save bandwidth, and/or decrease computation time, as a reduction in these factors invariably leads to power savings. These savings can directly translate into lowering greenhouse gas emissions and reduce computing's carbon footprint and its impact on the environment. Energy-awareness is critical in the networking infrastructure, especially in wireless 5G networks and beyond. Research on blockchain for 5G wireless networks is still in its infancy. But it is obvious that blockchain will significantly uplift the shape and experience of future mobile applications and services. Identifying the green networking analytics will lead to sustainable energy policy planning for the future.

INTRODUCTION

With the launch of fifth-generation (5G) network and its services in 2020, the traffic volumes are expected to increase 1000 times and the number of connected devices will be 10-100 times larger than before. Advances in technologies such as artificial intelligence, autonomous Internet of Things, big data analytics, blockchain, and augmented/virtual reality will play a major role in using the high-speed, low-latency, secure 5 G connectivity that is ubiquitous and reliable. Research is underway beyond the 5G networks, to support a greater number of users on higher transmission rate than 5G networks. Hence, the challenges for future networks are set to achieve 10 times the energy efficiency together with spectral efficiency and higher speed compared with today's 4G systems.

DOI: 10.4018/978-1-7998-6988-7.ch001

In this article, the concept of green networking is discussed and the objectives for energy saving in future high volume network traffic are clearly outlined. How the digital ledger technology can support the objectives of energy saving in various applications and services are discussed. Identifying green networking analytics for high quality of services and energy efficient solutions are discussed here. Green network analytics leverages the power of Machine. Learning and Machine Reasoning to provide accurate insights that are specific to network deployment and running.

Green Networking

There is an increase in the global use of Information and Communications Technology (ICT). The 2012 data points to 4.7% of the world-wide energy being consumed by ICT industry. A 2% increase discharges more than 830 million tons of CO2 every year. Mobile radio sector is responsible for 9% of that figure and expected to increase further due to exponential traffic increase in networks beyond 5G. Datacenters are the backbone of ICT networks and the second highest leading culprit in greenhouse gas emission, personal computers being the number one source. Every day there are over 2 Quintillion bytes of data created; these data come from unique sources and often need to be routed across the Internet. Currently with the COVID-19 pandemic situation, the traffic demand on wireless networks has reached a peak. Data networks consumed around 250 TWh in 2019, or about 1% of global electricity use, with mobile networks accounting for two-thirds. Based on current efficiency improvement trends, electricity consumption is projected to rise to around 270 TWh in 2022 [1-3].

The objective of green networking is to minimize greenhouse gas (GHG) emissions while maintaining the same level of performance. Green networking refers to all processes used to optimize networking and inter-networking functions, to make it more energy efficient. Green networking concepts can be extended to cover any method that reduces latency, saves bandwidth, and/or decreases computation time, as a reduction in these factors invariably leads to power savings. These savings can directly translate into lowering GHS emissions, and reduce computing's carbon footprint and its impact on the environment. Energy-awareness is critical in the networking infrastructure, especially in wireless 5G networks and beyond.

Blockchain

Blockchain technology has the potential to disrupt and revolutionize many businesses and professions. Blockchain-based cryptocurrency applications have been widely recognized and used, but blockchain applications have expanded to other fields. Many businesses appreciate it and have started to study its potentials. We can now see some blockchain use cases in different areas beyond finance and banking applications such as in supply chain management, advertising verification, energy-saving, and healthcare. In future, it is expected we will see more useful applications with the development of intuitive interfaces and more use cases. Access to information, data integrity, and operation resilience, among many other drivers, motivates businesses and industries to experiment and develop blockchain-based applications (Baoid et al., 2021).

Over the years, Blockchain has increased in popularity partly due to cryptocurrencies. But why is Blockchain such a hugely talked about area within cryptology? This is due to its revolutionary ability to store, validate, authorize, and move digital transactions across the Internet. One of the critical challenges on the Internet is trust. Are you doing business with the person that you think you are? Even with all the security mechanisms that we have, companies are still being hacked. To go to the next level with secure currencies, machine-to-machine transactions, we need Blockchain to create more security and trust.

Without Blockchain, a contract between two entities on the Internet requires one or more central authorities to validate the data. The central authority stores the data in their database. These databases are single points of failure and points where security attacks can provide a lot of damage. Besides, each central authority that is part of the transaction increases the cost and requires additional time to complete the transaction. A high level implementation of blockchain is shown in Figure 1.

Figure 1. A high-level diagram of the implementation of blockchain (Martin, n.d.)

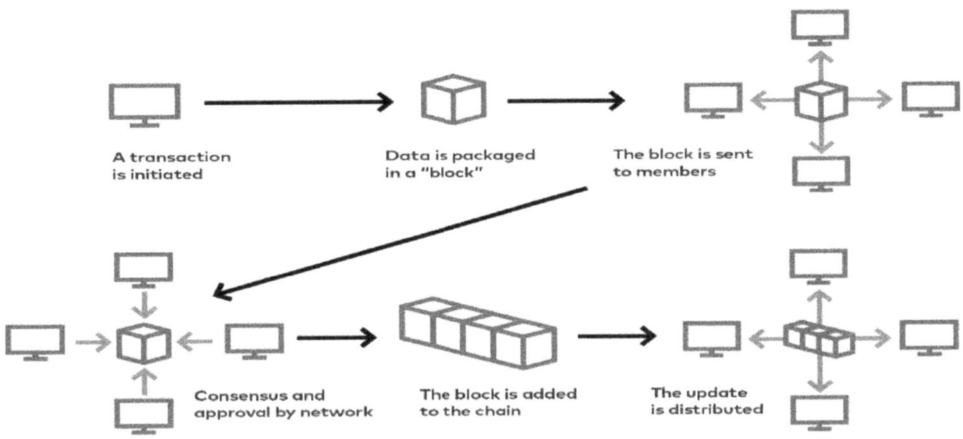

Blockchain removes the need to have a centralized authority, and it is more secure. The blockchain is a new type of database. A blockchain database is installed on individual computers used by the people who use the database. Every computer involved in the transaction has a copy of the database. The solution uses consensus-based permission in that all computers involved in the blockchain must reach an agreement before the change can happen. Every time a change is made to the record, a new block is added to the chain, and it is stored in the computer making the change. No record is ever deleted. All computers involved in the blockchain need to reach an agreement before the change can be made. This helps prevent a security attack in that a hacker would have to update every computer involved in the previous updates on the blockchain. Blockchain is sometimes called a distributed ledger, as it is like an accountant updating their books adding new rows each time a change occurs. Blockchain is a trust protocol, reducing central authorities and reducing costs.

Blockchain or distributed ledgers is an emerging technology that promises transparent, tamper-proof and secure systems for many businesses with smart contracts. Application of blockchain in healthcare domain is an ongoing research evolving rapidly. For instance, sharing electronic medical records, remote patient monitoring, drug supply chain, etc are several use cases of applying the blockchain technology to healthcare. Blockchain has been suggested as a way to solve critical challenges faced by healthcare, such as secured sharing of medical records and compliance with data privacy laws. The methodology used for reaching consensus in blockchain networks determines to a large extent key performance characteristic such as scalability, transaction speed, transaction finality, security and spending of resources such as electricity. Every method requires a procedure for generating and subsequently accepting a block. Minimizing resources or energy spent forms a significant criterion for blockchain performance. For example, the proof-of-work algorithms are known to be energy intensive as they spent significant amount of energy to validate a transaction.

Blockchain Enables 5G Technologies

The combination of blockchain and 5G is expected to pave the way for emerging mobile services. In fact, 5G is all about connecting heterogeneous devices and complex networks interconnecting more than 500 billion mobile devices by 2030. In such a context, the ultra-dense small cell networks, a fundamental component of 5G infrastructure, will provide connections and energy efficiencies of radio links with high data rates and low latencies. However, it introduces trust and secure interoperability concerns among complex sub-networks. Therefore, providing a reliable cooperation among heterogeneous devices is crucial for 5G mobile networks. In this regard, blockchain with its immutable and decentralized transaction ledgers can enable distributed massive communication with high security and trustworthiness. A big challenge for current 5G platforms is the need to guarantee an open, transparent, and secure system among the extraordinary number of resources and mobile users. Blockchain with its innovative concepts of decentralized operation can provide a high level of data privacy, security, transparency, immutability for storage of 5G heterogeneous data [6-9]. Blockchain is thus expected to be an indispensable tool to fulfill the performance expectations for 5G systems with minimal costs.

Blockchain has mainly cooperated with the key 5G enabling technologies including cloud computing, edge computing, Software Defined Networks, Network Function Virtualization, Network Slicing, and D2D communication. This enables the integration of blockchain and such 5G technologies. The benefits of blockchain for different 5G use cases and applications are limitless. Blockchain for cloud computing has drawn significant attention in the last decades thanks to its unlimited resources of storage and computation power, which can provide on-demand, powerful and efficient services with minimum management efforts. Cloud computing paradigms provide a number of technical solutions for realizing 5G services, such as optimizing the communications, processing and storage processes, 5G data content delivery and caching, resource allocation and data transmission management, and cloud-enabled small cell networking for 5G media services. Specially, in order to meet the ever-increasing demand of user association and resource allocation in cellular 5G networks, the architecture of cloud radio access networks (Cloud-RANs) is envisioned as an attractive model that manages the large number of small cells through the centralized cloud controller as baseband unit (BBU) pool. Cloud-RAN can offer high-speed interconnection and shared powerful processing to facilitate optimal multicell cooperation and collaborative radio, real-time cloud computing, which makes CloudRAN a promising candidate of next-generation 5G access networks. Table 1 shows the role of blockchain in every layer of a 5G network.

Table 1.

	Access Network	Edge Network	Core Network	Service Network
5G	Infrastructure management	Edge and access network management	Network Management	Mobile
Blockchain	Cloud-RAN	Virtualization	Traffic	Billing & payment

CHALLENGES AND OPPORTUNITIES

The existing cloud computing models remain unsolved challenges in terms of security, networking and computing performance degradation due to its centralized architecture. In the 5G era, the massive data traffic outsourced from IoT devices to the cloud has brought about a series of new security challenges, mainly including data availability, data privacy management, and data integrity.

- Data availability: In current cloud network architectures, cloud services are provided and managed centrally by the centralized authority. However, this configuration is vulnerable to single-point failures, which bring threats to the availability of cloud services for on-demand user access. A centralized cloud IoT system does not guarantee seamless provisions of IoT services when multiple users request simultaneously data or cloud servers are disrupted due to software bugs or cyberattacks.
- Data privacy management: Although the centralized cloud 5G networks can provide convenient services, this paradigm raises critical concerns related to user data privacy, considering a large amount of 5G heterogeneous data being collected, transferred, stored and used are on the dynamic cloud networks. In fact, IoT users often place their trust in cloud providers managing the applications while knowing very little about how data is transmitted and who is currently using their information (Yang, Margheri, Hu et al, 2018). In other words, by outsourcing data protection to the cloud, IoT data owners lose control over their data, which has also adverse impacts on the data ownership of individuals. Moreover, even in the distributed cloud IoT paradigms with multiple clouds, IoT data are not fully distributed but stored in some cloud data centres at high density. In this context, a massive amount of heterogeneous data may be leaked and user privacy is breached if one of the cloud servers is attacked.
- Data integrity: The storage and analysis of 5G data on clouds may give rise to integrity concerns. Having to place trust on the centralized cloud providers, outsourced data is put at risks of being modified or deleted by third parties without user consent. Moreover, adversaries can tamper with cloud data resources (Zhang et al., 2019), all of which can breach data integrity. For these reasons, many solutions have been applied to overcome the problem, by using public verification schemes where a third party auditor is needed to perform the integrity verification periodically. This scheme potentially raises several critical issues, including irresponsible verification to generate bias data integrity results or invalidated verification due to malicious auditors.
- Lack of immutability: The dynamic process of 5G data to clouds and data exchange between cloud providers and mobile users are vulnerable to information modifications and attacks caused by adversaries or third parties. Even entities within the network may be curious about transmitted data over the sharing and unauthorized, may obtain personal information (i.e. customer data of 5G smart grid or location information of vehicles in vehicular networks). These issues may lead to serious data leakage bottlenecks and consequently damage system immutability.
- Lack of transparency: In the conventional cloud systems, cloud resource providers have full control over outsourced network data (i.e. IoT data) while users are not aware of it and lacks the ability of tracking data after offloading to the cloud. This poses critical challenges on data users to perform verification and monitoring of data flows or usage, especially in the 5G scenarios where transparency among networks members is highly required to ensure fairness and openness, i.e.

cloud service providers and slice users in cloud-based network slicing, or between healthcare providers and patients in cloud e-health.

Recently, blockchains have been investigated and integrated in cloud computing to effectively address the above security challenges in the cloud-based 5G networks. A framework called BlockONet (Yang, Wu, Zhang et al, 2018) for 5G access scenarios, aims to improve the network credibility and security in 5G fronthaul using blockchain. Blockchain is employed to build a verification platform between IoT devices, BBU unit, and manufacturer, where user access information is stored immutably on the chain, while smart contracts are also leveraged to perform automatic user authentication.

The benefits from the use of blockchain in Cloud-RAN 5G networks are twofold. First, the concept of blockchain-based Cloud-RAN gets rid of centralized control at the core network and offers a decentralized fair agreement with blockchain consensus platform, which eliminates single point failure bottlenecks and improves significantly system trust. Second, by applying a decentralized blockchain without third parties, the blockchain-based cloud-RAN strategy can achieve optimal resource utilization and save a large amount of signalling and connection costs. Studies (Yang et al., 2017) are conducted to apply blockchain to build a trusted authentication architecture for cloud radio access network (Cloud-RAN) in the 5G era. They also show that the proposed schemes can address effectively network access authentication with trusted agreement among service providers and IoT users with reduced operation costs and improved spectrum usage over Cloud-RAN based mobile networks.

Blockchain is also integrated with cloud computing for 5G IoT networks. A cloud-centric IoT framework enabled by smart contracts and blockchain for secure data provenance is proposed (Ali et al., 2018) where blockchain incorporated in cloud computing, to build a comprehensive security network where IoT metadata (i.e. cryptographic hash) is stored in blockchain while actual data is kept in cloud storage, which makes it highly scalable for dense IoT deployments. In such a system, smart contracts with its autonomous, transparent and immutable properties are also adopted to ensure high cloud data validity. A secure data sharing architecture is introduced with attributed based-access to control cryptosystem. Its network model consists of four main components: IoT devices, a data owner, a blockchain network and a cloud computing platform. More specific, a permissioned blockchain model is adopted to manage IoT transactions and perform access control for device requests received by cloud, while cloud monitors closely the blockchain network. As a result, such a cloud blockchain integration brings a comprehensive security framework with enhanced privacy preservation, data ownership and secure data sharing. Similarly, a hierarchical access control structure for Cloud blockchain was investigated in (Ali et al., 2018) with a blockchain-based distributed key management. Especially, the blockchain network topology involves distributed side blockchains deployed at fog nodes and a multi blockchain operated in the cloud, which would speed up access verification offer flexible storage for scalable IoT networks. In addition, to protect cloud blockchain in security-critical applications, a forensic investigation framework is proposed using decentralized blockchain. Security issues from dynamic interactions between cloud service providers, clients, and IoT devices were considered and analysed with a tamper evident scheme. Blockchain is performed to audit evidence during the investigation of a criminal incident among cloud blockchain entities in a decentralized manner, and therefore avoiding single points of failure on the cloud storage and improving evidence availability. In addition, blockchain has also incorporated with the cloud federation architectures to further improve the performance of complex 5G-IoT networks in terms of transparent collaboration and interconnected services.

As an example, a blockchain framework was proposed on a joint cloud collaboration environment where multiple clouds are interconnected securely by peer-to-peer ledges (Chen et al., 2018). The proposed scheme contains three tiers with an IoT sensor network, a federation of multiple clouds, and a service platform. Typically, the blockchain platform can offer many advantages over the schemes based on a single cloud. For instance, since IoT data at each area is stored in a private local cloud in the multi-cloud network, its data security is significantly improved. Further, the single cloud can offer instant services for IoT users through the private blockchain network, which also mitigates risks of malicious attacks on cloud systems. Besides, a cloud blockchain model with micro-clouds was introduced in (Freitag, 2018) using blockchain-enabled distributed ledgers. A joint cloud-blockchain architecture to enable secure decentralized collaborative governance services, i.e. immutable data storage, transparent monitoring and resource management for suitable performance on lightweight computing nodes like IoT devices will be of high importance.

Blockchain for mobile edge computing: As an extension of cloud computing, mobile edge computing (MEC) has emerged as the promising technology to empower 5G services. Edge computing may have other names such as fog computing, mobile cloud or cloudlet. Similar to the cloud paradigm, edge computing can offer a series of computing services with capabilities of task processing, data storage, heterogeneity support and QoS improvements. In fact, edge servers are less powerful than remote clouds, but they are located at the edge of the network, with a close proximity to IoT devices, which enables highly efficient 5G data computation with much lower transmission delay, compared with the remote cloud (Taleb et al., 2017). As a result, edge computing can provide instant computing applications to IoT users with low latency and fast service response, which would be particularly useful in the next generation services (i.e. in 5G and beyond).

Resource Constraints and Allocation - Blockchain requires computation on the transaction before it is accepted or rejected. The consensus algorithms required for this purpose can be computationally intensive. Therefore, it is not feasible for all the nodes in the network to participate in the transaction validation process. For example, an MEC (or C-RAN) node may be already operating at full capacity providing services to the users and might not be able to perform the required computation for block validation in time hence resulting in delay. Such a situation may lead to bottleneck and network performance degradation as the required resource might not be provisioned in time. Due to resource constraint, an optimization framework that dynamically selects the mining node in a permissioned network is required for the 5G network. Therefore, resource provisioning for computing to support a Blockchain in a 5G network needs to be investigated. Besides, not all nodes are capable of running Blockchain, especially IoT devices. To overcome this challenge optimal placement of dedicated validating nodes may be needed in the network that needs to be investigated. Furthermore, the Blockchain requires broadcast of the transaction to be approved that may result in significant overhead adding to network traffic. Thus, reducing the overhead to minimize the storage and the processing burden brings additional challenges in adopting Blockchain in 5G network.

How Will Blockchain Benefit the Energy Industry?

Blockchain technology has the potential to transform the energy sector. The energy industry has been consistently catalyzed by innovations including rooftop solar, electric vehicles, and smart metering. Now, the Enterprise Ethereum blockchain presents itself as the next emerging technology to spur growth in the energy sector through its smart contracts and systems interoperability. Of the many use cases for blockchain, energy and sustainability are often less recognized. However, the World Economic Forum, Stanford Woods Institute for the Environment, and PwC released a joint report identifying more than 65 existing and emerging blockchain use-cases for the environment. These use cases include new business models for energy markets, real-time data management, and moving carbon credits or renewable energy certificates onto the blockchain. Distributed ledger technology has the potential to improve efficiencies for utility providers by tracking the chain of custody for grid materials. Beyond provenance tracking, blockchain offers unique solutions for renewable energy distribution.

Legacy energy sectors, such as oil and gas also stand to benefit from the implementation of Enterprise Ethereum solutions. Complex systems with multiple actors have the opportunity to benefit from blockchain technology. For example, petroleum is one of the most traded commodities and requires a network of refiners, tankers, jobbers, governments, and regulatory bodies. The complex network of participants suffers from siloed infrastructures and numerous process inefficiencies. Large scale oil and gas conglomerates are seeking to invest in and implement blockchain technology because of its ability to lower costs and reduce harmful environmental impacts.

Oil and gas companies are particularly concerned about privacy and trade secrets. These private blockchain networks offer data permissioning and selective consortium access to pre-approved parties. Private and consortium blockchains provide an interim solution until public blockchains can implement the necessary privacy features businesses demand.

Hence, the main benefits of blockchain in the energy sector are:

- Reduced costs
- Environmental sustainability
- Increased transparency for stakeholders while not compromising privacy

How can Blockchain Help With Renewable Energy?

This involves the use of Renewable Energy Certificates. When green energy is produced, many governments across the globe issue a credit certificate for the amount of electricity going into the public grid. These certificates track how much power is produced by green sources, but they are also tradable and salable between organizations. The present system of issuing and managing these certificates is relatively slow and quite awkward at times. The chances of error or misrepresentation may deter many investors in green energy technologies.

Recently, there are proposals to implement blockchain technology as a way to regulate certificates in the renewable energy industry. A peer-to-peer computer network issuing and tracking certificates could be ideal for securely managing these interactions without slowing processes or introducing human mistakes or dishonesty. Built-in functions for issuing, validating, and tracking certificate ownership would have to be introduced to the network in a highly scalable platform to accommodate both large and small energy producers.

How Blockchain be Incorporated Into a Power System?

With the technologies now employed by eco-friendly power plants, every time a given unit of energy is produced it is registered by a meter and logged in a spreadsheet. This in turn, is submitted to an approved registry system, which then creates a new certificate. Brokers are then called in to negotiate the buying and selling of certificates between organizations, at which time the certificates are verified again by an independent third party.

Because of all these steps and a dependency on human accuracy, there's room for error and little oversight for spotting mistakes. These inherent risks are one reason why the cost of managing certificates is fairly high. Blockchain technology could reduce or eliminate these issues. Through a 5G connection to a computer network, the meters which track energy production could write each unit directly into the blockchain. The information would then be validated and acknowledged by other nodes on the network.

The "consensus" problem is an existing challenge in blockchain technology. Most blockchain networks cannot support more than several hundred transactions per second. This is a problem because all servers must reach a consensus on each block before it becomes valid. Unstable networks and malicious software could slow things down even further. Low latency 5G network can help solve this problem.

Can Blockchain Transform Energy Grids

Adding blockchain technology to the power grid makes it possible to exchange energy certificates directly over a 5G network. Smart IoT devices can be used to ensure that all transactions are recorded instantly and automatically, Residential and business sources, as well as government regulators, would all have immediate access to the shared information (Chaudhry & Soptimizer, 2019; Dai et al., 2019; Nguyen et al., 2020; Tahir et al., 2020; Xiong et al., 2019). Those with solar panel arrays or wind turbine farms producing excess power could immediately sell their energy certificates to competing public buyers instead of to the local utility company as currently in place.

In many areas across the U.S. and the world, customers have only one power company, something the renewable energy industry is hoping to change. The International Energy Agency predicts solar power will be the most popular global energy source by 2050. Rising demand and falling prices suggest that solar energy prices will be cut in half by 2020. However, a central system for coordinating buyers and sellers would represent a high infrastructure investment. This is something small green energy producers may not be able to afford. A distributed and flexible blockchain system is much more efficient and affordable so that even very small sources could sell their excess electricity within their own community.

The global demand for sharing renewable energy relies on tracking energy certificates. Blockchain technology is poised to become the standard technology for managing these certificates in the near future over 5G network coverage. Transactions taking place over a peer-to-peer network mean reduced time and labor, fewer errors, greater security, and lower costs. There will be no need for intermediary brokers, while a decentralized system with no one point for failure means greater reliability for all concerned.

Blockchain Analytics

Blockchain analysis investigates classifying and monitoring blockchain addresses and transactions, to study the activities of various actors on the blockchain. Address classification is one of the main focus of blockchain analysis. Transaction monitoring monitor's every transaction related to an application. Risk analysis is based on the origin of the data, data flow and the history of the sender and the receiver. Compliance can analyze the risk associated with every transaction. Various rules and standards can be set to accomplish any regulatory obligations based on a jurisdiction. Surveillance and investigations enable law authorities to trace the data flow from source to destination in case of any misuse and criminal investigation.

GREEN NETWORKING IN 5G AND BEYOND

Several studies currently being conducted for green networking under hardware and software areas, mainly focus on network virtualization. Virtualization enables multiple physical machines to be merged on a single large physical machine, resulting in less energy being consumed at any instant of time. However, a virtualization solution designed explicitly to reduce network energy consumption is yet to appear. New research directions in both hardware and software green networking solutions are required as emerging networking technologies such as Fog and mobile Edge computing, Network virtualization, Network Function Virtualization(NFV) and Software Defined Networks (SDN) mature.

Virtualization can be applied to multiple kinds of resources, including network switches and links, storage devices, software resources, etc. A typical example of virtualization is the use of sharing multiple servers in data centers, thus reducing hardware costs, energy and cooling costs, ultimately reducing the carbon footprint of data centers. Large ICT companies such as Google and Amazon are also exploring geographical delocalization to reduce the energy consumption primarily focused to reduce the energy cost.

Energy Efficient Data Centers Beyond 5G

The development of "cloud computing" has a beneficial influence on energy consumption by sharing processors and other hardware, to avoid data centers being grossly underused. The ongoing research dedicated to energy optimization for computing is providing interesting early results. Note that computers are being put "to sleep" when they are not used, despite the fact that "going to sleep" and "waking up" are often computational and energy intensive. Moreover, Edge computing in datacenters provides execution resources (compute and storage) with sufficient connectivity (networking) at proximity to the data sources, typically within or at the boundary of access networks. The core benefits of edge solutions are low latency, high bandwidth, and trusted computing and storage.

Data traffic beyond 5G is expected to be time sensitive. Existing scheduling algorithms used for virtualization in data centers using edge network (Johnsson, n.d.) are not good enough for green networking. The focus is on improving the throughput and not necessarily reduction of energy consumption. New scheduling algorithms and energy friendly framework are needed for traffic classification and routing the traffic through edge/core to keep the overall energy consumption low Further, Fog & mobile Edge computing, NFV and SDN can be tuned to further reduce energy consumption by ICT.

Green Network Analytics

In the Data analysis process raw data obtained from various sources is subsequently converted into information useful for decision-making by users. *Data*, is collected and analyzed to answer questions, test hypotheses, or disprove theories. To identify these trends machine learning (ML) models can be built using supervised/unsupervised/semi-supervised/hybrid learning. In the **Data analysis** process of inspecting, cleansing, transforming and modeling data are done in stages with the goal of discovering useful information, informing conclusions, and supporting decision-making. Data analysis has multiple facets and approaches, encompassing diverse techniques under a variety of names, and is used in different business, science, and social science domains. In today's business world, data analysis plays a major role in making decisions more scientific and helping businesses operate more effectively.

Machine Learning (ML) which is a subset of Artificial Intelligence, can be used to predict events and reduce risks in any operation. To optimize performance, ML algorithms can be trained to predict risks by identifying patterns and surfacing risk (low, medium, high). The decision support systems can be designed through the ML models such that the response to such risks can be automated for the cases of non-urgent cases along with the identification of major trends of such events(Light et al., 2013; Mahasty et al., 2018; Muhammad et al., 2020). Regression Analysis is a well-known and well-understood algorithm in statistics and machine learning. It demonstrates how the dependent variable (Y) changes when one of the independent variables (X1/ X2......./ Xn) varies and enables to mathematically determine which of those variables really has an impact. Regression analysis can be linear, multi, and non-linear. Using this analysis, the Y value can be estimated over time based on given other input parameters. Other complex ML models like non-linear regression, support vector machine can be used for accurate results over the non-linear data. Also, Decision trees and Naive Bays methods can be used for the decision support systems based on history. The predictive analytics is important for any type of operations.

Data are collected from a variety of sources. Data, when initially obtained, must be processed or organized for analysis. Once processed and organized, the data may be incomplete, contain duplicates, or contain errors. The need for *data cleaning*, will arise from problems in the way that the datum are entered and stored. Data cleaning is the process of preventing and correcting these errors. Common tasks include record matching, identifying inaccuracy of data, overall quality of existing data, deduplication, and column segmentation.

Identifying the high-risk solutions at an early stages can be achieved by analyzing the trends based on specific parameters. In addition to these parameters, the historical data and test data may play a key role in identifying these trends. To identify these trends ML models can be built using supervised/un-supervised/hybrid learning algorithms for prediction/description. The ML process is shown in figure 2 showing the various stages of a machine learning process. Collected data is divided into Test Data and Training Data. Training data is used to train the Machine Learning Algorithms and then the Test data is applied to evaluate the system.

Figure 2.

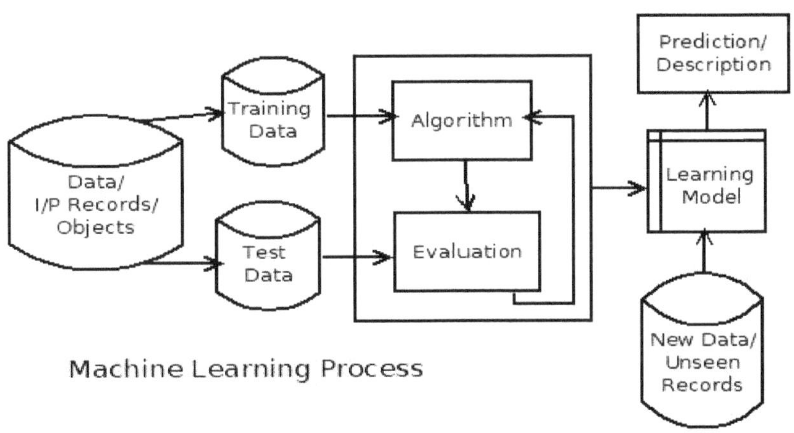

Supervised learning -Classification: In Classification, objects/data are assigned to well-known categories knows as class labels. These class labels are well defined for all the instances of the input; hence it is known as supervised learning. By classification, a model is built for the class label as a function of other attributes of the dataset. The model can be used as a tool to differentiate the objects of different classes (descriptive modelling) and to predict the class label of given records (predictive modelling).

Unsupervised learning – Clustering: In Cluster analysis (Han, 2012), objects are grouped based on their similarity and are commonly known as clusters. The clusters are formed in such a manner that the objects belonging to the same cluster are more similar to each other than the objects belonging to other clusters. Here, no class labels are defined. Hence it is an unsupervised learning. For the given data input of m attributes (x_{i1}, x_{i2}, x_{i3}, . . . x_{im}), i = 1, 2, . . . n, the key idea is to find out K clusters such that the intra-cluster distance $d(C_i) = \Sigma$ xÎ$C_{i,}$ yÎC_i distance(x, y) has to be minimum and the inter-cluster distance d(Ci, Cj) = Σ xÎCi,yÎCj distance(x, y) should be maximum. The distance (x, y) can be Euclidean distance or any other function which measures the distance between the two objects x, and y.

Machine Learning Model- Supervised: Decision tree is a technique widely used for pattern classification. The tree has three types of nodes- A root node that has no incoming edges and zero or more outgoing edges; Internal nodes, each of which has exactly one incoming edge and two or more outgoing edges. In a decision tree, each leaf node is assigned a class label. The non-terminal nodes, which include the root and other internal nodes, contain attribute test conditions to separate records that have different characteristics. Various split measures have been developed in the past for construction of decision trees. It is always preferred to construct a decision tree that is smaller in size and lesser in height and a good split measure helps in achieving this.

Machine Learning Model- Unsupervised: The analysis of the data can further be done using Cluster Analysis. Hierarchical clustering will enable us to do the comparative analysis of individual event with a cluster of events. Also, one cluster of events can be compared with another cluster of events.

Energy Profiling for Green Networking

A Taxonomy of green networking criteria for off-line solutions and on-line solutions are given in (Bianzino et al., n.d.). The criteria specific are also listed and they all are related to network specific parameters. The list of energy profiling listed here are general for any applications running on ICT infrastructure using 5G and beyond networks;

- Traffic prediction, resource pre-allocation and healing
- Energy usage for servers (database servers, web servers, data servers)
- Energy usage for cloud
- Energy usage by ICT infrastructures (switches, routers)
- Profiling energy usage by nodes/network capacity
- Profiling usage by IoT devices

The list can be exhaustive. As newer technology evolves, the applications can focus on some of the above analytics that would bring cost effective and environment friendly solutions.

Once the data are analyzed, it may be reported in many formats to the users of the analysis to support their requirements. The users may have feedback, which results in additional analysis. As such, much of the analytical cycle is iterative.

GREEN NETWORKING ANALYTICS

Green network analytics leverages the power of Machine. Learning and Machine Reasoning to provide accurate insights that are specific to network deployment and running. Quick troubleshoot of issues can be conducted from the insights obtained in real-time as well as non-real-time. Based on the discussions about issues with blockchain and green networking in 5G and beyond, following metrics are identified as indicators of performance and optimized operations.

- Data integrity
- availability
- Storage management
- Scalability
- Spectrum management
- Interference management
- Energy saving according to cluster size
- Trust and Security levels
- Privacy management
- Transparency
- Power sufficiency
- Reliability
- Economy
- Environment effects
- Natural resource usage
- Green cost efficiency

Analytics can be of any one of the four types: prescriptive, predictive, diagnostic, and descriptive as shown in Figure 3.

Choice of the type of data analytics outcome is entirely dependent on the criteria for network operations and demands of the environment.

CONCLUSION

Blockchain is an emerging technology that has drawn significant attention recently and is recognized as one of the key enablers for 5G networks thanks to its unique role to security assurance and network performance improvements. This chapter describes how blockchain supports 5G technologies and presents some challenges of energy saving for the future mobile and wireless network beyond 5 G. Energy saving results calculated from theoretical models and algorithms are valuable to develop energy efficient communication systems of the future beyond 5 G. Autonomous and parameter-free base stations can learn from the environment and activate the relevant energy saving features when needed with the optimized parameter settings. Solutions for control procedures that optimize networks sites power consumption and power usage depending on a variety of parameters that could be non-correlated. AI-based network energy efficiency, spectral efficiency assessment and optimization including the point-of-presence, the access and the core networks.

Figure 3.

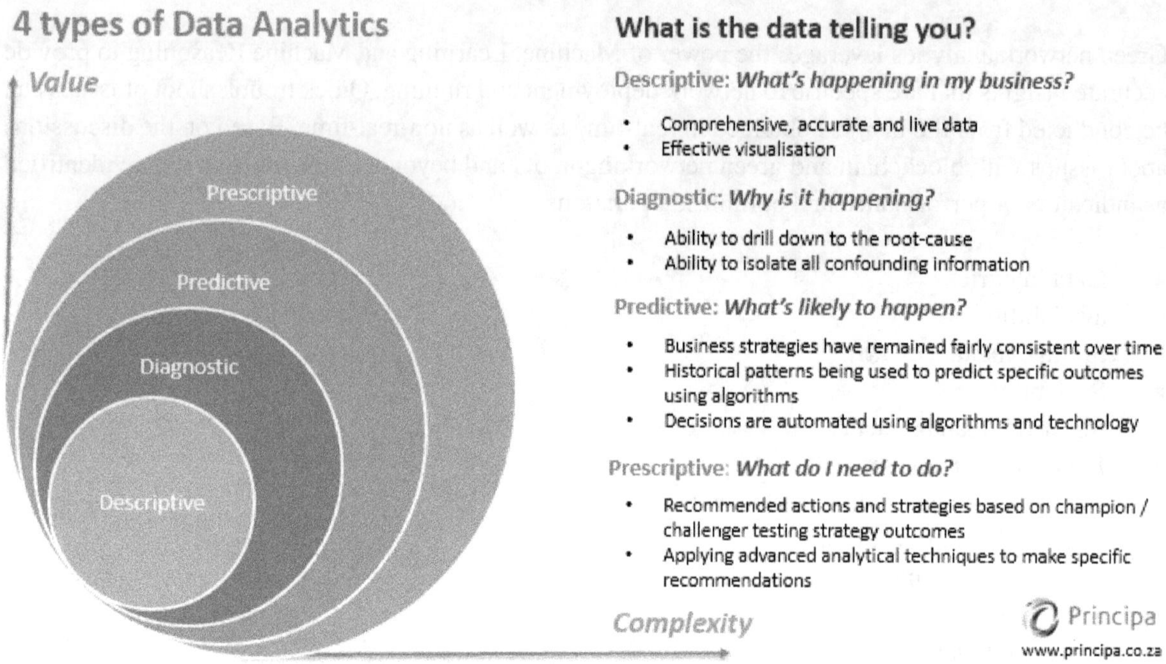

By 2023, 5G will make up around one-fifth of all mobile data traffic, with subscription uptake forecast to reach 1 billion. 4G LTE has also exploded since its launch in 2010, accounting for some 50 percent of total traffic in 2017 and still growing. Distributed cloud computing is paving the way for the future of network communications and, rather than waiting for 5G, operators should build tomorrow's networks today. This chapter discusses some energy efficient network functions across multiple layers of communication.

FUTURE DIRECTIONS

Beyond 5G, some fundamental issues that need to be addressed are higher system capacity, higher data rate, lower latency, higher security, and improved QoS compared to the 5G system (Chowdhury et al., 2020; Loven et al., 2019; Prabadevi et al., 2021; Rappaport et al., 2019; Stoica & Abreu, 2019). Technologies such as artificial intelligence, terahertz communications, wireless optical technology, free-space optical network, blockchain, three-dimensional networking, quantum communications, unmanned aerial vehicles, cell-free communications, integration of wireless information and energy transfer, integrated sensing and communication, integrated access-backhaul networks, dynamic network slicing, holographic beamforming, backscatter communication, intelligent reflecting surface, proactive caching, and big data analytics that can assist the 6G architecture development in guaranteeing the QoS. The key factors that will characterize the 6G communication system will be AI integrated communication, tactile Internet, high energy efficiency, low backhaul and access network congestion, and enhanced data security. Advancements in machine learning create more intelligent networks for real-time communications in 6G. Using numerous analytics, it can be determined how a complex target job is performed energy efficient.

With these emerging technologies, models can be developed to see business end tools running on these platforms that are energy efficient and providing high QoS.

REFERENCES

5G and blockchain: The building blocks of the shared economy. (n.d.). Available: https://www.ericsson.com/en/blog/2019/ 10/5G-blockchain-shared-economy

Al-Dunainawi. (2018, October). Green Network Costs of 5G and Beyond, Expectations Vs Reality. *IEEE Access: Practical Innovations, Open Solutions.*

Ali, Wang, Bhuiyan, & Jiang. (2018). Secure data provenance in cloud-centric internet of things via blockchain smart contracts. *2018 IEEE SmartWorld, Ubiquitous Intelligence & Computing, Advanced & Trusted Computing, Scalable Computing & Communications, Cloud & Big Data Computing, Internet of People and Smart City Innovation (SmartWorld/SCALCOM/UIC/ATC/CBDCom/IOP/SCI),* 991–998.

Baoid, Light, & Mahanti. (2021). Blockchain Technology and its Applications Across Multiple Domains: A Survey. *The Journal of International Technology and Information Management.*

Bianzino, Chaudet, Rossi, & Rougier. (n.d.). *A Survey of Green Networking Research.* Institut TELECOM, TELECOM ParisTech, CNRS LTCI UMR 5141.

Bilal, Khan, & Zomaya. (2013). Green Data Center Networks: Challenges and Opportunities. *11th International Conference on Frontiers of Information Technology.*

Blockchain: A key enabler for 5G. (n.d.). Available: https://www.standardsuniversity.org/e-magazine/may-2019-volume-9- issue-1-blockchain-standards/blockchain-a-key-enabler-for-5G /

Chaer, A., Salah, K., Lima, C., Ray, P. P., & Sheltami, T. (2019). Blockchain for 5G: Opportunities and challenges. *Proc. IEEE Globecom Workshops (GC Wkshps),* 1–6. 10.1109/GCWkshps45667.2019.9024627

Chaudhry, M. A. R., & Soptimizer, Z. A. (2019). Blockchain: A key enabler for 5G. *IEEE Standards Univ., 10*(1). Available: https://www.standardsuniversity.org/e-magazine/may-2019-volume-9- issue-1-blockchain-standards/blockchain-a-key-enabler-for-5g/

Chen, W., Ma, M., Ye, Y., Zheng, Z., & Zhou, Y. (2018). IoT service based on jointcloud blockchain: The case study of smart traveling. *2018 IEEE Symposium on Service-Oriented System Engineering (SOSE),* 216–221. 10.1109/SOSE.2018.00036

Chowdhury, Shahjalal, Ahmed, & Jang. (2020). 6G Wireless Communication Systems: Applications, Requirements, Technologies, Challenges, and Research Directions. *IEEE Open Journal of the Communication Society.*

Dai, Y., Xu, D., Maharjan, S., Chen, Z., He, Q., & Zhang, Y. (2019, May). Blockchain and deep reinforcement learning empowered intelligent 5G beyond. *IEEE Network, 33*(3), 10–17. doi:10.1109/MNET.2019.1800376

Freitag, F. (2018). On the collaborative governance of decentralized edge microclouds with blockchain-based distributed ledgers. *2018 IEEE/WIC/ACM International Conference on Web Intelligence (WI)*, 709–712. 10.1109/WI.2018.000-7

Han, J. (2012). *Concepts and Techniques*. Morgan Kaufmann.

Johnsson, L. (n.d.). *Overview of data centers energy efficiency evolution*. Available: https://pdfs.semanticscholar.org/559f/5b4bb297999ed00d4a787cf9317ec515afa1.pdf

Light, Selvi, Li, & Malali. (2013). Fall Pattern Classification from Brain Signals using Machine Learning Models. *Journal of Selected Areas in Health Informatics (JSHI), in the Cyber Journals: Multidisciplinary Journals in Science and Technology, 3*(12).

Light, J. (2020). Green Networking: A Simulation of Energy Efficient Methods. *Procedia Computer Science, 171*, 1489–1497. doi:10.1016/j.procs.2020.04.159

Loven, Peltonen, Leppänen, & Partala. (2019). Edge AI: A vision for distributed, edge-native artificial intelligence in future 6G networks. *Proc. 6G Wireless Summit*, 1-2.

Mahasty, Thompson, & Soleimani. (2018). A Concise Temporal Data Representation Model for Prediction in Biomedical Wearable Devices. *IEEE Internet of Things Journal, 6*(2), 1438 - 1445.

Manashty, A., & Light, J. (2019, March). Life Model: A novel representation of life-long temporal sequences in health predictive analytics. *Future Generation Computer Systems, 92*, 141–156. doi:10.1016/j.future.2018.09.033

Martin, T. (n.d.). *How blockchain will disrupt your industry*. https://www.slalom.com/in-sights/how-blockchain-will-disrupt-your-industry

Muhammad, Algehyne, Usman, Ahmad, Chakraborty, & Mohammed. (2020). *Supervised Machine Learning Models for Prediction of COVID-19 Infection using Epidemiology Dataset.* Springer Nature Singapore Pte Ltd.

Nguyen, Pathirana, & Seneviratne. (2020). Blockchain for 5G and Beyond Networks: A State of the Art Survey. *Computer Science, Engineering, Mathematics, J. Netw. Comput. Appl.*

Nguyen, D. C., Pathirana, P. N., Ding, M., & Seneviratne, A. (2019). *Blockchain for 5G and beyond networks: A state of the art survey*. https://arxiv.org/abs/1912.05062

Prabadevi, Deepa, Pham, Nguyen, Reddy, Reddy, Pathirana, & Dobre. (2021). Towards Blockchain for Edge-of-Things: A New Paradigm, Opportunities, and Future Directions. *IEEE Internet of Things Magazine*.

Rappaport, T. S., Xing, Y., Kanhere, O., Ju, S., Madanayake, A., Mandal, S., Alkhateeb, A., & Trichopoulos, G. C. (2019). Wireless communications and applications above 100 GHz: Opportunities and challenges for 6G and beyond. *IEEE Access: Practical Innovations, Open Solutions, 7*, 78729–78757. doi:10.1109/ACCESS.2019.2921522

Stoica, R. A., & Abreu, G. T. F. (2019). *6G: The wireless communications network for collaborative and AI applications*. Available: arXiv:1904.03413.

Tahir, O., Habebi, M. H., Dabbagh, M., Mugheesi, A., Ahad, A., & Ahmed, K. I. (2020, July). A Review on Application of Blockchain in 5G and Beyond Networks: Taxonomy, Field-Trials, Challenges and Opportunities. *IEEE Access: Practical Innovations, Open Solutions*, *8*, 115876–115904. doi:10.1109/ACCESS.2020.3003020

Taleb, T., Samdanis, K., Mada, B., Flinck, H., Dutta, S., & Sabella, D. (2017). On multi-access edge computing: A survey of the emerging 5G network edge cloud architecture and orchestration. *IEEE Communications Surveys and Tutorials*, *19*(3), 1657–1681. doi:10.1109/COMST.2017.2705720

Xiong, Z., Feng, S., Wang, W., Niyato, D., Wang, P., & Han, Z. (2019, June). Cloud/fog computing resource management and pricing for blockchain networks. *IEEE Internet of Things Journal*, *6*(3), 4585–4600. doi:10.1109/JIOT.2018.2871706

Yang, H., Wu, Y., Zhang, J., Zheng, H., Ji, Y., & Lee, Y. (2018). Blockonet: blockchain-based trusted cloud radio over optical fiber network for 5G fronthaul. In *Optical Fiber Communication Conference*. Optical Society of America. 10.1364/OFC.2018.W2A.25

Yang, H., Zheng, H., Zhang, J., Wu, Y., Lee, Y., & Ji, Y. (2017). Blockchainbased trusted authentication in cloud radio over fiber network for 5G. *2017 16th International Conference on Optical Communications and Networks (ICOCN)*, 1–3.

Yang, M., Margheri, A., Hu, R., & Sassone, V. (2018). Differentially private data sharing in a cloud federation with blockchain. *IEEE Cloud Computing*, *5*(6), 69–79. doi:10.1109/MCC.2018.064181122

Zhang, Y., Xu, C., Lin, X., & Shen, X. S. (2019). *Blockchain-based public integrity verification for cloud storage against procrastinating auditors. IEEE Transactions on Cloud Computing*. doi:10.1109/TCC.2019.2908400

Chapter 2
Improving Water Efficiency in the Beverage Industry With the Internet of Things

Sandeep Jagtap
https://orcid.org/0000-0001-5271-0192
Cranfield University, UK

George Skouteris
Helmholtz-Zentrum Dresden-Rossendorf, Germany

Vilendra Choudhari
Jubilant FoodWorks Limited, India

Shahin Rahimifard
Loughborough University, UK

ABSTRACT

The food and beverage industry is one of the most water-intensive industries, with water required for various processes (e.g., washing, cooking, cleaning) at almost every stage of the production, as well as being a key constituent in many food and drink products. Therefore, a real-time efficient water management strategy is imperative, and the novel internet of things (IoT)-based technologies can be of significant help in developing it. This chapter presents the architecture of an IoT-based water-monitoring system followed by the demonstration of a case study of a beverage factory wherein the monitoring system helped understand the detailed water usage as well as finding solutions and addressing overconsumption of water during the manufacturing processes. The successful deployment of IoT helped reduce the annual water consumption by 6.7%, monitor water usage in real-time, and improve it.

DOI: 10.4018/978-1-7998-6988-7.ch002

INTRODUCTION

Background

Out of the total existing water on earth's surface 97% is ocean water, 2.5% is entrapped in glaciers and ice and only 0.5% is accessible as freshwater (Mullen, 2012). From the abstracted freshwater about 70% is used for irrigation, 19% is used for manufacturing purposes and the 11% left is consumed for domestic usage (Luckmann, Grethe, McDonald, Orlov, & Siddig, 2014). With respect to manufacturing, water is extensively used in the food industry (Poretti, 1990). In more detail, it is estimated that food and drink industries consumption of water in 2010 was between 185.5-195.7% (Bromley-Challenor, Kowalski, Barnard, & Lynn, 2013) as shown in Table 1. With the rising world population, which is set to reach 9.7 billion by 2050 (United Nations, 2017), an increasing amount of freshwater will be needed for drinking, food production, hygiene and sanitation. This increases pressure on water resources and exposes the food industry vulnerability to water scarcity.

Problem Overview and Scope of Work

Water is an essential resource for food and drink sector as it is embedded within the food product, is needed for processing or for cleaning purposes (Casani, Rouhany, & Knøchel, 2005). This sector considers water efficiency and sustainability as the topmost priority in decision-making processes for Food Supply Chain (FSC) stakeholders (Jagtap, 2019; Jagtap, & Rahimifard, 2018). To meet both demand and supply for freshwater, a well-aware and responsive water management system is required. Only through communication, collaboration, and collective actions of all the stakeholders within the FSC, water efficient practices can be implemented (Skouteris et al. 2018; Webb, Skouteris, & Rahimifard, 2018). Hence, a real-time water consumption tracking system is needed through which a detailed information on water usage activities can be monitored to identify wastage and find the opportunities to reduce the consumption.

Table 1. Food and drink sector water use

Food and Drink Industry	Total Water Use (million m^3/annum)	
	2007	2010
Food and Drink Manufacturing	230.9 (56.1%)	185.5-195.7 (53.4%)
Retail	10.1 (2.5%)	6.9-10.1 (2.0-2.8%)
Wholesale	1.6 (0.4%)	1.1-1.7 (0.3-0.5%)
Hospitality and Food Service	169.0 (41.1%)	153.7-158.8 (44.3 -43.3%)

Source: (Bromley-Challenor, et al., 2013)

The Internet of Things (IoT) is accepted as one of the most important areas of future technology and is gaining careful attention from a wide range of industries (Lee & Lee, 2015). The IoT concept, which aims to support the transparency and visibility, could be utilized to provide detailed information on water consumption in FSC through smart sensors and meters from each machine component to whole of the supply chain (Jagtap, et al., 2021a; Jagtap, Garcia-Garcia, & Rahimifard, 2021b). Thus, real-time water consumption data from food manufacturing processes can be gathered seamlessly and then analyzed, to increase water-aware decision-making.

This paper provides an understanding of water efficient practices that are undertaken through the application of IoT and addresses the benefits of adopting such management practices. Furthermore, a framework is introduced to support the incorporation of collected water data into an FSC's planning tools and information technology platforms. The final goal of the framework is to highlight how decision-making processes based on such data could support and enhance water efficiency and thereby increase the effectiveness of FSC. Finally, the case study results back the adoption of IoT in value-based manner and water management practices that are more in line with FSC development.

IOT-BASED WATER MONITORING SYSTEM

IoT Based Water Monitoring Architecture

Figure 1 shows a detailed IoT architecture for water management. The bottom right quadrant is termed as Sensing layer and its primary function is to acquire data and information on water flowrate and quality in real-time. This data is collected using a number of sensors, such as pressure transducers, flow meters and water quality sensors. The bottom left quadrant is called as Networking layer which follows certain procedures for reading sensors and devices. It executes basic functions of linking up of sensing layer to database systems and software platforms. It uses short-range wireless networks such as WiFi, Bluetooth, RFID and ZigBee. The upper left quadrant is the Service layer, and it involves management of data and information, software applications and platforms. Service layer is responsible for collecting the data from all IoT-gateways. It processes an excessive amount of data and sorts it before storing it in a data warehouse. The stored data is then made available for data mining and analyzing by applications running in a cloud to extract useful information. The Application layer which is the upper right quadrant generates real-time water data analysis and water trend reports and presents information to user over the Internet via HTTP. The web application is powered by ASP, .NET, HTML5 and supports user-friendly functionalities such as diverting water from certain food production processes to other secondary processes depending upon the water quality, checks the water usage, sends alerts and allows to view historical data.

IoT-based Water Efficiency System

Figure 1 illustrates the process of IoT-based water monitoring system deployed in the case company. The system consisted of pressure sensors, flowmeters and water quality sensors to sense various water parameters which are crucial for food production. The system recorded data for water flowrate, water pressure for identifying leakages and water quality parameters such as pH, temperature, chlorine, electrical conductivity, dissolved oxygen and oxidation/reduction potential for possible contamination. The data collected is stored on the secured cloud server and is made available in real-time irrespective of their location. It uses a specially designed software suite that mines for the suitable water data and recognizes water usage patterns. It further uses mathematical and statistical algorithms to capture behavioral changes and variations in pressure, flow and quality of water. It distinguishes between abnormal events when compared to the standard operating procedures and statistically filters out false alerts. As shown in Figure 2, the Water monitoring system can identify and measure water loss in real-time and detect

Figure 1. IoT Architecture for Water Monitoring

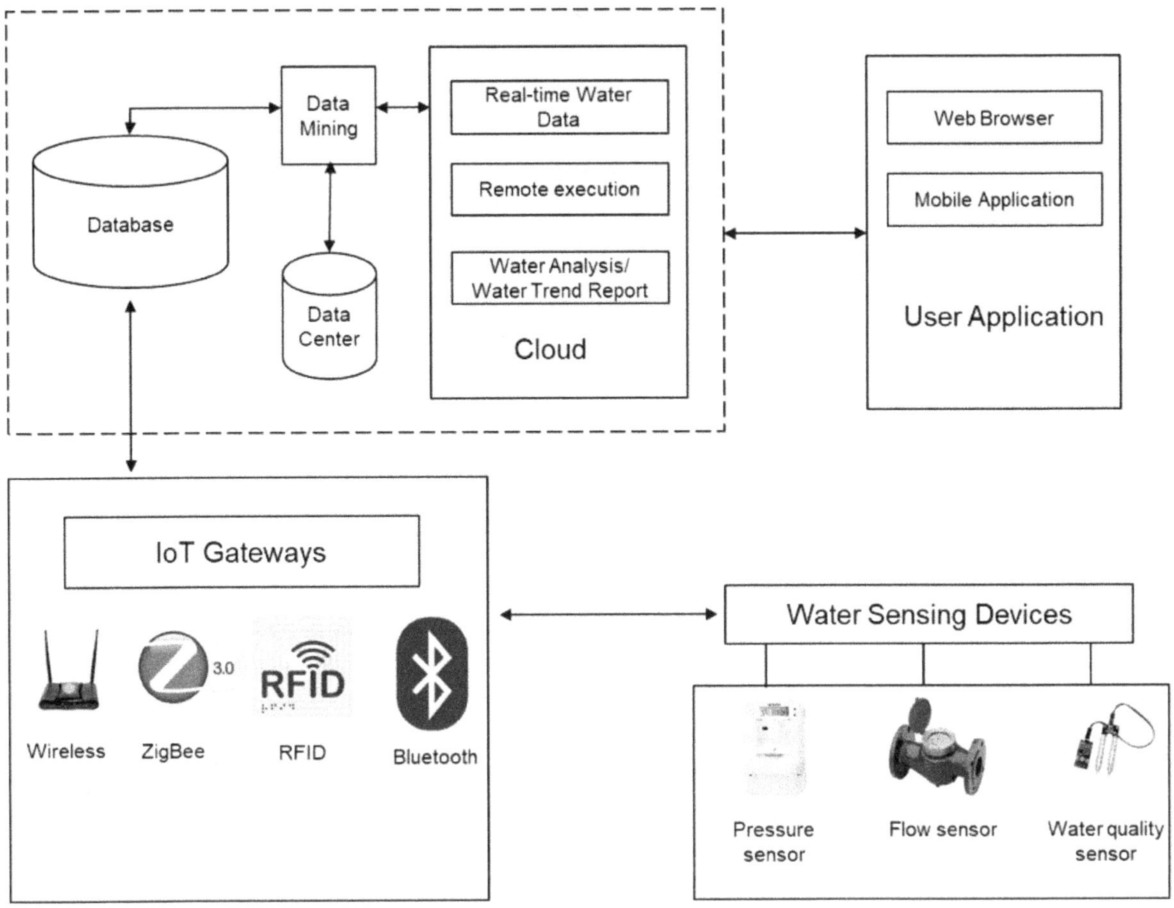

leaks in the system as well as detecting contamination. This allows the users to contain the wastage of water immediately as soon as it is detected and reinstate the water usage back to normal.

CASE STUDY

Beverage industry is one of the largest users of water resources and is highly relevant as it consumes between 89-99% of potable water (Beverage Industry Environmental Roundtable, 2011). Traditionally, water usage in the beverage industry has been quantified on a total volume or normalized volume (volume water used per volume product packaged). This ratio is typically well known and has become the standard for measuring water use efficiency in the beverage sector. As per Statista 2018, production volume of aerated and soft drinks across India from 2015 to 2017 (in million litres) has dropped as shown in Figure 3. But, still 2644.56 million liters for the year 2017 is quite significant because every 1 litre of beverage production consume anywhere between 1.7 - 4.2 liters of water and in some cases even more than that (Beverage Industry Environmental Roundtable, 2011).

Figure 2. Pathway for IoT-based Water Efficiency

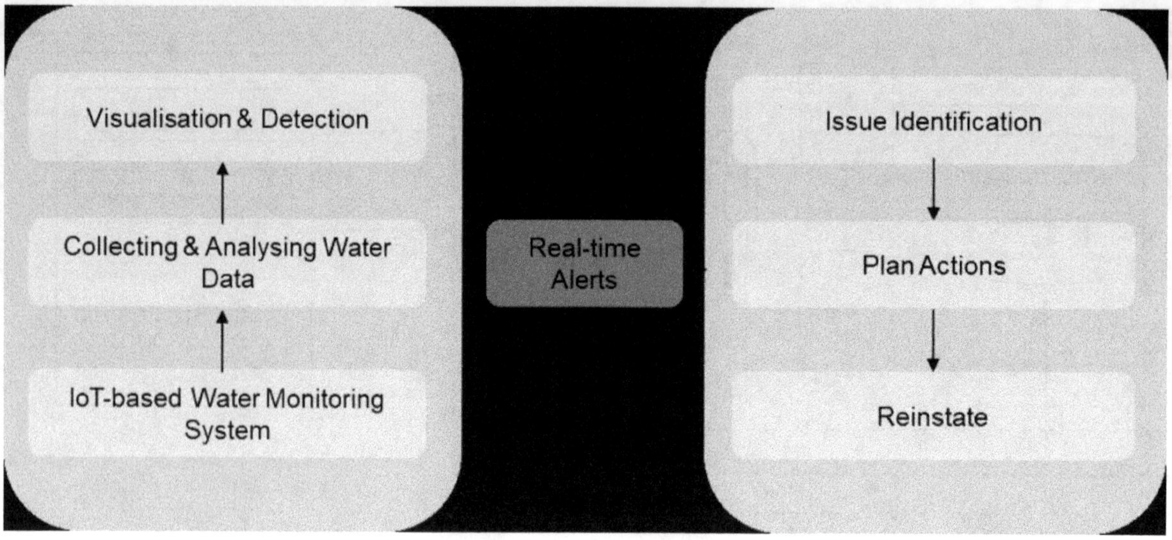

Manufacturing Process

Figure 3. Production Volume of Aerated & Soft drinks – India (Source: Statista 2018)

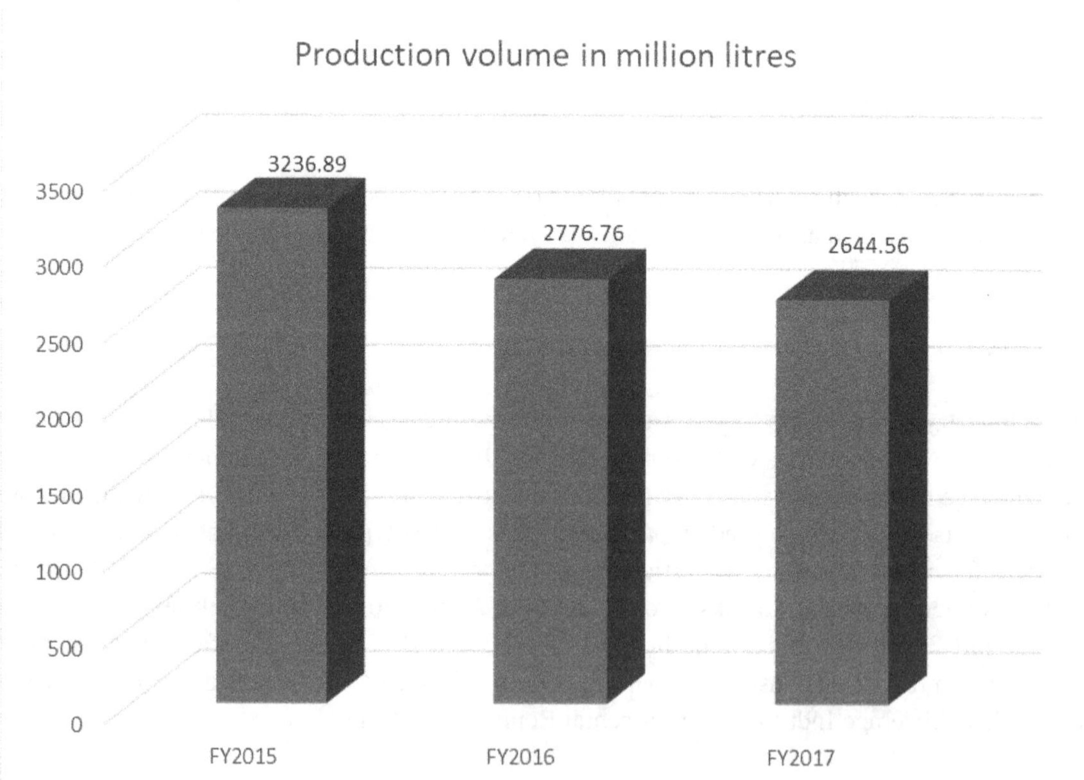

Figure 4 shows the soft drink production process. The raw water drawn from the water-wells goes through various purification and filtration processes until it reaches the blenders. Once inside the blenders, flavorings and sugar are added to the water as per the recipe and mixed thoroughly. The mix is sent to the filling machines wherein CO_2 gas is added while filling the bottles.

Figure 4. Beverage Production Flowchart

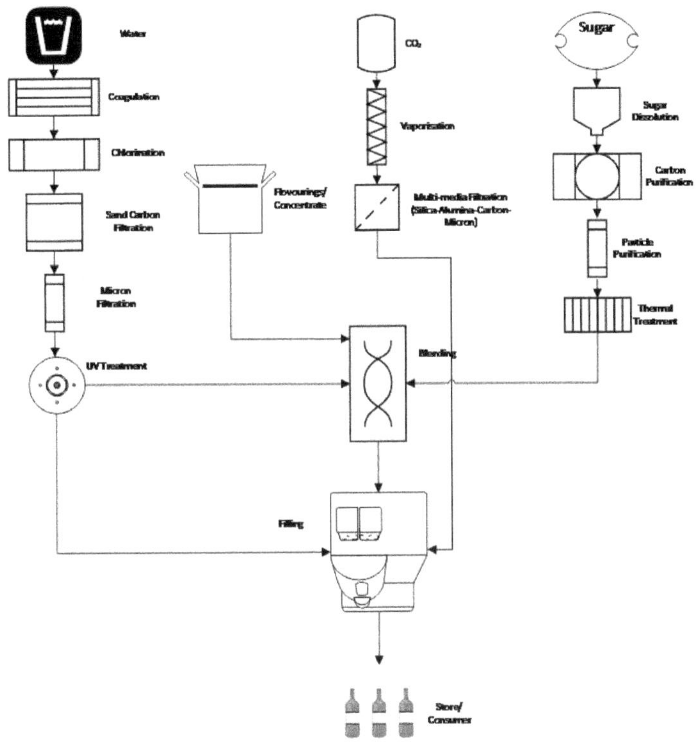

Water Consumption in Factory

The case company consumes 127,500 m^3/month of potable water and through IoT-based water monitoring system identified the five major areas where water is consumed extensively: 1) Cooling, 2) Bottle cleaning/filling, 3) Plant cleaning, 4) Utilities and 5) Raw material washing. As illustrated in Figure 3, cooling is the largest water consuming water activity taking up 57% of total water which equates to 72,675 m^3/month. The other losses which accounted to 29% was due to evaporation, leaks and due to other reasons.

IoT-System Implementation

The beverage company before installation of the IoT system were not aware of their in-detail water usage and its breakdown. For the year 2017, the water usage was not uniform (as shown in Table 2) and the company used mass-balance calculations to estimate the water consumption. The company was aware of the raw water intake, the volume (in litres) of finished beverages and the amount of water discharged

Figure 5. Breakdown of Water Consumption in Plant

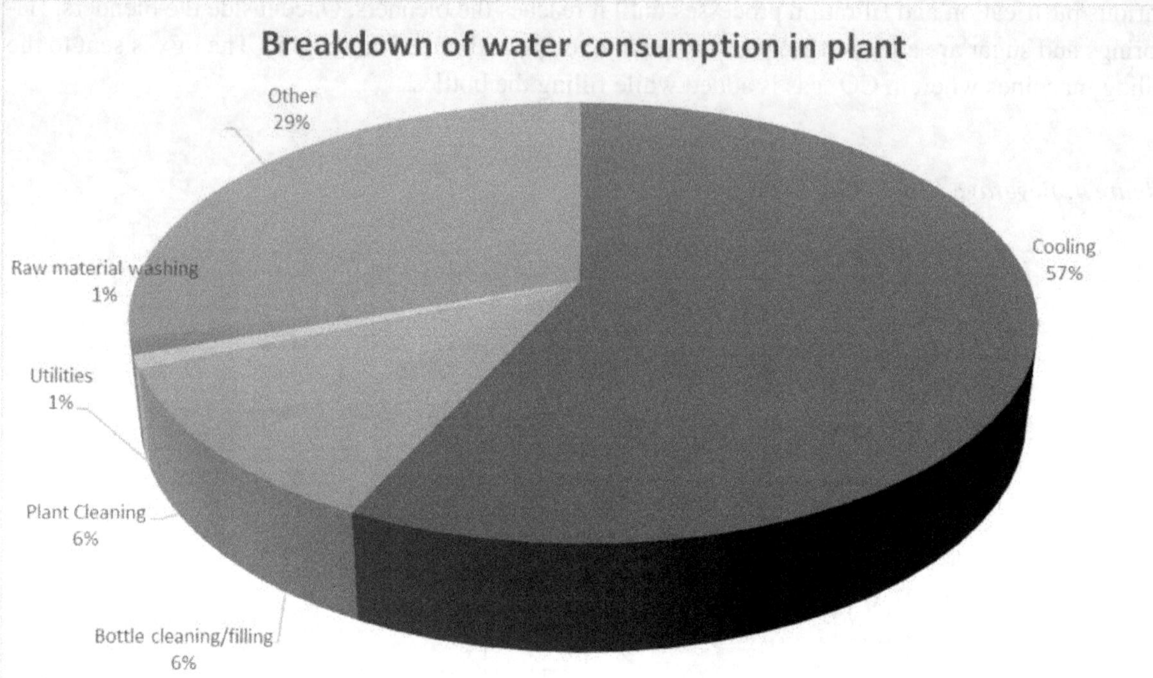

to the effluent treatment plant (ETP) and by working out them they estimated the water losses. This practice never allowed them to look for the root cause or reasons behind the water losses.

The company undertook the project to install the real-time IoT-based water monitoring system by the end of December 2017 and started monitoring water consumption from January 2018 onwards. They realized the various water saving opportunities and focused on the issues which did not require extra investments and can be addressed with the resources at their disposal (as shown in Table 3). The outcome of that was a continuous reduction in the water usage from January 2018 to June 2018.

Table 2. Total water in litres to make one liter of finished beverage

2018	**JAN**	**FEB**	**MAR**	**APR**	**MAY**	**JUN**
Total Water Lt/ Lt beverage	2.03	2.01	1.99	1.98	1.92	1.90
2017	**JAN**	**FEB**	**MAR**	**APR**	**MAY**	**JUN**
Total Water Lt/ Lt beverage	1.74	2.11	2.19	1.99	2.14	2.49

CONCLUSION

The objective of this case study was to analyze water-intensive processes and identify opportunities to reduce the water consumption of the beverage company. Through IoT-based water monitoring system, processes and areas where significant amount of potential water savings was achieved are listed in the Table 3.

Table 3. Water-saving opportunities

1. Filler bowl gasket damage leading to water leakage
2. Filler valve vent tube issue, dispensing liquid even when bottle is absent
3. Rinser auto system not working properly leading to excessive water usage
4. Recovery tank overflowing pump, float and pipe line changing for water losses
5. Filler room, ozone tester, rinse water diversion to recovery tank
6. CIP system validation for water reduction
7. Final syrup room water recovery system to be install
8. Production line vacuum water to be circulated or diverted to raw water tank

Due to the adoption of the IoT-based water monitoring system, the total water (in liters) required to make one liter of beverage was reduced from 2.10 liters to 1.96 liters on an average. They managed to reduce the overall annual consumption of water by 6.66% which was equivalent to reducing 8491.5 m^3 per month of total water intake. The company also realized that by utilizing the water used for cooling towers can be reutilized and diverted to recovery tank with some adjustments in the piping designs.

The results from this case study shows that there is a tremendous potential with the real-time IoT-based monitoring system to measure and reduce the water consumption of various processes and this can be replicated into other food factories.

ACKNOWLEDGMENT

This work was supported by the Engineering and Physical Sciences Research Council (EPSRC) Centre for Innovative Manufacturing in Food [Reference: EP/K030957/1].

REFERENCES

Beverage Industry Environmental Roundtable. (2011, December). *A Practical Perspective on Water Accounting in the Beverage Sector.* Retrieved from Water footprint: https://www.waterfootprint.org/media/downloads/BIER-2011-WaterAccountingSectorPerspective.pdf

Bromley-Challenor, K., Kowalski, M., Barnard, R., & Lynn, S. (2013). *Water use in the UK food and drink industry - A review of water use in the food and drink industry in 2007 and 2010, by sub-sector and UK nations.* Banbury: WRAP.

Casani, S., Rouhany, M., & Knøchel, S. (2005). A discussion paper on challenges and limitations to water reuse and hygiene in the food industry. *Water Research, 39*(6), 1134–1146. doi:10.1016/j.watres.2004.12.015 PMID:15766968

Jagtap, S. (2019). *Utilising the internet of things concepts to improve the resource efficiency of food manufacturing* (Doctoral dissertation). Loughborough University.

Jagtap, S., Bader, F., Garcia-Garcia, G., Trollman, H., Fadiji, T., & Salonitis, K. (2021a). Food logistics 4.0: Opportunities and challenges. *Logistics*, *5*(1), 2. doi:10.3390/logistics5010002

Jagtap, S., Garcia-Garcia, G., & Rahimifard, S. (2021b). Optimisation of the resource efficiency of food manufacturing via the Internet of Things. *Computers in Industry*, *127*, 103397. doi:10.1016/j.compind.2021.103397

Jagtap, S., & Rahimifard, S. (2018). Real-time data collection to improve energy efficiency in food manufacturing. In *International Congress on Organizational Management, Energy Efficiency and Occupational Health and Safety in Agrifood Industry (+ AGRO 2018), Castelo Branco, Portugal* (pp. 3-4). Academic Press.

Lee, I., & Lee, K. (2015). The Internet of Things (IoT): Applications, investments, and challenges for enterprises. *Business Horizons*, *58*(4), 431–440. doi:10.1016/j.bushor.2015.03.008

Luckmann, J., Grethe, H., McDonald, S., Orlov, A., & Siddig, K. (2014). An integrated economic model of multiple types and uses of water. *Water Resources Research*, *50*(5), 3875–3892. doi:10.1002/2013WR014750

Mullen, K. (2012). *Information on Earth's water*. Retrieved September 11, 2017, from https://www.ngwa.org/Fundamentals/teachers/Pages/information-on-earth-water.aspx

Poretti, M. (1990). Quality control of water as a raw material in the food industry. *Food Control*, *1*(2), 79–83. doi:10.1016/0956-7135(90)90089-U

Skouteris, G., Webb, D. P., Shin, K. L. F., & Rahimifard, S. (2018). Assessment of the capability of an optical sensor for in-line real-time wastewater quality analysis in food manufacturing. *Water Resources and Industry*, *20*, 75–81. doi:10.1016/j.wri.2018.10.002

Statista. (2018). *Production volume of aerated and soft drinks across India from FY 2015 to FY 2018 (in million liters)*. Retrieved from Statista: https://www.statista.com/statistics/762413/india-aerated-and-soft-drinks-production-volume/

United Nations. (2017). *World Population Prospects 2017*. Retrieved June 29, 2017, from https://esa.un.org/unpd/wpp/DataQuery/

Webb, D. P., Skouteris, G., & Rahimifard, S. (2018). In-plant real-time manufacturing water content characterisation. *Water Resources and Industry*, *20*, 37–45. doi:10.1016/j.wri.2018.08.003

Chapter 3
Network Intrusion Detection Using Linear and Ensemble ML Modeling

Shilpi Hiteshkumar Parikh
U and P U. Patel Department of Computer Engineering, CSPIT, Charotar University of Science and Technology (CHARUSAT), Changa, India

Anushka Gaurang Sandesara
U and P U. Patel Department of Computer Engineering, CSPIT, Charotar University of Science and Technology (CHARUSAT), Changa, India

Chintan Bhatt
U and P U. Patel Department of Computer Engineering, CSPIT, Charotar University of Science and Technology (CHARUSAT), Changa, India

ABSTRACT

Network attacks are continuously surging, and attackers keep on changing their ways in penetrating a system. A network intrusion detection system is created to monitor traffic in the network and to warn regarding the breach in security by invading foreign entities in the network. Specific experiments have been performed on the NSL-KDD dataset instead of the KDD dataset because it does not have redundant data so the output produced from classifiers will not be biased. The main types of attacks are divided into four categories: denial of service (DoS), probe attack, user to root attack (U2R), remote to local attack (R2L). Overall, this chapter proposes an intense study on linear and ensemble models such as logistic regression, stochastic gradient descent (SGD), naïve bayes, light GBM (LGBM), and XGBoost. Lastly, a stacked model is developed that is trained on the above-mentioned classifiers, and it is applied to detect intrusion in networks. From the plethora of approaches taken into consideration, the authors have found maximum accuracy (98.6%) from stacked model and XGBoost.

DOI: 10.4018/978-1-7998-6988-7.ch003

INTRODUCTION

Currently, we are thriving in a world that is limitless and with no boundaries. With the augment in advances technologically and scientifically there are high chances of attacks, breaches, and other vulnerabilities in the network. Besides this, the surge of internet facilities and online utilities available in a fraction of seconds have resulted in high cases of cyber-crime. Before, two decades the detection of breaches and attacks were carried independently by users without any intervention from the machine. But nowadays due to the high-amount of cyber crimes and intrusions in networks, it is not possible to solve the crime manually and hence it is more efficient with the machine learning and deep learning methods available. Still, there is a huge demand for a novel technique that predicts the intrusions as well as guides the users of the network on how to resolve them.

When we talk about data in wireless networks, different types of data in structure, dimension, size come into picture. According to the authors (Yuanwei et al., 2019), big data resources are utilized by analytical and statistical machine learning tools to support new intelligent applications which are proposed in wireless networks. Accordingly, the prevailing variant types of data can be categorized into major three forms: Wireless Data, Social Data and Cloud Data. The most notable challenge to perform data analytics in wireless networks is to accurately predict user preference distribution. Performing data analytics on wireless networks can also help to look into the odd behaviour of some data and help to figure out the malicious activity taking place inside the wireless networks.

Basically, an intrusion detection system can be of two types-software or hardware. It is the choice of the manufacturer to select a software or hardware system and the system can be attached to the different network domains such as Ethernet, FDDI, or any other. The IDS system continuously inspects the traffic from the original point of installation and performs an observation of passing traffic on the entire subnet and matches the traffic that is passed on the subnets to the collection of known attacks. Such advanced systems are not possible to attack by invaders because any malevolent activity is directly reported to the administrator. The IDS system developed by the researchers here supervises both inbound and outbound traffic on the network, as well as data traversing between systems within the network.

Most present-day businesses require top-tier safety to protect their credentials for work. Even though there are conventional techniques such as authentication and authorization (Xu et al., 2014) but they are not able to ensure complete security in the systems, Intrusion Detection System on the other hand provides a great level of safeguard for protecting the system from attacks and other threats. One important advantage of the IDS system is that it provides an immediate alert to the administrator about the prevailing attacks on the network so that the administrator is at least aware that the network has been infected. Being aware of future possible attacks and breaches, an IT person can take appropriate steps to stop the attacker or prevent it from happening. So, the basic step of any IDS system is to detect the type of attack that would be taking place. Although the system is not able to resolve the attack, perceiving the intrusion will benefit the security officials and hence Intrusion Detection (ID) is the first and foremost step.

This chapter takes into consideration four basic attacks. Amongst them, the DOS attack (Alharbi et al., 2018) is the most hazardous because it generates a lot of traffic making it so full of memory and extraneous resources that the system fails to recognize legal user requests. The main purpose of the R2L attack is to get an illicit permit to the system's resources and the privacy of the whole network is disrupted. The U2R attack gives access to the attacker as a root user so confidentiality of data is again lost. The Probe attack is where the attacker investigates the network for weaknesses that can prove useful to recognize services that can be executed (Chao-yang, 2011).

LITERATURE REVIEW

A wide spectrum of wireless communication systems and wireless devices have emerged in the present era. This led to the advent of a Big Data era in wireless communication networks. In large-scale wireless networks, Big Data has the significant features of huge variety, the high volume of network data, real-time data and velocity. According to the survey performed by (Hong-Ning Dai et al., 2019), the BDA in wireless networks can be divided into 4 parts, 1st being Data Acquisition, 2nd Data Preprocessing, 3rd Data Storage and the last being the Data Analytics. The data provided to the Data Acquisition storage can be data from mobile communication networks, vehicular networks, Mobile social networks, IOT or any other relevant wireless network data. The primary goal of data analytics was to extract meaningful data from wireless networks. BDA in large scale wireless networks becomes quite challenging because of the huge volume of the networks plus due to the heterogeneous data structure in networks along with the wide data diversity and dimension. If one wants to perform data analytics for wireless networks 4 challenges that will be encountered will be Temporal-spatial correlation, Privacy, Efficiency and Real-time. Privacy gets hampered the most when we observe intrusions in large-scale wireless networks.

Adhering to the complexity of handling and understanding the data obtained in large scale wireless networks, Machine Learning concepts can prove to be of great significance. (Mirza et al., 2018) explained the benefits Machine Learning and AI can provide in decreasing the complexity of next-generation wireless networks. Machine Learning and AI have always proved beneficial and efficient tools in handling a large amount of data specifically for giving suggestions and making predictions all based on the type of data set provided. To manage a wireless network that eventually keeps on growing in size comes up with a challenge in itself since there is a constant need to integrate the new elements. Machine Learning techniques can be useful for performing analytics on large-scale data because it helps to extract valuable and meaningful information from the raw data and helps to better visualize the data to provide a better insight of the data. Machine Learning and AI are found to be useful for networks even to address the new areas and where there is no historical data available.

Various researchers are currently working on finding different machine learning and deep learning classifiers that accurately predict the intrusions into the system by selecting appropriate features. The most important step of the detection system is feature engineering because the extracted features are responsible for the prime detection rate and the least false positive metric.

The existing work on intrusion detection in (Sharmila & Nagapadma, 2019) takes into consideration the Naive Bayes Algorithm- using the sklearn library and in the absence of it. In the original dataset, there are various attributes so PCA is implemented to reduce extraneous features. PCA is selected because it transmutes a large vector of dimensions into a small set that certainly reduces computational time and enhances accuracy. The system was prepared on both the normal and PCA-based Naive Bayes algorithm and the results clearly depicted that PCA implementation for 10 prime features achieves high accuracy in comparison to the conventional technique.

The proposed technique in (Amato et al., 2017) implements IDS using an artificial feed-forward neural network called multilayer perceptron (MLP). Due to many features and features being of different categories such as continuous, separate, and symbolic the authors have selected important features by two different algorithms. Greedy StepWise is more advantageous because it uses nearly 28 seconds to compute and extract 11 features, while on the other hand Best-first is akin to the Greedy method and extracts 11 features, but the only difference is it uses backtracking and takes more time to compute nearly

32 seconds. The main benefit of this method is it does not require the whole network to be retrained when a new attack is added instead only the series of layers consisting of a new attack as feed in.

Another approach that has been widely used for the implementation of IDS is by using the Random Forest classifier. It is an ensemble classifier. In the research paper (Jha & Ragha, 2013) the authors have made use of a random forest classifier for intrusion detection in the network. A noteworthy advantage of random forest classifiers is that the generated forest can be used for future reference and it overcomes the problem of overfitting. In this presented paper, the authors focused on 4 attacks- DOS, U2R, Probe, and R2L. In this methodology, the authors used the NSL-KDD dataset as input, pre-processed the data then classified the features based on the type of attack. After the features have been classified, a set of features were selected using the feature subset selection measure Symmetrical Uncertainty and on those set of features, a Random Forest algorithm was applied. The proposed method produced a high Detection Rate and low False Alarm Rate to classify the attacks. Overall, the use of symmetrical uncertainty reduced the problems of information gain and helped the system to achieve higher accuracy.

A few years earlier, SVM was known to be the best algorithm that was used for the implementation of IDS in the network. This notion was prevalent due to the noticeable generalization nature and ability to overcome dimensionality. Still, SVM came along with few restrictions, to overcome these limitations, the authors (Farnaaz & Jabbar, 2016) proposed a methodology that used SVM along with the K-Means algorithm. They used featured weights while training the SVM model. They used a reduced NSL-KDD dataset to improve the performance of the SVM model. This eventually helped to decrease the train and test time of the model proposed in this methodology.

The rapid growth of technological advances finds ways to ease the life of people but on the contrary side it breaches the security of the data. (T.Saranya et al., 2020) proposed a comparative analysis of various machine learning algorithms that can be used for implementing intrusion detection systems for detection of anomalies in wireless networks. According to the methodology proposed by them and the comparative analysis done, the detection rate, false positive rate and the accuracy of the system not only depends on the type of algorithm used but also on the area of application.

With the evolving development of Big Data and computing, deep learning algorithms have also been used for the implementation of IDS in networks. Various research is already taking place in improving the accuracy of IDS using deep learning approaches. In the paper (Yin et al., 2017), the authors have proposed a three-layered RNN structure. This architecture takes into consideration 41 features as input to the system and 4 intrusion types as output from the system. The experimental results reveal that the performance of RNN-IDS is better than the traditional classifiers. In the present methodology, the RNN architecture consists mainly of 2 parts. The first part is Forward Propagation which is used for the calculation of output values and the second part is Backward Propagation which is responsible for passing the residuals that were accumulated for updating weights. The RNN does not only provide high modeling for IDS but it provides high accuracy for both multiclass and binary classification.

METHODOLOGY

It is very common sometimes to consider false alarms as a threat to the network. This eventually leads to neglect of the actual attack on the network. When there is a high noise rate in the network, the overall intrusion detection rate gets badly affected. As the intrusion detection system monitors the whole network it might happen that they become vulnerable for the same types of attacks that are found in the networks and as a result sometimes, some protocol-based attacks can fail the entire IDS. To add on, traditional network intrusion detection systems create a bottleneck, as all the inbound and outbound traffic passes through it. But when Machine Learning got introduced in network intrusion detection, it solved several types of problems which were encountered in the traditional network systems. Machine Learning and Deep Learning techniques hold the strength to learn through the pattern and detect an anomaly as well as detect any new attack on the network. The below section gives the overview of various machine learning techniques used to solve the problems of traditional IDS.

Overview of NSL-KDD Dataset

NSL-KDD dataset is the new version of the KDD'99 dataset. It was designed to solve the problems encountered in the KDD'99 dataset. This dataset is broadly used for comparing different intrusion detection systems and types of attacks. The size of this new version of the dataset is smaller than the KDD'99 dataset. Every entry/record in the dataset consists of 43 unique attributes. The dataset basically has one file for Train data and another for Test data. There are major 4 types of attacks, DoS, R2L, Probe and U2R. The train set contains 23 unique types of attacks and the test set contains another 17 different types of attack. Summarizing the total number of attacks in both the train and test set, there are a total of 148,516 attacks in the dataset. Table 1 below shows the detailed information regarding the type of attacks in the train and test data set as well as the number of attacks pertaining to individual train and test data set of the NSL-KDD dataset.

Table 1. Dataset Overview

Major Attack Category	**Attack Type**		**Quantity of Attack (Training Data)**	**Quantity of Attack (Testing Data)**
	Train data set	**Test data set**		
DoS	land, neptune, teardrop, pod, back, smurf	land, neptune, teardrop, pod, back, smurf, udpstorm, process table, mail bomb, apache2	45927 (37%)	7458 (33%)
R2L	spy, phf, multihop, imap, ftp_write, guess_psswd, warezmaster	spy, phf, multihop, imap, ftp_write, guess_psswd, warezmaster, xlock, named, sendmail, httptunnel, snoop, snmpgetattack	995 (0.85%)	2754 (12.1%)
U2R	perl, rootkit, loadmodule, buffer_overflow	perl, rootkit, loadmodule, buffer_ overflow, worm, ps, snmpguess, xterm, sqlattack	52 (0.04%)	200 (0.9%)
Probe	ipsweep, nmap, Satan, portsweep	ipsweep, nmap, Satan, portsweep, saint, mscan	11656 (9.11%)	2421 (11%)
Normal			67343 (53%)	9711 (43%)
Total			**125973**	**22544**

The feature types in the NSL-KDD dataset can be divided into 4 major categories- Categorical, Binary, Discrete and Continuous. Table 2 shown provides an overview of the 4 Categorical features present. The attributes which are categorical depict that the values are discrete and belong to a specific finite set of categories.

Table 2. Categorical Features

Feature No.	Feature Name	Description	Type	Value Type
F1	Protocol Type	Protocol used in the connection	Categorical	Strings
F2	Service	Service used at destination network	Categorical	Strings
F3	Flag	Connection Status – Normal or Error	Categorical	Strings
F4	Label	Classification of the traffic input	Categorical	Strings

Table 3 shown provides an overview of the 5 Binary features present. These attributes have only two values which are either 0 or 1 where 0 means attribute is absent and 1 means attribute is present and takes the value.

Table 3. Binary Features

Feature No.	Feature Name	Description	Type	Value Type
F5	Land	If source and destination IP addresses and port numbers are equal then, this variable takes value 1 otherwise 0	Binary	Integers
F6	Logged In	Displays login status, 1 for successful login else 0	Binary	Integers
F7	Root shell	Root shell obtained:1 else 0	Binary	Integers
F8	Is Hot Logins	Belonging to 'hot' list then 1 else 0	Binary	Integers
F9	Is Guest Login	Belonging to 'guest' list then 1 else 0	Binary	Integers

Table 4 shown provides an overview of 11 Continuous features present. These attributes have very minimal differences between one value and the next value.

Table 5 shown provides an overview of the 23 Discrete features present. These attributes can have numeric values for the binary features as well as have integer and floating-point values. The difference between discrete and continuous attributes is that continuous attributes come from the infinite set while discrete attributes come from finite or countably infinite.

Feature Engineering

This step is the most crucial step to perform before building the model. To obtain the best accuracy of the detection methods implemented, it is essential to extract the most correlated features. Before extracting the necessary features, the Pearson correlation graph is visualised as shown in Figure 1 to make extractions more robust.

The NSL-KDD dataset used here consists of 42 features so the extraction of features will reduce computational time and enhance the accuracy of the model. For this step, the authors have taken into

Table 4. Continuous Features

Feature No.	*Feature Name*	*Description*	*Type*	*Value Type*
F10	Duration	Length of Connection time duration	Continuous	Integers
F11	Src Bytes	Number of data bytes transferred from source to destination in single connection	Continuous	Integers
F12	Dst Bytes	Number of data bytes transferred from destination to source in single connection	Continuous	Integers
F13	Hot	Number of "hot" indicators in the content such as: entering a system directory, creating programs and executing programs	Continuous	Integers
F14	Num Failed Logins	Count of failed login attempts	Continuous	Integers
F15	Num Compromised	Number of "compromised" conditions	Continuous	Integers
F16	Num Root	Number of "root" accesses or number of operations performed as a root in the connection	Continuous	Integers
F17	Num File Creations	Number of file creation operations in the connection	Continuous	Integers
F18	Num Shells	Number of shell prompts	Continuous	Integers
F19	Num Access Files	Number of operations on access control files	Continuous	Integers
F20	Num Outbound Cmds	Number of outbound commands in an ftp session	Continuous	Integers

consideration the Random forest classifier because it delivers a good predictive rate with less over-fitting. Feature pruning using this method falls under the group of Embedded methods that uses a mixture of filter and wrapper procedures. The method calculates the importance of features depending on their usefulness to anticipate the target value using the inbuilt functions as shown in Figure 2.

Based on the insights obtained from the values of importance and the maximum score obtained, the important features are extracted. The scores associated with each column depict the pertinence in predicting the outcome. The lowest value of any feature shows that they are least relevant for predicting the value. The scores obtained are useful for a variety of reasons such as- improved understanding of the model and input data. Therefore, all the irrelevant attributes that are not related to the prediction of the target variable are removed and the selected features are only 10 as shown in Figure 3 that represents the highest correlation with the prediction column.

ARCHITECTURE

In the present system, first of all, the NSL-KDD dataset is loaded into the environment. Stopwords are also been downloaded into the system. Once the dataset is ready to use, Data PreProcessing is performed on the dataset. Mapping is performed for the attack class. For encoding of the data, OneHotEncoder() is used. After performing the preprocessing on the dataset, to get more insight into the dataset, Exploratory Data Analysis (EDA) is conducted on the data. With the help of StanderScaler() and LabelEncoder(), feature engineering is performed on the dataset. The next step is feature selection from the features that have been administered. For the selection of valuable features, RandomForest Classifier is taken into consideration. The system generates 10 features that will help in producing the most favourable result. Later, the dataset is split into Train and Test set and fitted. For classification of the types of attack, two different types of classifiers are considered, Linear Models and Ensemble Models as shown in Figure 4.

Table 5. Discrete Features

Feature No.	Feature Name	Description	Type	Value Type
F21	Wrong Fragment	Total number of wrong fragments in this connection	Discrete	Integers
F22	Urgent	Number of urgent packets in this connection	Discrete	Integers
F23	Su Attempted	Su root command done: 1 else 0	Discrete	Integers
F24	Count	No. of connections to the same destination host as the current connection in the past 2 seconds	Discrete	Integers
F25	Srv Count	No. of connections to the same service (port number) as the current connection in the past 2 seconds	Discrete	Integers
F26	Serror Rate	The % of connections that have activated the flag – s0, s1, s2 or s3, among the connections aggregated in count	Discrete	Float
F27	Srv Serror Rate	The % of connections that have activated the flag – s0, s1, s2 or s3, among the connections aggregated in srv_count	Discrete	Float
F28	Rerror Rate	The % of connections that have activated the flag REJ, among the connections aggregated in count	Discrete	Float
F29	Srv Rerror Rate	The % of connections that have activated the flag REJ, among the connections aggregated in srv_count	Discrete	Float
F30	Same Srv Rate	The % of connections that were to the same service, among the connections aggregated in count	Discrete	Float
F31	Diff Srv Rate	The % of connections that were to different service, among the connections aggregated in count	Discrete	Float
F32	Srv Diff Host Rate	The % of connections that were to different destination machines, among the connections aggregatedrv_ in scount	Discrete	Float
F33	Dst Host Count	No. of connections having the same destination host IP address	Discrete	Integers
F34	Dst Host Srv Count	No. of connections having the same port number	Discrete	Integers
F35	Dst Host Same Srv Rate	The % of connections that were to same service, among the connections aggregated in dst_host_count	Discrete	Float
F36	Dst Host Diff Srv Rate	The % of connections that were to different service, among the connections aggregated in dst_host_count	Discrete	Float
F37	Dst Host Same Src Port Rate	The % of connections that were to the same source port, among the connections aggregated in dst_host_srv_count	Discrete	Float
F38	Dst Host Srv Diff Host Rate	The % of connections that were to different destinations machines, among the connections aggregated in dst_host_ srv_count	Discrete	Float
F39	Dst Host Serror Rate	The % of connections that have activated the flag – s0, s1, s2 or s3, among the connections aggregated in dst_host_count	Discrete	Float
F40	Dst Host Srv Serror Rate	The % of connections that have activated the flag – s0, s1, s2 or s3, among the connections aggregated in dst_host_srv_ count	Discrete	Float
F41	Dst Host Rerror Rate	The % of connections that have activated the flag REJ, among the connections aggregated in dst_host_Count	Discrete	Float
F42	Dst Host Srv Rerror rate	The % of connections that have activated the flagREJ, among the connections aggregated in dst_host_srv_count	Discrete	Float
F43	Score	Difficulty Level	Discrete	Integers

Figure 1. Pearson Correlation Analysis

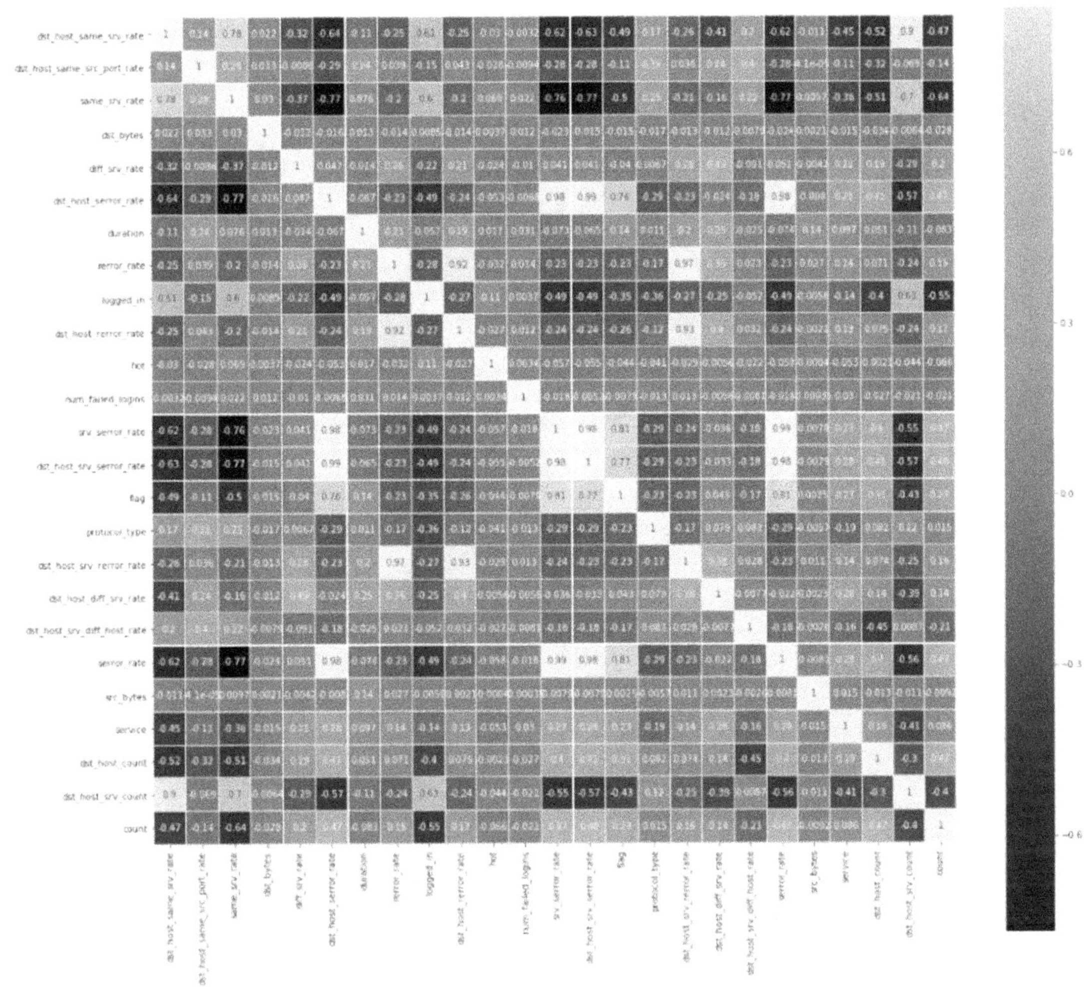

Figure 2. Importance Value of Features

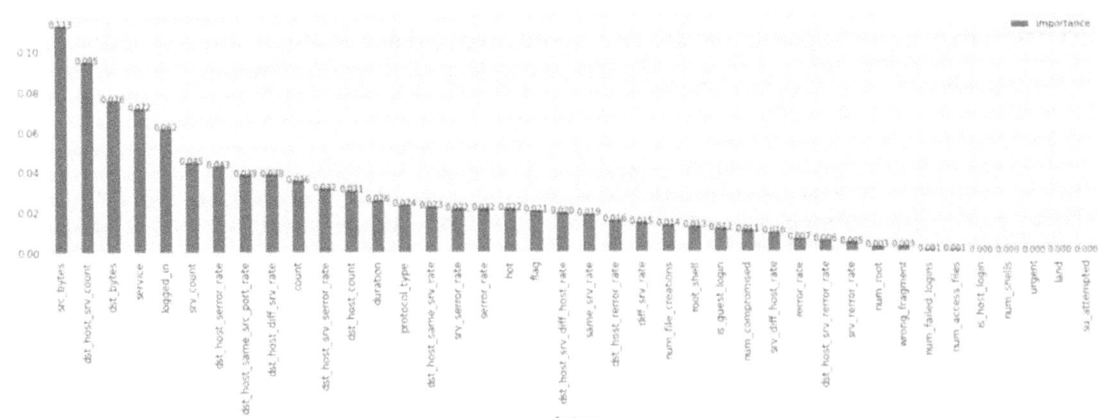

Figure 3. Selected Features

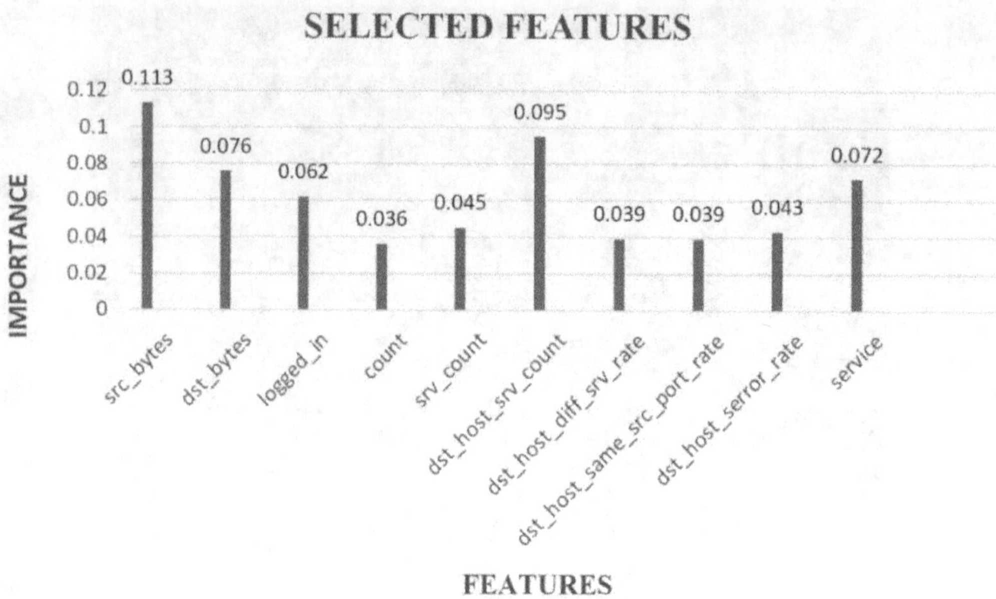

Linear Models

There are various machine learning models available which are continuously used in the prediction task to understand the problem and provide solutions likewise. For linear modeling, the researchers have taken into consideration three models- Logistic Regression (LR), Stochastic Gradient Descent (SGD) and Naïve Bayes Model. These models are generally referred to as "old-school" but they are relatively faster to train and output can be easily interpreted in comparison to the ensemble models.

Logistic Regression (LR)

This linear model is referred to as a logarithm linear model as is easy to implement as well as it can be trained rapidly. The major disadvantage of working with this model is it is prone to overfitting and it doesn't work precisely with non-linear data. This algorithm calculates the possibilities of various classes by parametric logistic distribution. The formula implemented for that is as follows-

Table 6. List of Classifiers Implemented

Linear Model	Ensemble Model
1. Logistic Regression	1. LGBM Model
2. SGD Classifier	2. XGBoost Classifier
3. Naïve Bayes	3. Stacked Model

Figure 4. Architecture of the proposed system

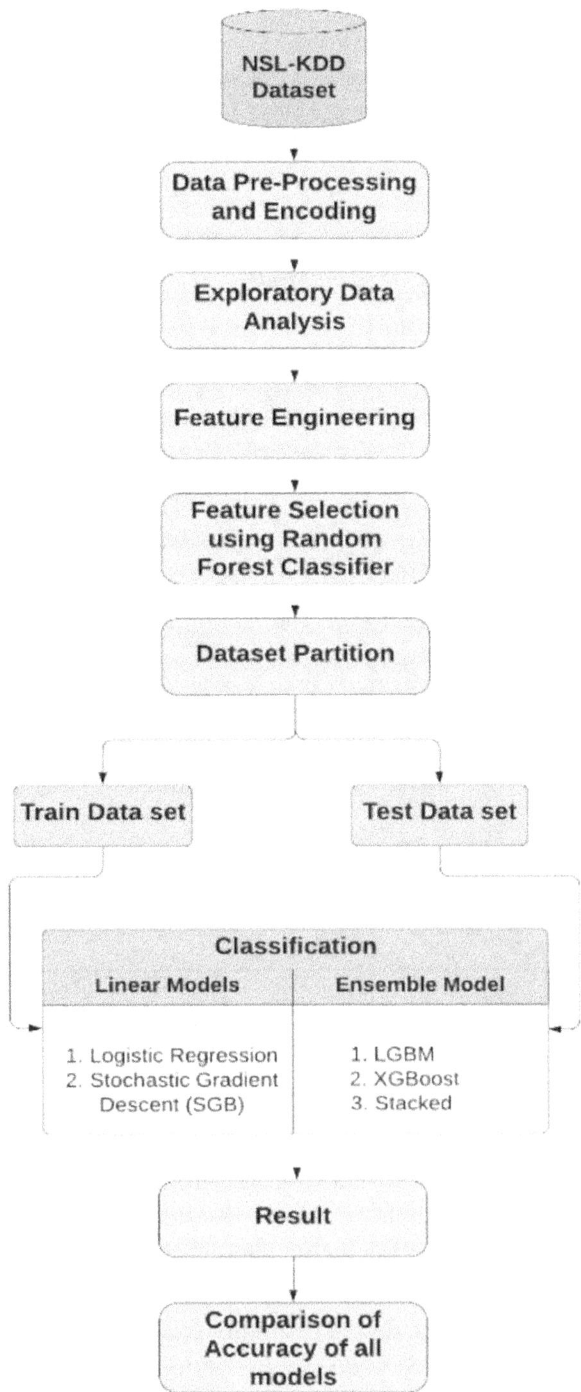

$$P(Y = k \mid x) = \frac{e^{\omega_k * x}}{1 + \sum_k^{k-1} e^{\omega_k * x}}$$

From the confusion matrix shown in Figure 5, it is clear that the wrongly classified outcomes are 2580 from a total of 1,34,686. In addition to the correct and wrong predictions, we have also visualised the precision, recall, F1-score and support for all the classifiers.

Figure 5. Confusion Matrix of Logistic Regression Model

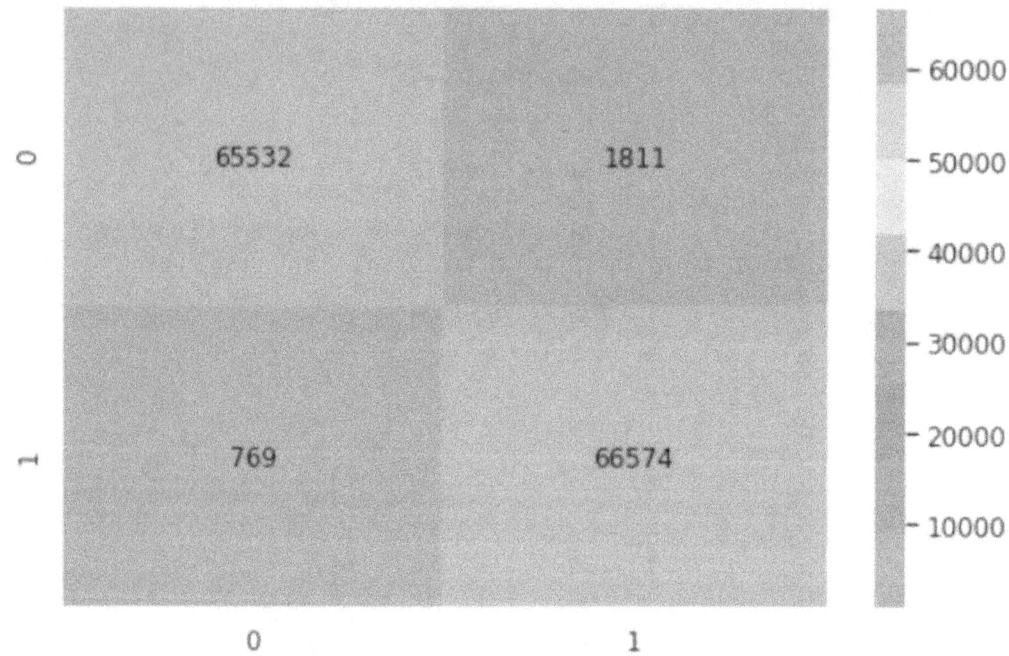

Stochastic Gradient Descent (SGD)

It is a recursive procedure that is used to calculate derivate from all training data samples and append it immediately. In this classifier, some samples are selected instead of taking the whole dataset as input for each iteration. A sum of samples from the whole dataset is used to perform calculations for each iteration and that is called a batch. The entire dataset can be taken as input but the disadvantage occurs when the dataset is huge so the sample is always randomly selected from the entire dataset to reduce the computational time.

for *i* in *range* (*m*):

$$\theta_j = \theta_j - \alpha\left(\hat{y}^i - y^i\right)x_j^i$$

From the confusion matrix shown in Figure 6, it shows that the wrongly classified items are 2565 and the drawbacks of this classifier include it needs various hyperparameters and repetitions to train and it is tactful towards feature pruning and scaling so attention is needed during that phase.

Figure 6: Confusion Matrix of Stochastic Gradient Descent Model

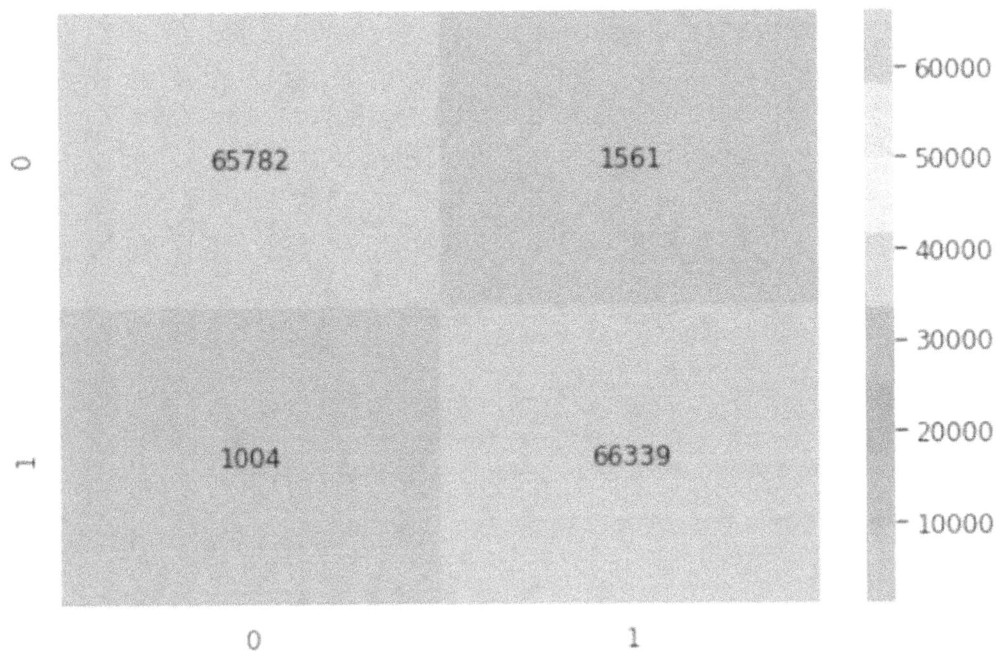

Naïve Bayes

It is a probabilistic ML model that is used widely for classification tasks. Using this theorem, it can find the probability of an attack taking place given that previously the attack has already taken place. In simple terms, the model predicts the outcome of A event happening given that event B has already taken place before. The algorithm takes one assumption into account that all predictors and features are independent so the presence of any one feature does not particularly affect the other feature. The calculation for each class is done differently and the conditional probability class is classified then into maximum probability class. For the calculation of each class, conditional probability formula is used which is as follows

$$P\left(X = x \mid Y = c_k\right) = \prod_{i=1}^{n} P\left(X^{(i)} = x^{(i)} \mid Y = c_k\right)$$

From the confusion matrix shown in Figure 7, it shows that the wrongly classified outcomes are 3533 which is greater than LR and SGD models. So, this classifier works poorly in comparison to the other two because the data we are taking as input is attribute related data hence it is clear that because the classifier relies solely on the discrete data, it is not an empirical approach for intrusion detection.

Figure 7. Confusion Matrix of Naive Bayes Model

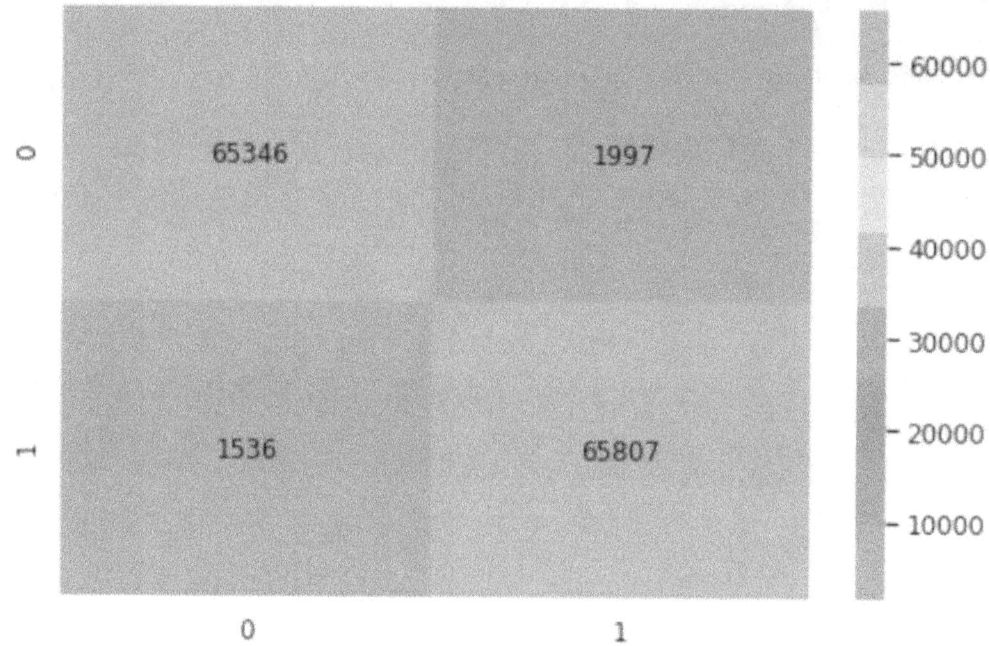

Ensemble Models

Combining different linear or non-linear models together to enhance the overall performance of the system, is the ultimate goal of Ensemble Modeling. More precision of the model can be obtained by the use of ensemble modeling. There can be two types of ensemble models:1) Homogeneous and 2) Heterogeneous. In the Homogeneous Ensemble model, the base learners are of a similar type whereas the Heterogeneous Ensemble model uses different types of base learners. To implement these ensemble models, different types of techniques are used. Two techniques that will be covered in this chapter is 1) Boosting and 2) Stacking

Boosting Technique

Boosting is a sequential technique wherein, models are added at each stage/level of the ensemble modeling. The role of each model at every stage is to solve the errors of the previous stage models and eventually enhance the performance of previous models at the next stage. The succeeding models depend on the preceding models. The main advantage of boosting is that any weak learning model can be made strong by placing the weak learner into the boosting techniques. If these weak learning models are tested alone on the entire dataset then they would not give such good output. Hence boosting algorithms combines different weak learners to form a strong learner.

Light Gradient Boosting Machine (LGBM)

When it comes to dealing with large datasets, all the classifiers such as SVM, GBDT, Logistic regression and many such would end up becoming very slow in the training phase. In this scenario, LGBM classifier is the most suitable one. The standout features of LGBM are higher speed in training the dataset along with higher accuracy in predicting the outcomes and it also supports the GPU and distributed learning. LGBM is a gradient boosting technique that uses a tree-based algorithm. Rather than following a level approach, LGBM follows a leaf-wise approach in training the data and predicting the outcome.

In the experimental setup to detect intrusion in the network, the NSL-KDD dataset was trained on the LGBM Classifier. It uses Gradient-based One-sided sampling in order to filter out the samples. The total number of leaves set for the classifier are 4 and 42 random states were used. Figure 8 shows the confusion matrix developed from the experimental results.

Figure 8. Confusion Matrix of Light Gradient Boosting Machine Model

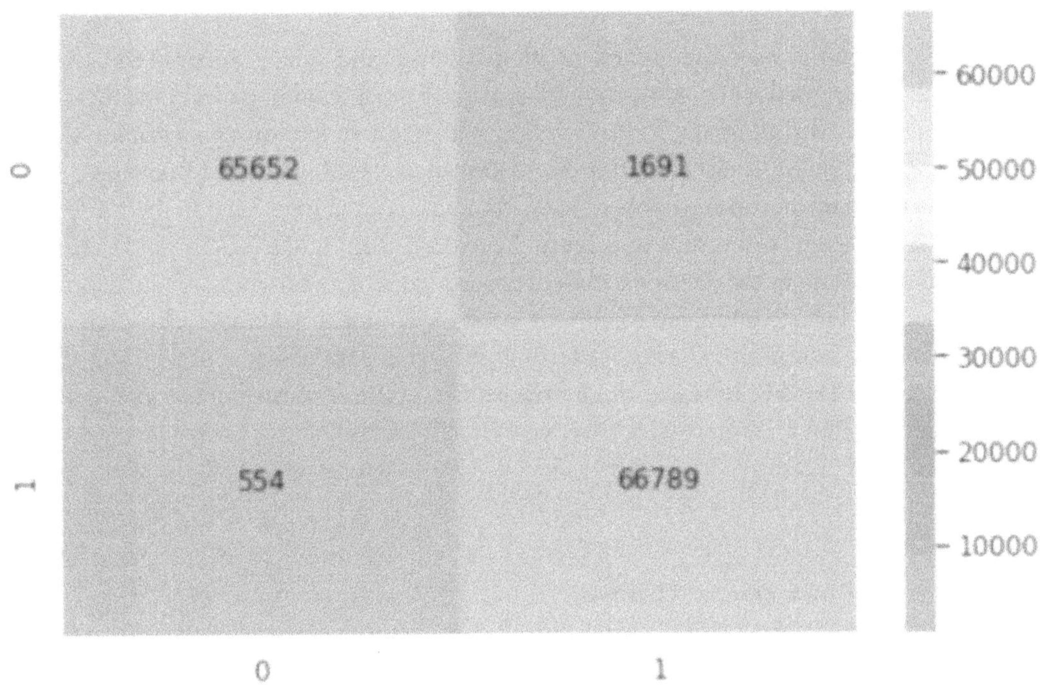

It can be observed from the confusion matrix shown in Figure 8 that there are a total of 2245 wrongly classified outcomes. Hence, this classifier has shown better performance than the other models and also less learning time.

In the testing phase of the LGBM classifier, the parameters that are shown in Table 7 were used. Total 4 number of leaves were used for the LGBM classifier and a random state is assigned 42 so every time it would generate the same output. These set of parameters proved to give better results than the linear models.

Table 7. Optimal Hyperparameters for LGBM

Hyperparameters	Optimal Value
num_leaves	4
max_depth	100000
subsample	0.4
min_child_samples	10
learning_rate	0.001
n_estimators	200
random_state	42

Extreme Gradient Boosting (XGBoost)

XGBoost classifier is said to have the highest predictive power and it is considered to be almost 10 times faster than the other gradient boosting technique. It has a great feature to overcome the overfitting problem. To remove the overfitting, the XGBoost classifier uses a variety of regularization techniques and for the same reason, the XGBoost classifier is also known as the 'Regularized Boosting' technique. This technique also implements an ensemble of decision trees. All the trees used in the XGBoost classifier are aggregated through their results to achieve the overall score of the XGBoost classifier.

In the experimental setup of the XGBoost classifier used in the network intrusion detection system, 10 boost rounds are used and 10 stopping rounds are used for its implementation. This classifier gives the best score than any other model. During the testing phase, the parameters applied for the XGBoost classifier are as listed in Table 8. By using these parameters, XGBoost turned out to give the best performance compared to other classifiers which were previously used.

Table 8. Optimal Hyperparameters for XGBoost

Hyperparameters	Optimal Value
nthread	1
eta	0.01
max_depth	3
min_child_weight	1
subsample	0.1
colsample_bytree	0.1

Stacking Technique

Compared to boosting, in Stacking technique we can use different base models to create one new model which gives overall better accuracy. Initially all the individual base models are trained separately on the available dataset. And then to make the final prediction, the predictions of the base classifier models are used as input to the new combined model classifier.

Stacked Model

To create one mixed classifier, Logistic Regression classifier, SGD classifier, Naïve Bayes classifier, LGBM classifier and XGBoost classifier are used as the base classifiers. The data trained from all these classifiers are used as input to the mixed Stacked model. As an outcome, the overall score of the stacked model increased than any other model.

RESULTS

In this chapter, the presented system has been explained from the initial point (data-set) to the final point (the result of each model). The steps of implementation of the project are:

- **Data-set:** NSL-KDD dataset that is an improved version of the KDD dataset and does not contain redundant attributes.

Intrusion Detection

In the dataset, there were 43 features initially. Among them, some columns contained all null values and could not be useful to detect intrusion, so researchers performed feature engineering by first converting the attributes to numerical before feeding it to the model, one-hot encoding and label encoding were used to implement it. After converting the attributes, visualization of Pearson Correlation is performed and then Random Forest classifier is implemented to obtain the necessary features that can be beneficial to implement the model. This classifier is frequently implemented in Data Science because the tree-based developments used by random forest genuinely rank high on how they enhance the purity of the node. These 12 features were selected for the final implementation of the model. ['src_bytes', 'dst_bytes', 'logged_in', 'count', 'srv_count', 'dst_host_srv_count', 'dst_host_diff_srv_rate', 'dst_host_same_src_port_rate', 'dst_host_serror_rate', 'service']

Performance Metrics: The performance for each model can be known by classification report and confusion matrix. A classification report is created for each model to better understand the quality of predictions the model is generating. Specifically, it provides clear insights on True Positive (TP), True Negative (TN), False Positive (FP) and False Negative (FN). Confusion Matrix is a table as shown in Table 9 that is used to understand the performance of a model on a set of test data for which the true values are already known.

1. **TN / True Negative:** This represents the case for which the classifier has predicted data as 'normal data' and the data are actually 'normal'

Table 9. Confusion Matrix (also called Error Matrix)

		Predicted Class	
		Class = Yes	**Class = No**
Actual Class	**Class = Yes**	True Positive	False Negative
	Class = No	False Positive	True Negative

2. **TP / True Positive:** This represents the case for which the classifier has predicted data as 'attack data' and the data is actually 'attack data'.
3. **FN / False Negative:** This represents the case for which classifier has predicted data as 'normal data' but the data is actually 'attack data'
4. **FP / False Positive:** This represents the case for which classifier has predicted data as 'attack data' but the data is actually 'normal'

The Classification report generated for each model consists of the following metrics: -

1. Accuracy.
2. Recall.
3. Precision.
4. F1-Score.
5. Support

Accuracy: - It is the ratio of accurately predicted instances to the total instances as. It is written mathematically as: Accurately classified samples/ Total number of samples.

So, = TP+TN/ TP+TN+FP+FN

Recall: - It is also referred to as sensitivity. It is the ratio of accurately predicted positive instances to the total instances in a class. In simple terms, it is the fraction of positives that were correctly identified.

So, Recall = TP/TP+FN

Precision: -It is one of the basic performance indicators obtained from the classification report. It presents the complete number of instances that are correctly classified as attack divided by a total number of records classified as an attack. The precision can be obtained mathematically from the below equation.

So, Precision = TP/TP+FP

F1-Score: - It is considered a weighted harmonic mean of precision and recall taking into consideration the best score as 1.0 and worst as 0.0. If there is a class distribution and that is not even then in such circumstances the value of F1 would be greater than the accuracy. In some cases, if the values of FP and FN are too far apart or very different, it is better to look at the values obtained from Precision and Recall.

So, F1 Score = 2*(Recall * Precision) / (Recall + Precision)

Support: - It is the number of samples of the true response that lie in a specific class. Disparity obtained in this score during training indicates that there is some weakness in the score reported by the respective classifier and it shows that there is a need to perform stratified sampling.

Table 10 shows the classifiers trained and the precision, recall and F1-score for all the classifiers. In general, the authors observed the highest TP rate in the classification report for XGBoost and Stacked Model. Also, the lowest TP rate was noted in Naïve Bayes Model so it indicates that the classifier suffers from a high FP value and hence a huge number of normal packets that are not generating intrusion are classified as attack packets. Overall, it can be said that besides accuracy value TP, FP, FN, TN and classification report holds an important place to understand the performance of the model.

Table 10. Performance Metrics of Models Implemented

Column	ML Classifier	Precision	Recall	F1- Score
A	Logistic Regression (LR)	98%	97%	98%
B	Stochastic Gradient Descent (SGD)	98%	98%	98%
C	Naïve Bayes	98%	97%	97%
D	LGBM	99%	97%	98%
E	XGBoost	99%	99%	99%
A+B+C+D+E	Stacked Model	99%	99%	99%

But taking into consideration the recent innovations in network intrusion detection using machine learning approaches, it shows that the most crucial parameters to evaluate the model are FP and FN. Any classifier implemented must possess minimal value for these parameters. The researchers have worked on minimizing the values to obtain the best accuracy of the classifier.

As Figure 9 shows, the FP and FN values of the classifiers implemented, it is clear that XGBoost and Stacked Model have the least FP and FN values so they can be considered more accurate than other classifiers. Moreover, the Naïve Bayes classifier shows the highest amount of FP and FN values so it can be considered least accurate because there is an exorbitant amount of wrongly classified instances.

Table 11 summarizes the accurately classified instances and incorrectly classified instances for each model. It shows that XGBoost and Stacked Model works best in comparison to other classifiers because there are the least number of wrongly classified instances with the smallest FP and FN values. All the ML classifiers implemented showed notable accuracy and precision in detecting normal packets. Both the Stacked and XGBoost model outperforms the accuracy obtained from other models. Out of all six models, it is observed that the Naïve Bayes classifier works poorly with the highest FP and FN rates. Figure 10 shows the graphical comparison of the performance of the six models. It can be said that the better performance of ensemble models over linear models is primarily due to their power for parallel and distributed processing and efficiency to get trained over large datasets. Also stacked model takes less computational time than linear models, so there is the optimal usage of hardware and memory.

Figure 9. FP and FN Values

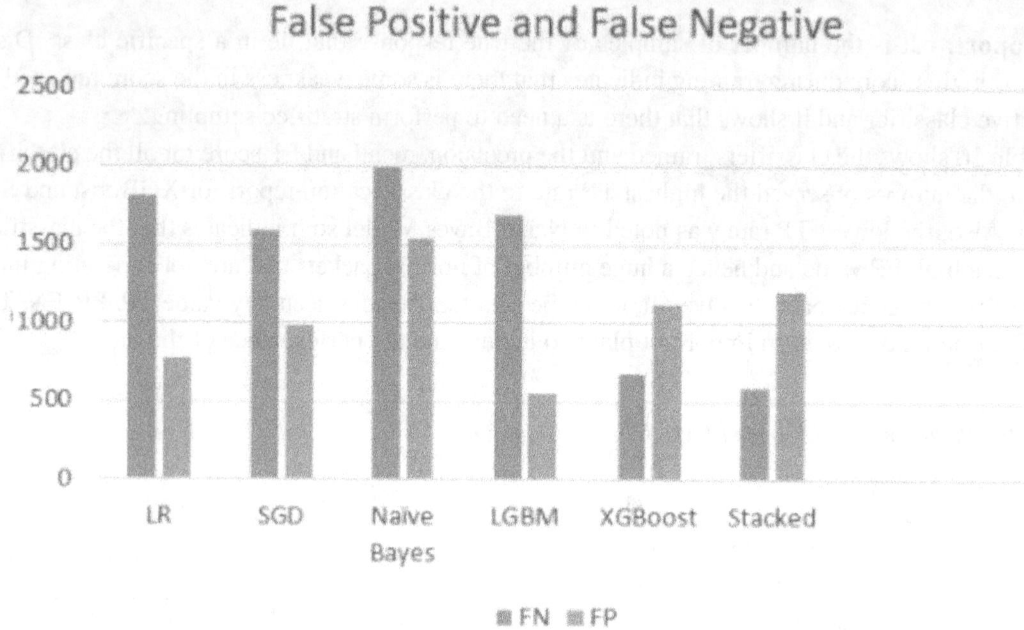

Pros and Cons of Metrics Measurement

There are mainly two downsides of only considering Accuracy for all the models and not diving into other statistics. Firstly, when data has two or more classes then it may already give the accuracy of 80% but then it is not possible to judge that all classes are predicted properly or the model is trained with bias and is able to detect only two classes out of four. Second, if the dataset is unbalanced such as there are no even number of classes then also there is a high probability of the model achieving high accuracy by always predicting the most common class. So, besides accuracy, it is necessary to consider other parameters too such as Confusion Matrix and metrics (Precision, Recall, F1 score) based on Confusion Matrix to do a comprehensive study on which model is performing accurately. By using these parameters to evaluate the model it accesses the problem without any bias and even handles the problem of

Table 11. Accuracy of Classifiers

ML Classifier	Correctly Classified Instances	Wrongly Classified Instances	Accuracy
Logistic Regression (LR)	132106	2580	98.08%
Stochastic Gradient Descent (SGD)	132121	2565	98.098%
Naïve Bayes	131153	3533	97.37%
Light Gradient Boosting Machine (LGBM)	132441	2245	98.3%
XGBoost	132885	1801	98.6%
Stacked Model	132883	1803	98.6%

Figure 10. Graphical Comparison of Accuracy of all Classifiers

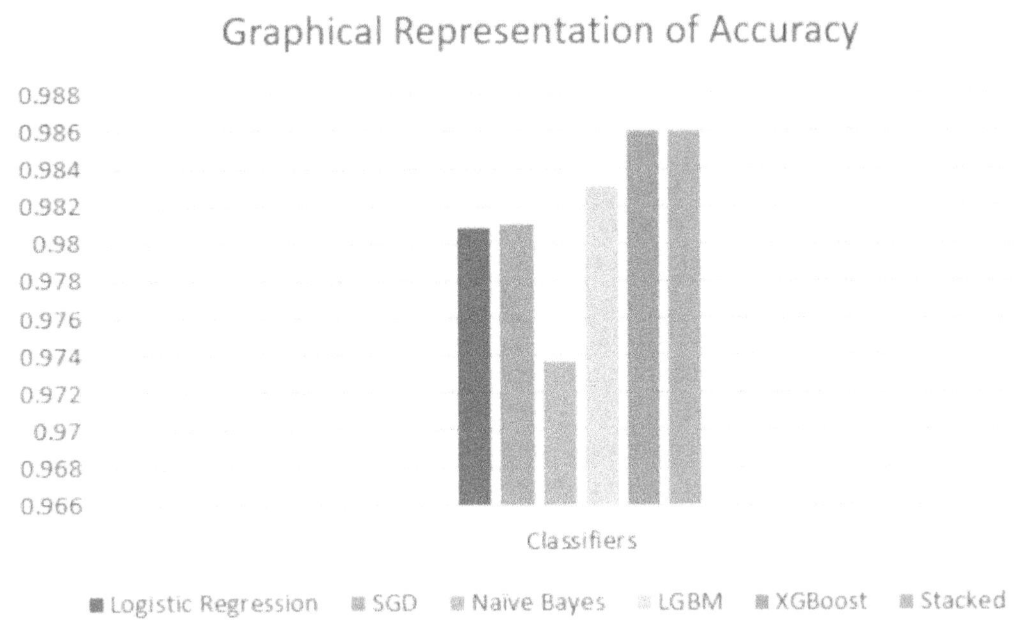

the unbalanced dataset. Accuracy is preferred in cases where there is a balanced dataset with even class distribution while an F1 score is more preferable where there is uneven class distribution. Furthermore, for detecting the intrusions and wrongly classified instances it is vital to understand FP and FN rate and F1 score is more preferable when FP and FN rate are considered while on the contrary if TP and TN rate are more important then Accuracy metric is taken into consideration.

Comparison of various ML Algorithms used for IDS

The XGBoost and Stacked Model achieved the highest accuracy rate 98.6% with the smallest RMSE value and false positive rate and regarding the average accuracy rate, there is no huge difference between LR and SGD classifier. The Naive Bayes classifier did not reach optimum accuracy but it has a low demand for training the model. So from the statistics and comprehensive study, it can be deduced that the Stacked and XGBoost model presents an acceptable accuracy with the lowest FN rate which augments the confidentiality of network resources taken into consideration. In order to remove irrelevant features from the dataset and preprocess it efficiently, a stacked model has been created which helps to train the model better and detect the intrusion taking place in the network. To better analyse the models, computational time and performed statistics have been calculated by calculating Accuracy, Precision, Recall and F1 score along with FP, FN, TP and TN rate. So, the model with the least accuracy Naive Bayes takes less computational time in comparison to other classifiers but the wrong classified instances also get increased. Similarly XGBoost and Stacked Model required a higher amount of time to complete computations but it performed effectively with the least false-negative and wrong classified instances.

DISCUSSION

Cybercrime is increasing each and every day so it is very important to protect computer systems and other wireless devices by using advanced IDS systems to reduce security breaches. In order to develop an accurate IDS system, it is essential to study in-depth the data analysis on wireless communication to deploy a model with precision. Every day billions of data is generated by mankind and wireless devices and so it is very difficult to maintain these large data pools without leveraging analytics. Data Analysis for intrusion detection is gaining more attraction because it promotes deep study of large unstructured data from multiple sources, overcoming cyberattacks and detecting anomalies in networks. Besides this the analysis on data obtained from wireless networks is beneficial to understand network traffic and any virus signature. It can even calculate the underlying correlations of attributes from heterogeneous sources and enhance security by monitoring. In today's era data analysis has become an unavoidable step and is becoming the most agile way to gain useful knowledge and improve the security in wireless networks. Performing data analysis before deploying IDS can help to retain the value of the system for longer duration and is flexible to changes and other improvements whenever needed because data might keep changing or augmenting as time passes.s

CONCLUSION

This chapter provides a thorough study of linear and ensemble classifiers that on implementation give better solutions to intrusion detection in the network. It is clear that security breaches are increasing day by day so to prevent cyber crimes and intrusions it is necessary to deploy the intrusion detection system. Detecting attacks into the network has become a crucial part of any firm that is handling an enormous amount of data online. The chapter focuses on linear models -Logistic Regression, Stochastic Gradient Descent, Naive Bayes and ensemble models- Light Gradient Boosting Machine, XGBoost and Stacked Model that is a mixture of linear and ensemble models proposed.

Overall, XGBoost and Stacked Model outperformed other models discussed by achieving the highest accuracy rate of 98.6% for detecting DOS, Probe, U2R and R2L attack categories and achieves the lowest FN rate and Naive Bayes classifier achieves the lowest accuracy and maximum FN and FP rate for the detection of intrusions. Therefore, it can be concluded that XGBoost and Stacked Model can be practically taken into consideration for the detection of intrusion in the network. Using these models, better analytics can be performed on wireless networks.

REFERENCES

Ahmad, Z., Khan, A. S., Shiang, C. W., Abdullah, J., & Ahmad, F. (2020). *Network intrusion detection system: A systematic study of machine learning and deep learning approaches*. https://onlinelibrary.wiley.com/doi/full/10.1002/ett.4150

Ahmed, M., Mahmood, A., & Hu, J. (2016). *A survey of network anomaly detection techniques*. https://www.sciencedirect.com/science/article/abs/pii/S1084804515002891

Alharbi, A., Alhaidari, S., & Zohdy, M. (2018). *Denial-of-Service, Probing, User to Root (U2R) & Remote to User (R2L) Attack Detection using Hidden Markov Models*. https://www.ijcit.com/archives/volume7/issue5/IJCIT070501.pdf

Amato, F., Mazzocca, N., Moscato, F., & Vivenzio. (2017). *Multilayer Perceptron: An Intelligent Model for Classification and Intrusion Detection*. https://ieeexplore.ieee.org/document/7929765

Britel, M. (2018). *Big Data Analytic for Intrusion Detection System*. https://ieeexplore.ieee.org/document/8610578

Canêdo, D., & Romariz, A. (2019). *Data Analysis of Wireless Networks Using Classification Techniques*. https://arxiv.org/abs/1908.07329 doi:10.5121/csit.2019.90905

Chao-yang, Z. (2011, August 1). *DOS Attack Analysis and Study of New Measures to Prevent*. https://ieeexplore.ieee.org/document/5997473

Dai, H., Wong, R., Wang, H., Zheng, Z., & Vasilakos, A. (2019). *Big Data Analytics for Large-scale Wireless Networks: Challenges and Opportunities*. https://dl.acm.org/doi/fullHtml/10.1145/3337065

Fan, Y., & Zhang, R. (2014). *Research on Network Security and Identity Authentication*. https://www.scientific.net/AMR.926-930.2046

Farnaaz, N., & Jabbar, M. (2016). *Random Forest Modeling for Network Intrusion Detection System*. doi:10.1016/j.procs.2016.06.047

Hamid, Y., Sugumaran, M., & Journaux, L. (2016). *A Comparative Analysis. Machine Learning Techniques for Intrusion Detection*. https://dl.acm.org/doi/10.1145/2980258.2980378

Jha, J., & Ragha, L. (2013). *Intrusion Detection System using Support Vector Machine*. https://research.ijais.org/icwac/number3/icwac1342.pdf

Kibria, M., Nguyen, K., Villardi, G., Zhao, O., Ishizu, K., & Kojima, F. (2018). *Big Data Analytics, Machine Learning, and Artificial Intelligence in Next-Generation Wireless Networks*. https://ieeexplore.ieee.org/document/8360430

Krishna, P., Yenduri, S., & Ariwa, E. (2020). *Data analytics in wireless systems and IoT issues and challenges*. https://onlinelibrary.wiley.com/doi/full/10.1002/dac.4522

Kumar, S., Viinikainen, A., & Hamalainen, T. (2016). *Machine learning classification model for Network based Intrusion Detection System*. https://ieeexplore.ieee.org/document/7856705

Lahre, M. K., Diwan, M. T., Kashyap, S., & Agrawal, P. (2013). *Analyze Different approaches for IDS using KDD 99 Data Set*. https://www.academia.edu/4823609/Analyze_Different_approaches_for_IDS_using_KDD_99_Data_Set

Othman, S. M., Ba-Alwi, F. M., Alsohybe, N. T., & Al-Hashida, A. Y. (2018). *Intrusion detection model using machine learning algorithm on Big Data environment*. doi:10.1186/s40537-018-0145-4

Rai, M., & Mandoria, H. (2019). *Network Intrusion Detection: A comparative study using state-of-the-art machine learning methods*. https://ieeexplore.ieee.org/document/8977679

Sarumi, O., Adetunmbi, A., & Adetoye, F. (2020). *Discovering computer networks intrusion using data analytics and machine intelligence*. https://www.sciencedirect.com/science/article/pii/S2468227620302386

Sharmila, B., & Nagapadma, R. (2019). *Intrusion Detection System using Naive Bayes algorithm*. https://ieeexplore.ieee.org/document/9019921

Tavallaee, M., Bagheri, E., Lu, W., & Ghorbani, A. (2009). *A detailed analysis of the KDD CUP 99 data set*. https://ieeexplore.ieee.org/document/5356528

Yin, C., Zhu, Y., Fei, J., & He, X. (2017). *A Deep Learning Approach for Intrusion Detection Using Recurrent Neural Networks*. https://ieeexplore.ieee.org/document/8066291

Chapter 4
5G in Healthcare:
Features, Advantages, Limitations, and Applications

Vijay Prakash
Thapar Institute of Engineering and Technology, India

Lalit Garg
https://orcid.org/0000-0002-3868-0481
University of Malta, Malta

Luke Camilleri
University of Malta, Malta

Joseph Curmi
University of Malta, Malta

Darren Camilleri
University of Malta, Malta

ABSTRACT

5G is a new universal wireless standard, a new form of mobile network engineered to bring everyone and everything virtually together. 5G is not only for mobile phones, but it is also the foundation for virtual reality (VR), the internet of things (IoT), and autonomous driving, connecting many electronic devices to the internet. Having good healthcare is very important as it affects all parts of human life and social well-being. Moreover, it is crucial to have a great healthcare system if we want economic growth, workforce productivity, and society to advance. Despite all the hard work done by scientists and medical professionals, today's healthcare is mainly inefficient, and a significant overhaul is required. This chapter discusses the primary advantages, including the 5G's main features in healthcare and their limitations and probable solutions and applications to the latest scenario.

DOI: 10.4018/978-1-7998-6988-7.ch004

INTRODUCTION

5G is defined as the 5th generation technology for mobile networks. 5G is a new universal wireless standard, a new mobile network engineered to bring virtually everyone and everything together. This infrastructure is designed to bring consumers higher data rates of multi-gigabits speed, low latency, better network capability, enhance connectivity, and better reliability. A few decades ago, in the 1980s, the first generation of mobile wireless communications was initiated, and from that day, it continued to evolve gradually approximately every decade (Hossain, 2013; Jain, 2016).

The first generation of commercial cellular networks used analogue signals. However, it faced numerous issues as at that time, mobile phones had poor battery life and were more prominent in size and were inconvenient. 2G emerged in the early 1990s and used digital signals instead of analogue and introduced a new digital tool called Global System for Mobile (GSM) (Peersman, Cvetkovic, Griffiths, & Spear, 2000). GSM supports features like conference calls and SMS etc. In the early 2000s, 3G started to take over the scenario. 3G was developed to improve frequency capacity and data transmissions. With 3G features like sharing photos, engaging on social media platforms, downloading video, and many more became possible (Korhonen, 2003). 4G was introduced at the end of 2010, and with it came the introduction of Long-Term Evolution (LTE). Standards included in this generation are faster speeds and higher quality, better security, and lower costs for data usage (Cox, 2012). In 2018, 5G started to penetrate the market to improve standards and address the growing Internet of Things where networks can serve their communication needs for the billions of devices connected (Eze, N. O. Sadiku, & M. Musa, 2018). As more people are getting connected to the internet, 4G is starting to reach its peak, and it will not be able to handle a sizable quantity of users connected all at once. This technology will be capable of handling 1000 times more traffic and up to 10 times faster. 5G is not only for mobile phones, but it is also the foundation for VR, IoT, and autonomous driving, making many electronic devices connected to the Internet (Schulz et al., 2017).

Emerging technologies (Paramita, Bebartta, & Pattanayak, 2021) act as the backbone of 5G technology, including small cell, millimetre waves and Beamforming (Agyapong, P. K., Iwamura, M., Staehle, D., Kiess, W., & Benjebbour, 2014). With 5G, the spectrum is opened to shorter waves, known as millimetre waves which will support faster data speeds to more users at the same time due to the high-frequency waves. This will enable more data bandwidth, with a higher performance where people can send and receive concurrently immeasurable data volume. 5G waves are of shorter wavelength and, therefore, high frequency. These millimetre waves cannot travel very far or even through obstacles, and thus there is the need for small cell networks. These small cell networks use lower power small base stations instead of large high-power towers to cover large distances. These are closer together, transmitting signals like a relay team around obstacles (Qiao et al., 2015). The user's device can switch to different base stations closest to his device, which will enable him to keep the connection.

Moreover, Multiple-Input and Multiple-Output (MIMO) will have an increased capacity of around a hundred ports which will broadcast information in every direction at once. This could lead to interference. Hence, to avoid disruption, 5G uses beamforming technology to efficiently aim and focus precisely streams of data transmissions rather than transmitting signals everywhere. Base stations will be able to support more data at once. For this technology to succeed, devices would need a 5G radio chip to connect to the 5G network. Major Smartphone developers plan or have already released devices that support 5G connectivity (Lee et al., 2018). MIMO has become a promising technology for 5G systems because of its enormous spectrum capacity and low power consumption. On the other hand, Pilot contamination

(PC) severely inhibits the performance of MIMO systems. As a result, two pilot scheduling strategies have been presented, such as Fractional Pilot Reuse (FPR) and Asynchronous Fractional Pilot Scheduling Scheme (AFPS). These scheduling strategies are beneficial to minimize significantly the number of PCs utilized by PCs in uplink time division duplex (TDD) gigantic MIMO systems. Customers are assigned to the central cell and edge cell depending on their signal-to-interference-plus-noise ratio (SINR) (Zahoor et al., 2020). 5G isn't just about speed; it has the potential to transform the whole healthcare system by overcoming all of the limits of today's wireless networking technologies while also providing a fantastic user experience (Dananjayan & Raj, 2020).

Having good health care is very important as it affects all parts of human life and social well-being. Moreover, it is crucial to have a great healthcare system if we want economic growth, workforce productivity, and society to advance. Unfortunately, despite all the hard work done by scientists and medical professionals, today's healthcare is mainly inefficient, and a significant overhaul is required (Latif, Qadir, Farooq, & Imran, 2017). The current healthcare system has four main flaws. Firstly, it is not patient-centric; the lack of convenience for patients could neglect the attention given to routine and appropriate health checks from practitioners. Secondly, the system is not personalized in line with the individual patient, where doctors administer medicines based on population averages instead of human characteristics. Thirdly, it is not equitably accessible, and this access constraint to primary healthcare can cause unfortunate health outcomes to patients. Lastly, it is not data-driven, resulting in errors that could have been prevented (Haskell, 2020). The main research contribution to this chapter is the primary advantages, including the main features of 5G in healthcare and limitations and applications considered to the current scenario.

LITERATURE REVIEW

Nowadays, Information and Communication Technologies have evolved rapidly, and it has introduced a lot of disruptive and emerging technologies. These are technologies that are still being developed and are disrupting industries. Emerging technologies are evolving every day, and one place where they are being implemented is in healthcare. The following library databases were accessed: ACM digital library, Springer, IEEE Xplore digital library, Science Direct, Scopus, Wiley Online Library and Metalib. Google Scholar is used to pointing out the specific journals databases not previously mentioned. The current literature also searched theses on the topic "Healthcare" and "Introduction to 5G." The searching keywords used were "Healthcare", "5G", and "Benefits of Good Healthcare", "5G in Healthcare", "Advantages of 5G in Healthcare", and "Limitations of 5G in Healthcare".

One of the most innovative and game-changing emerging technologies is 5G. This technology is the fifth invention of mobile communications technology; it is a breakthrough innovation that can provide 1000 times' increased capacity and 10–100 times higher data rate, with low latency than preceding technologies (Sharma et al., 2018). Furthermore, it can also support many connected devices compared to today's mobile technology systems. Google Trends has shown the increasing popularity of 5G in healthcare during the last years ("Google Trends," 2021) (see Figure 2), during the past three months (see Figure 3), and during the past seven days (see Figure 4).

This will bring new healthcare opportunities as medical professionals require high-speed internet connectivity to video call patients for telemedicine (Baker & Stanley, 2018). Nonetheless, high speeds are not always accessible due to the inaccessibility of current frequency resources caused by the present

Figure 1. PRISMA flow diagram for the related survey articles

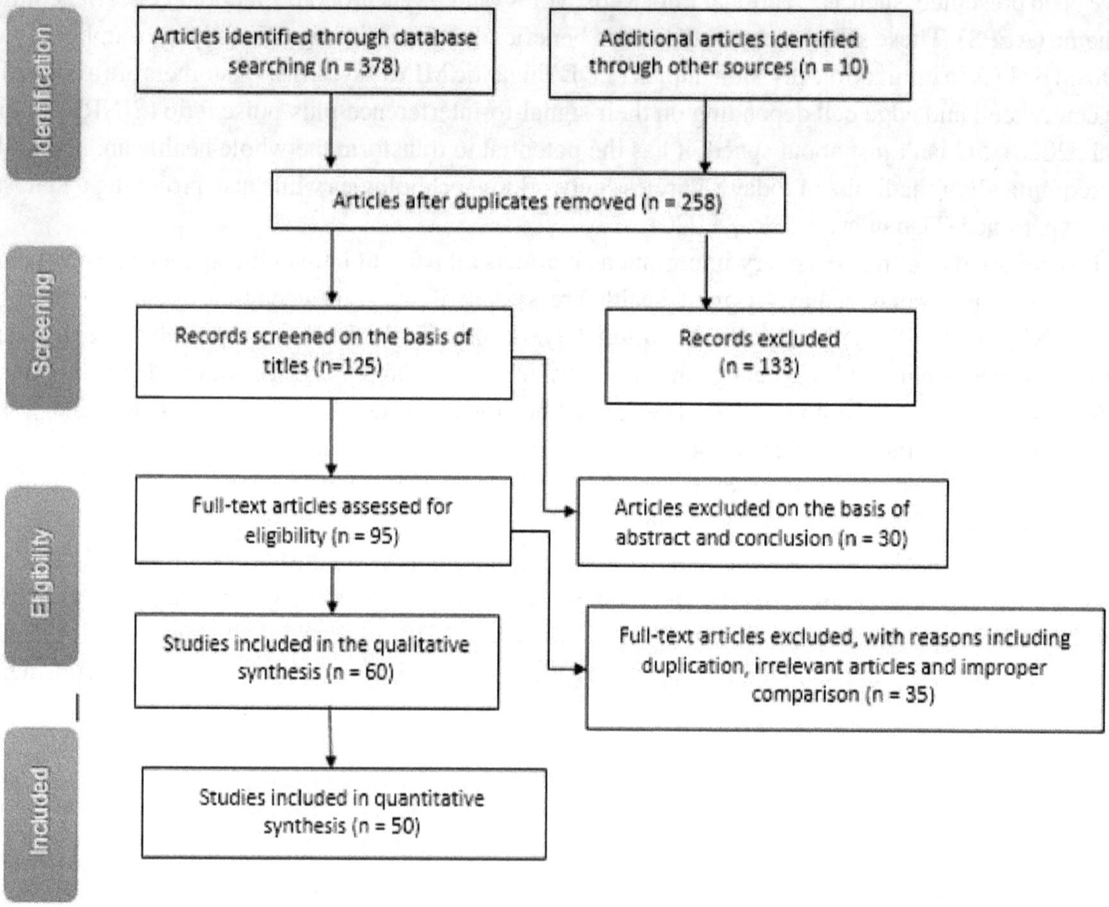

fixed allocation policy and fragmentation of the spectrum (Marković, 2017). 5G uses high carrier frequencies with massive bandwidths. Therefore they will require a vast number of base stations and antennae (Andrews, J. G., Buzzi, S., Choi, W., Hanly, S. V., Lozano, A., Soong, A. C., & Zhang, 2014; Zaidi et al., 2016). A popular way 5G increases speed is due to MIMO (multiple in multiple out) technologies (Bogale & Le, 2016). This technology today is still in its early stages, and many prospects lie ahead. In healthcare, 5G is changing how physicians meet patients, make a diagnosis and provide treatment.

Additionally, this technology offers personalized and preventive healthcare. Hence, this high-speed network is solving many problems which patients suffer when they want medical care. In the future of medicine, the distinctive characteristics of 5G are the guiding force (Jia, Gu, Guo, Xiang, & Zhang, 2016). One problem 5G is solving is hospital visits for patients who live in rural areas (Magsi et al., 2018). The introduction of remote home monitoring systems gave patients the ability to receive the required care from their homes. However, patients who live in rural areas do not have access to high-speed broadband using fibre optic networks, and 3G and 4G networks would not work due to their higher latency and network traffic (Oleshchuk & Fensli, 2011).

Another way 5G is improving the healthcare industry is by transmitting large imaging files quickly. Usually, MRIs and other image machine files are pretty significant. Sometimes they can even go up to a

Figure 2. Google Trends for 5G in healthcare in past five years ("Google Trends," 2021)

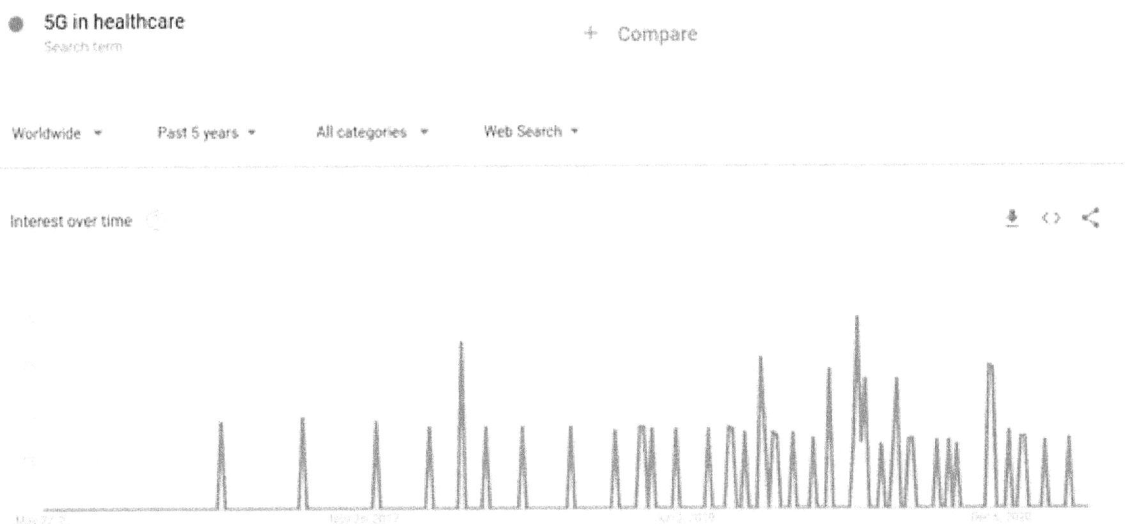

gigabyte of information per patient (Böhle, Eitel, Weygandt, & Ritter, 2019). With 5G, sending large files will only take seconds, unlike minutes with 4G or a copper wired connection. Likewise, remote monitoring of patients using wearable sensors, communication equipment, and the internet of things requires high internet speeds. Medical professionals need to correct the patient's (Oleshchuk & Fensli, 2011).

With 5G, patients with tele-homecare scenarios give the ability for a medical specialist to continuously monitor for irregular body health signs like irregular heart rhythm (arterial fibrillation), high blood pressure or high blood glucose level (Dias & Cunha, 2018). All of this from their wearables or smart devices

Figure 3. Google Trends for 5G in healthcare in past three months ("Google Trends," 2021)

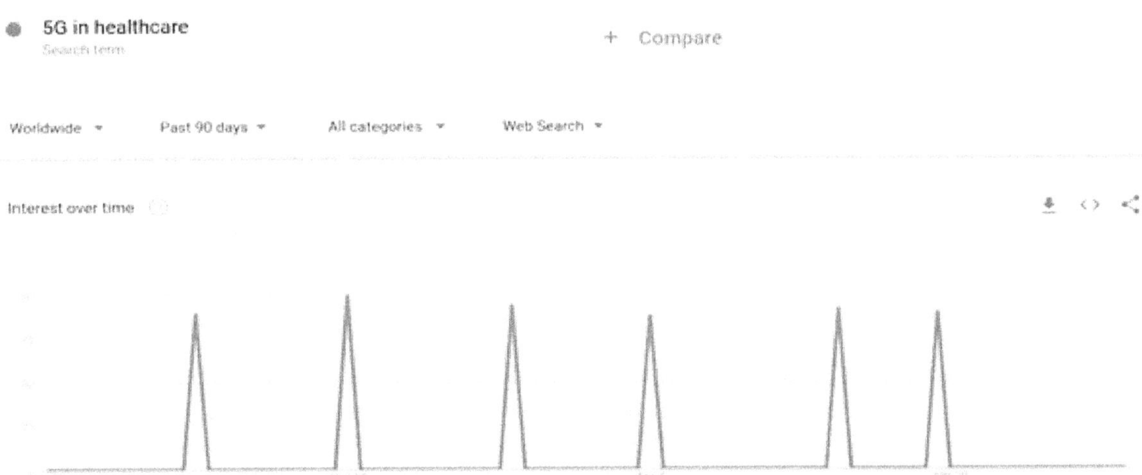

Figure 4. Google Trends for 5G in healthcare in past seven days ("Google Trends," 2021)

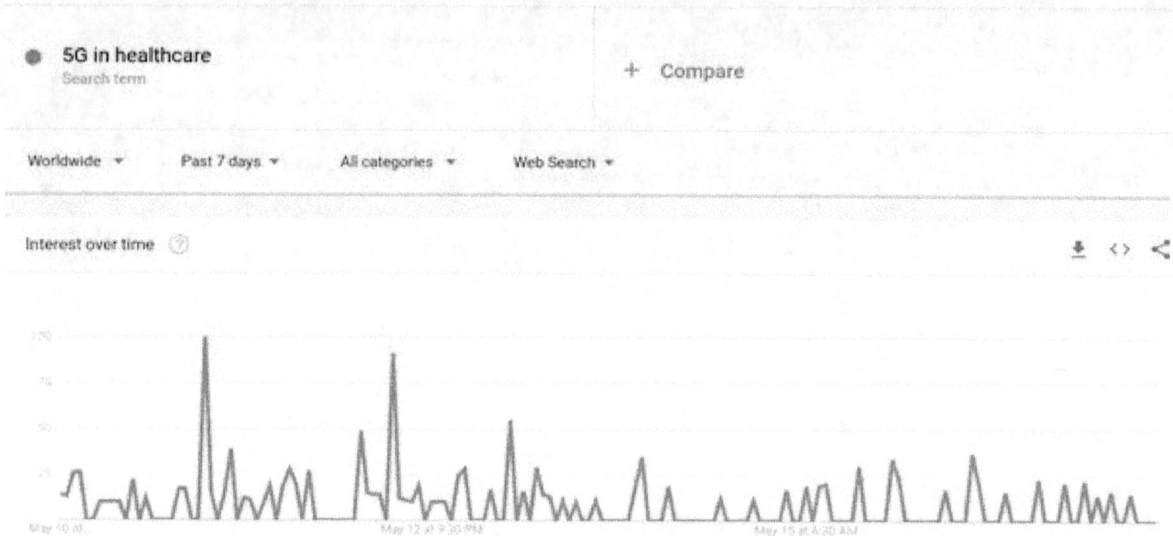

connected to the state-of-the-art 5G network. As the elderly population is increasing rapidly due to many people living longer, many patients who are not in a worse condition will only visit the hospital for a short stay. This is done to leave beds for more urgent or severe cases. By using 5G with telemedicine, doctors could know instantaneously with notifications if a patient is improving or not without utilizing a hospital bed (Jagadeeswari, Subramaniyaswamy, Logesh, & Vijayakumar, 2018).

Ambulance services can also be upgraded with 5G. 5G-powered sensors can be fitted to the ambulance to instantly transfer the patient's vitals and without any network issues. In addition, the medical care professionals at the hospital can help or instruct the ambulance crew. Another problem that can evade is using real-time vehicle parts' tracking can avoid mechanical failures, and patients arrive safely and without any unnecessary delay to the hospital (Nayak & Patgiri, 2020).

5G can also be used for surgery. Surgery is a medical practice where a treatment or injury is completed by cutting open the body and eliminating or repairing a broken part. One primary use of 5G in surgery is Wireless Tele Surgery (WTS) (Soldani et al., 2017). Wireless Service Robots (WSR) are used to link medical experts and patients who are far away from each other (Choi, P. J., Oskouian, R. J., & Tubbs, 2018; Soldani et al., 2017). With telesurgery, high speed, consistent and low latency internet connection is vital. Even a couple of seconds delay can make the surgery unsuccessful. Nonetheless, with 5G, issues like these are solved, and patients who cannot travel far will get the treatment they need. Advancements in technology are constantly evolving; this operation is becoming more viable (Choi, P. J., Oskouian, R. J., & Tubbs, 2018).

Another way 5G is transforming healthcare is cost reduction. With 5G, many costs both for the patient and for the medical workforce are reduced heavily. Through 5G high-speed connectivity, many patients and medical staff with home care and telesurgery will receive better healthcare and save money from reducing travelling and hospital resources. Although 5G technology is not cheap, it will surely be rewarding in the long run. 5G integrates well with other Internet of Things (IoT) devices; it will improve efficiency, save time, and reduce paperwork, reducing costs (Palattella et al., 2016). These intelligent gadgets will gather data efficiently, and with this technology, will connect every gadget instantaneously.

Hence, data can be compiled to be analyzed when it is collected by the sensor (Ahad, Tahir, & Yau, 2019). Whether the gadget is connected to a hospital patient or a patient recovering at home will notify the medical staff without interruptions or time consumption.

Additionally, 5G enhances teamwork in the healthcare industry, improving the service patients receive from their medical experts (Sigwele et al., 2018). Specialists and patients will be able to identify any illnesses or diseases early by using this technology. Patients will be able to see all of their past visits, surgeries and tests from the comfort of their home. The moment data is collected from the smart devices, and the 5G cellular network sends the data to the cloud-based server, where data is stored for future use (Ullah et al., 2019). Moreover, this data is visible to both the medical professionals and the patient. The patient could use this past data for future hospital visits and self-monitoring. This way, medical professionals do not need to type in any data or use traditional files to record diagnoses and treatments. The paperwork and manual work is reduced, errors are therefore reduced, and time is also saved.

5G is not all glitter; it also brings about obstacles that need to be addressed before it can be implemented successfully, the first of which is security and privacy (Khan, Kumar, Jayakody, & Liyanage, 2020). Through many devices and sensors, the network will connect every part of human life to the internet. This would likely lead to different hazards to the protection and confidentiality of data. In healthcare networks, 5G security risks are more severe as malware attacks can harm society. IoT devices would be more vulnerable. This is because small sensors and electronics with low processing power are helpless and unable to manage complex encryption being encoded. Powerful tools are needed to protect such exposed systems (Zhou, Cao, Dong, & Vasilakos, 2017). There are also cost barriers for the 5G implementation. There have been several proposals and solutions to achieve the proposed 5G architecture objectives. Such solutions can include approaches to spectrum allocation, large MIMO, the heterogeneous architecture of cells, content caching or NFV. Each approach should accomplish many objectives to solve a variety of issues. These suggestions would entail further modifications in the telecommunications groundwork to function. Changes such as supplementary micro-cell installations, copper replacement with fibre installation and core network enhancements require change that adds up to enormous expenditures (Agyapong, P. K., Iwamura, M., Staehle, D., Kiess, W., & Benjebbour, 2014).

The day 5G with high-speed connectivity becomes widely available, a broader network of devices, including the most miniature sensors and cloud-based storage, would communicate for the same (Akpakwu, Silva, Hancke, & Abu-Mahfouz, 2017). Furthermore, practical computation based on the cloud would establish a range of opportunities. The examples stated above are early examples of what to come in the healthcare sector. Furthermore, the persistent issues in the world, such as the unequal distribution of healthcare services, the inequalities in healthcare, the rise in chronic illnesses, and the rise in medical costs, will be significantly reduced with the convergence of innovations of emerging technologies.

Diabetes is a prevalent chronic disease from which nearly 8.5 per cent of the world population suffer; 422 million people worldwide have to struggle with diabetes. It is crucial to note that type 2 diabetes mellitus makes up about 90 per cent of the cases. More critically, the situation will be worse, as reported in, with more teenagers and youth becoming susceptible to diabetes as well (Chen et al., 2018).

The number of people who are developing diabetes has been on the rise in the last few decades. Diabetes can result in several complications. The most popular include heart disease and high blood pressure. Therefore, developing new wearable devices for diabetics can help people manage them better. In addition, health care professions can immediately access the data gathered from the devices to better monitor patients (Alfian et al., 2018). These new 5G devices need to achieve five goals: comfortability, personalization, smartness, cost-effectiveness and sustainability.

Two aspects need to be tackled to achieve cost-effectiveness. The first is to reduce the number of people developing diabetes. This can be done by keeping users in a healthy lifestyle. The second cost reduction can be made by out-of-hospital treatments, which will reduce cost compared to the patient's short/long-term hospitalization. For patients to stay comfortable, these devices will need to avoid affecting patients' daily activities as much as possible. Standardized care was good to scale up service, but now we can move beyond it. Personalizing care through machine learning and cognitive computing algorithms (data made available through the use of 5G) can enable personalized treatments and solutions for every patient (Habibzadeh et al., 2020). Treatment can be more sustainable because data is collected continuously and can see improvements/ decline. Thus new strategies can be made more easily and quickly. Last, these 5G capable devices can also detect symptoms earlier and prevent diabetes through personalized treatment (Chen et al., 2018).

The authors (Yu, Lin, Alazab, Tan, & Gu, 2020) have studied Intelligent Transportation Systems (ITS). Autonomous vehicles face poor intention detection rates and real-time performance when predicting driving direction as ITS systems become more complicated. These concerns can have a significant impact on the safety and comfort of mixed-traffic systems. As a result, autonomous cars' capacity to forecast driving direction in real-time based on the surrounding traffic situation should be enhanced. In 5G-enabled ITS, the authors developed a deep learning-based traffic safety solution for a mix of autonomous and manual vehicles. The driving path datasets and the natural-driving datasets for long-term memory networks in 5G-enabled ITS will be employed as network inputs in the proposed system.

Automotive communication augmented reality (AR), virtual reality (VR), Industry 4.0 and remote healthcare are among the vertical sectors and new services that the 5G communications infrastructure is expected to support. The authors (Abidi et al., 2021) have worked on the optimal 5G network slicing approach and proposed a glowworm swarm-based deer hunting optimization algorithm (GS-DHOA). The three main stages of the proposed approach are data collection, optimal weighted feature extraction (OWFE) and different slicing classification for 5G networks. The proposed approach works effectively for network slicing. Although, the developed method need improvements to solve more complex problems.

The authors (Bera et al., 2020) discussed the challenges and issues that apply to blockchain on the Internet of Drones (IoD) environment launched by the 5G-based Internet of Things (IoT). Furthermore, the authors proposed and explored the new blockchain-based secure framework for data management of IoD communication between entities. This plan can prevent many of the attacks required in an IoT-enabled IoD environment. Furthermore, the proposed scheme provides better protection, operational requirements, minimal communication and computational overhead than other related schemes.

Diabetes is one of the most common illnesses, impacting 400 million people globally. Regrettably, over half of them reside in rural regions and are ignorant of the disease's seriousness. Diabetes treatment is feasible, but it is difficult and costly. The authors (Rajput et al., 2021) create a reference model for rural diabetics. It assists rural Indians in detecting diabetes two sufferers at an early stage. In addition, this paradigm facilitates patient-physician contact and engagement. The purpose of their research study's investigation is to compile a list of existing risk variables and the relationships between them.

FEATURES OF 5G IN HEALTHCARE

5G wireless network connectivity is set to transform nearly every industry; it will change how citizens learn, work, communicate and commute. It will make companies more productive, manufacturers more competitive, and healthcare better and accessible. In our case, we will look at how 5G is transforming the healthcare sector and the industry that comes with it (Latif et al., 2017). 5G could aid healthcare in significant ways by providing telehealth, remote surgery, real-time monitoring, and delivering treatment information and support to patients continually because of the reliable networking, speed, and scale (Darrell M. West, 2009). The healthcare system is showing how fragile and how much it can be affected when facing natural disasters. For example, with the COVID-19 pandemic we are going through, what is taking place right now puts unprecedented pressure on the healthcare industry throughout the world. With 5G applications, healthcare companies and healthcare technology providers can improve and develop new opportunities. That could transform the care of patients while bringing forth the emergence of a new healthcare ecosystem, which, compared to the current one, will be more connected and more intelligent. This ecosystem can be seen to be aligned with the 4P medicine, that of being predictive, preventative, personalized and participatory (M, G, K, N.d, & L, 2013) (Sagner, Michael, Amy McNeil, Pekka Puska, Charles Auffray, Nathan D. Price, Leroy Hood, 2017).

1. **Predictive:** The new health system will have the ability to forecast patient's threats as healthcare providers are warned of the patients' problems due to the constant flow of data on their lifestyle behaviours. This will help healthcare practitioners tackle the issue before it escalated further.
2. **Preventative:** The ability to act by tracking and trace and having early detection; an example is a pandemic. In this contagion, geolocation data is combined with diagnostic profiles and tests to pinpoint who is at risk of spreading and passing the illness, which will then initiate alerts and intervention to prevent the spread.
3. **Personalized:** This creates many opportunities for people to personalize their healthcare and interventions because of the real-time health monitoring over the 5g networks. An example is well-being advice that is tailored and delivered to the individual.
4. **Participatory:** 5G health ecosystem enables patients to be involved and participate in their care, increasing the efficiency of medicine. This creates an approach where patients are empowered, which can benefit them by driving their outcome. At this stage, Patients are considered part of the care team and have complete transparency by allowing them to view their laboratory results and others online.

5G IN HEALTHCARE: ADVANTAGES AND APPLICATIONS

5G in health care can provide many advantages discusses by authors (Chamola, Hassija, Gupta, & Guizani, 2020; Gupta, N., Juneja, P. K., Sharma, S., & Garg, 2021; Ramaraju, 2020; Ren, Shen, Tang, & Feng, 2020) both to the patient and to the medical professionals. Firstly, today's healthcare system is not patient-centric as patients have to go to a physician's office or hospital even if they have a minor injury. This is not convenient, and for people with mobility problems and carers would be a hassle for them to visit the doctor for a check-up. Moreover, patients will need to stick with their appointment else they will lose their place. With 5G, this problem could be solved, as patients could easily communicate with their doctors via their devices such as a smartphone or tablet. With this technology, patients can get the care they want from anywhere. Secondly, the current healthcare infrastructure is not equally accessible due to geographic and accessibility reasons (Latif et al., 2017). 5G could solve this problem as everywhere there is access to mobile network connectivity, patients could access the healthcare services they want. Thirdly, patient's data can be processed, transmitted and shared between hospital departments and medical professionals with ease. This way, doctors could get their hands on the patient's examinations quicker which reduce time wastage. 5G is solving problems that previous technologies did not. If we take previous mobile networks like 3G and 4G, they could not handle the data traffic nowadays mobile users are demanding. Moreover, 5G has a higher network capacity due to massive MIMO and carrier aggregation. Some 5G networks are built using multiple bands from the network spectrum. These are the low-band, mid-band, and high-band. The low band provides slower speeds than the other bands, but then it is the most reachable from users far from the antennae as these types of frequencies can pass through building and obstacles. Nonetheless, the high band can provide super speeds, but they cannot penetrate through buildings and barriers (Sharma et al., 2018) (Wang et al., 2014).

Applications

5G will be perfectly applied to the healthcare sector. This technology will change healthcare for many people, and as it is an emerging technology, who knows what the future will bring. 5G could be applied in different areas of healthcare (Liu, Effiok, & Hitchcock, 2020). This including telemedicine, remote surgery, wearable notifications and data transfer (Latif et al., 2017). Telemedicine involves communication from the patient to the medical professional through wireless technologies. With telemedicine, many patients are now getting the care they need without travelling and waiting in line. Doctors could set virtual appointments and connect with patients from their clinic (Yellowlees et al., 2020). This way, the patient will not disrupt his daily lifestyle. Remote surgery or telesurgery involves operating on a patient where the surgeon and the patient are in a different location (Vatandsoost & Litkouhi, 2019). Remote surgery has been growing throughout this decade as surgeons find it easier to operate using robotic arms and magnified views. Wearable notifications are notices received on the patient's/physician's smartwatch (Latif et al., 2017). Last but not least, data transfer gives the ability for medical staff to share and analyze data with ease. As we can see, 5G can be used in various settings in the healthcare industry. With this technology, healthcare could leap forward. Furthermore, we can notice that 5G is not a standalone technology but a backbone for other evolving technologies such as robotics in telesurgery. The Internet of Things (IoT) with wearables, high-speed video call using cloud computing and artificial intelligence (Naveen & Kounte, 2020).

Example

An example of 5G in healthcare would be the first remote robotic surgery in China. On June 27 2019, the first robotic surgery was performed in Beijing Jishuitan Hospital. 5G technology from Huawei was used to connect Beijing Jishuitan Hospital with the Yantai Shan Hospital of Shandong and the Jinxing No. 2 Hospital of Zhejiang. The patient in Jinxing had a fractured lumbar vertebra (bone in the lower spine collapses), and the patient in Yantai had a fractured thoracic vertebra (bone in the mid spine collapses). Traditional surgery methods cannot be performed correctly on patients with such issues. This was the first time that two surgeries were carried out concurrently by remotely controlling orthopaedic surgery robots. According to Tian Wei, a professor and chief scientist of Tianjin orthopaedic surgery robots, it is crucial to improve the quality of surgeries and ensure that everyone has equal access to good healthcare. With 5G, this is possible since there is no signal freezing, feedback latency and untimely data processing. This type of surgery brings the start of a new era where robotic surgery could become part of almost every surgery ("China Telecom's 5G Helped Beijing Jishuitan Hospital Perform the World's First Remote Robotic Surgery," 2019).

LIMITATIONS AND PROBABLE SOLUTIONS

5G is projected to bring around a revolution in communication. This will enable healthcare systems to provide a more dynamically and enhanced service. However, 5G cannot do everything as, like all other technologies, it has its limitations. Apart from this, some theories noted in academic articles suggested that 5G creates health risks due to 5G's higher radiofrequency (Dananjayan & Raj, 2020). The scientific evidence, however, does not support any of these claims. This is agreed upon by worldwide known NGOs such as the National Cancer Society, WHO, FCC and others.

The first drawback 5G has is the small range compared to previous generations since 5G uses higher frequency bands. Furthermore, the connection is easily disrupted when there is an obstacle between the device and the antennae. This limitation will make it easier to lose access to 5G and switch back to other networks (Boccardi, Heath, R. W., Marzetta, Popovski, & Lozano Solsona, 2013).

In some places where 5G antennas have already been installed, some residents are resisting these installations. This is because these antennae are being installed near homes, presenting two main concerns for residents. Some complained about the aesthetics of their antennas, while some are worried about harmful electromagnetic waves being emitted near where they live. The reason for installing them near homes is due to the short effective range of this technology. These cellular radios are installed as close as hundreds of feet apart (Mercola, 2020).

Another significant limitation is the costs of developing 5G and other related infrastructure. That is just the upfront cost. There are also increased upholding costs which is a big issue. New strategies have to be adopted to offset the vast costs as only some of these costs will be put on the customers. Standard measures such as cost-saving efforts may not be enough to bear the cost, so that some alternative approaches may be needed. These may include network sharing between the service providers and finding new revenue models (Mercola, 2020).

One of the most significant issues for 5G is rural access. Some of the population may not even have a cellular connection, let alone providing them with 5G access. As many people live in cities rather than scattered evenly around the country, it might be easier to offer 5G access. However, a percentage of the population may and will be left behind in advancing this technology (Mercola, 2020). Rural areas lack access to the best healthcare; eliminating this problem would be great, but it is impractical from an economic point of view. It is proposed to get rural areas a form of 5G called "low-band", which has less capacity, it still has lower latency which is very important for healthcare with 5G. Speeds, however, may not even be equal to the average of 4G speeds we are used to today (Akpakwu et al., 2017).

The probable solutions to various limitations in healthcare sectors such as home care and remote patient monitoring, ambulance services, virtual consultation, real-time maintenance, augmented reality, and virtual reality and Robotics can be taken care of using the advanced features of 5G technology. Several sensors may be used to monitor the essential information of patients, thanks to advancements in technology and smartphones—the so-called Internet of Medical Things (IoMT). A high-bandwidth 5G mobile network is necessary to link many IoMT devices and share enormous amounts of data remotely. Through 5G-powered wearables or other remote monitoring and rehabilitative equipment, these sensors detect the ECG signal, blood pressure, body temperature, glucose level, and other parameters.

Wearables with various sensors can also help patients with chronic illnesses such as cancer, stroke, spinal cord injuries, and chronic pulmonary diseases track their physical activity and provide in-home rehabilitation. Wearable smart gadgets for non-interventional blood sample analysis (such as haemoglobin concentration) and seizure prediction are also developed. The IoMT may also be used to track medicine consumption in patients (through ingestible sensors), allowing for remote monitoring of pharmacotherapy compliance. Patient's vitals are relayed in real-time to doctors via 5G-powered sensors fitted to ambulances and mobile health monitoring systems, with no network issues. Therefore, necessary treatments can be given to patients on the go with the help of ambulance crews, paving the way for smart ambulance services.

Doctors may attend to several patients without having to meet in person, thanks to High Definition streaming. Doctors may prescribe without delay depending on patient input and retrieve vitals from the real-time remote monitoring system. Patients may contact doctors from the comfort of their own homes, saving money on travel expenses. Outpatient hospital overcrowding can be avoided if only those patients who require emergency medical attention are seen in the hospital.

All types of medical equipment are equipped with sensors that transmit the state of each component in real-time to the maintenance team, which not only aids in the prediction of hardware failure but also reduces the time it takes to fix or replace it. Medical robotic-assisted surgery has the potential to improve accuracy and aid treating doctors in improving postoperative outcomes. 5G-enabled telesurgery improves the outreach to marginalized populations and the much-needed real-time cooperation among surgeons across many facilities. For remote processing, large files may be sent around the world in a short period. This allows several researchers to access data in real-time, which helps them conduct scientific investigations with better data analysis.

The significant attributes like problems in current healthcare, 5G applications, healthcare requirements, 5G specifications and 5G obstacle, along with 5G limitations discussed in the above sections, are briefly summarised in Table 1.

CONCLUSION AND FUTURE PERSPECTIVES

In this paper, the generation of communication networks has been discussed and the recent trends of 5G in Healthcare. Various emerging technologies found in the literature like big data, cloud computing, internet of things, machine and deep learning, blockchain and artificial intelligence. The various limitations of the healthcare system can be sorted out using these different technologies. Further, it is safe to assume that with 5G and the confluence of these emerging technologies, the healthcare industry will change drastically as these technologies are used to augment human capacity and enhance the effectiveness of human potential.

In the future, the high demand for 5G will increase. To perform mission-critical medical functions, you need to have access over 90% of the time to 5G. Otherwise, other people's lives may be at risk, and this new technology will be counterproductive. 5G will enable the healthcare industry to transform and reimagine itself to provide better service and healthcare access to more people. 5G will allow new

Table 1. Major attributes of 5G in healthcare

Attribute	Description	Reference
Problems in the current healthcare system	• The current system is not patient-centric.	(Latif et al., 2017)
5G applications	• 5G can be applied in various healthcare sectors, including telemedicine, remote surgery and wearable notifications. • 5G network sends data to the based cloud server. • 5G improves collaboration between healthcare systems. • Connecting physicians/surgeons in different hospitals. • Telesurgery becoming more feasible. • Connecting physicians/surgeons in different hospitals. • Monitoring of irregular health signs • Better ambulances with a real-time connection with the hospital • 5G will bring around a revolution in healthcare.	Latif et al., 2017), (Ullah et al., 2019), (Sigwele et al., 2018); (Soldani et al., 2017); (Choi, P. J., Oskouian, R. J., & Tubbs, 2018); (Oleshchuk & Fensli, 2011) (Dananjayan & Raj, 2020)
Healthcare requirements	• High Internet speeds to make a correct diagnosis. • Tele homecare scenarios	(Oleshchuk & Fensli, 2011)
5G Specifications	• MIMO technology • High carrier frequencies, many base stations and antennae • Better capacity, low latency, higher data rates and a higher number of linked devices • Low-band, mid-band, and high-band • 5G devices need to be effective, comfortable, personalized, sustainable and smart	(Bogale & Le, 2016); (Andrews, J. G., Buzzi, S., Choi, W., Hanly, S. V., Lozano, A., Soong, A. C., & Zhang, 2014); (Sharma et al., 2018); (Wang et al., 2014); (Yang et al., 2015)
5G obstacles	• Obstacles that need to be addressed before 5G can be implemented successfully • Looking into privacy and security	(Agyapong, P. K., Iwamura, M., Staehle, D., Kiess, W., & Benjebbour, 2014); (Zhou et al., 2017)
5G limitations	• 5G still has limitations mainly due to short-range, which leads to many other issues. • Higher band provides faster speeds nearby, but obstacles can interfere, while lower bands provide slower speeds at a distance with little interference.	(Yang et al., 2015)

devices which can gather data regularly rather than at a particular time. This massive amount of data can be used to form patterns and identify issues earlier and accurately.

New wearables, which are devices capable of gathering and sharing medical data, can identify problems proactively and warn the patient to get help before any visible symptoms (patient still thinks he/she is healthy). Proactive action will enable users to tackle any health problems before it might be too late. Other wearables can be designed for patients with chronic conditions, enabling healthcare professionals to collect and analyze data continuously. An example of this is a blood glucose monitor, which will monitor the blood sugar level. This will help patients who do not know that they are pre-diabetic or diabetic discover their condition and get the medical help they need.

Another technology in healthcare that is starting to be used is AI. This technology will help medical specialist's access up-to-date information on patients' diagnosis and treatment. AI requires high-bandwidth and low latency networks, which is precisely what 5G promises to bring around. Remote medicine will enable highly experienced doctors to offer their specialized knowledge and service, not only to urban areas but also to rural areas if there are enough bandwidth and low latency. These doctors may offer consultations online, send and receive medical images in real-time and surgeons will even perform surgeries through tools such as virtual or augmented reality and robotics. Robotics are already being used in surgeries but operating them through a network have mainly been used in a few pilot trials.

REFERENCES

Abidi, M. H., Alkhalefah, H., Moiduddin, K., Alazab, M., Mohammed, M. K., Ameen, W., & Gadekallu, T. R. (2021). Optimal 5G network slicing using machine learning and deep learning concepts. *Computer Standards & Interfaces*, *76*, 103518. Advance online publication. doi:10.1016/j.csi.2021.103518

Agyapong, P. K., Iwamura, M., Staehle, D., Kiess, W., & Benjebbour, A. (2014). Design Considerations for a 5G Network Architecture. *IEEE Communications Magazine*, *52*(11), 65–65. doi:10.1109/MCOM.2014.6957145

Ahad, A., Tahir, M., & Yau, K. L. A. (2019). 5G-based smart healthcare network: Architecture, taxonomy, challenges and future research directions. *IEEE Access: Practical Innovations, Open Solutions*, *7*, 100747–100762. doi:10.1109/ACCESS.2019.2930628

Akpakwu, G. A., Silva, B. J., Hancke, G. P., & Abu-Mahfouz, A. M. (2017). A Survey on 5G Networks for the Internet of Things: Communication Technologies and Challenges. *IEEE Access: Practical Innovations, Open Solutions*, *6*, 3619–3647. doi:10.1109/ACCESS.2017.2779844

Alfian, G., Syafrudin, M., Ijaz, M. F., Syaekhoni, M. A., Fitriyani, N. L., & Rhee, J. (2018). A personalized healthcare monitoring system for diabetic patients by utilizing BLE-based sensors and real-time data processing. *Sensors (Switzerland)*, *18*(7), 2183. Advance online publication. doi:10.339018072183 PMID:29986473

Andrews, J. G., Buzzi, S., Choi, W., Hanly, S. V., Lozano, A., Soong, A. C., & Zhang, J. C. (2014). What Will 5G Be? *IEEE Journal on Selected Areas in Communications*, *32*(6), 1065–1082. doi:10.1109/JSAC.2014.2328098

Baker, J., & Stanley, A. (2018). Telemedicine Technology: A Review of Services, Equipment, and Other Aspects. *Current Allergy and Asthma Reports*, *18*(11), 60. Advance online publication. doi:10.100711882-018-0814-6 PMID:30259201

Bera, B., Saha, S., Das, A. K., Kumar, N., Lorenz, P., & Alazab, M. (2020). Blockchain-Envisioned Secure Data Delivery and Collection Scheme for 5G-Based IoT-Enabled Internet of Drones Environment. *IEEE Transactions on Vehicular Technology*, *69*(8), 9097–9111. doi:10.1109/TVT.2020.3000576

Boccardi, F., & Heath, R. W. J., Marzetta, T. L., Popovski, P., & Lozano Solsona, A. (2013). Five Disruptive Technology Directions for 5G. *IEEE Communications Magazine*, (February), 74–80.

Bogale, T. E., & Le, L. B. (2016). Massive MIMO and mmWave for 5G Wireless HetNet: Potential Benefits and Challenges. *IEEE Vehicular Technology Magazine*, *11*(1), 64–75. doi:10.1109/MVT.2015.2496240

Böhle, M., Eitel, F., Weygandt, M., & Ritter, K. (2019). Layer-wise relevance propagation for explaining deep neural network decisions in MRI-based Alzheimer's disease classification. *Frontiers in Aging Neuroscience*, *10*(JUL), 194. Advance online publication. doi:10.3389/fnagi.2019.00194 PMID:31417397

Chamola, V., Hassija, V., Gupta, V., & Guizani, M. (2020). A Comprehensive Review of the COVID-19 Pandemic and the Role of IoT, Drones, AI, Blockchain, and 5G in Managing its Impact. *IEEE Access: Practical Innovations, Open Solutions*, *8*, 90225–90265. doi:10.1109/ACCESS.2020.2992341

Chen, M., Yang, J., Zhou, J., Hao, Y., Zhang, J., & Youn, C. H. (2018). 5G-Smart Diabetes: Toward Personalized Diabetes Diagnosis with Healthcare Big Data Clouds. *IEEE Communications Magazine*, *56*(4), 16–23. doi:10.1109/MCOM.2018.1700788

China Telecom's 5G Helped Beijing Jishuitan Hospital Perform the World's First Remote Robotic Surgery. (2019). Retrieved May 17, 2021, from https://carrier.huawei.com/en/success-stories/Industries-5G/Medical/beijing

Choi, P. J., Oskouian, R. J., & Tubbs, R. S. (2018). Telesurgery: Past, Present, and Future. *Cureus*, *10*(5). https://www.ncbi.nlm.nih.gov/pmc/articles/PMC6067812/ PMID:30079282

Cox, C. (2012). An Introduction to LTE: LTE, LTE-Advanced, SAE and 4G Mobile Communications. In An Introduction to LTE: LTE, LTE-Advanced, SAE and 4G Mobile Communications. doi:10.1002/9781119942825

Dananjayan, S., & Raj, G. M. (2020). 5G in healthcare: How fast will be the transformation? *Irish Journal of Medical Science*. Advance online publication. doi:10.100711845-020-02329-w PMID:32737688

Darrell, M. W. (2009). How 5G technology enables the health Internet of Things. *Cyber Resilience of Systems and Networks*, (July), 1–150. Retrieved from https://link.springer.com/10.1007/978-3-319-77492-3_16

Dias, D., & Cunha, J. P. S. (2018). Wearable health devices—Vital sign monitoring, systems and technologies. *Sensors (Switzerland)*, *18*(8), 2414. Advance online publication. doi:10.339018082414 PMID:30044415

Gupta, N., Juneja, P. K., Sharma, S., & Garg, U. (2021). Future Aspect of 5G-IoT Architecture in Smart Healthcare System. *2021 5th International Conference on Intelligent Computing and Control Systems (ICICCS)*, 406–411.

Habibzadeh, H., Dinesh, K., Rajabi Shishvan, O., Boggio-Dandry, A., Sharma, G., & Soyata, T. (2020). A Survey of Healthcare Internet of Things (HIoT): A Clinical Perspective. *IEEE Internet of Things Journal*, *7*(1), 53–71. doi:10.1109/JIOT.2019.2946359 PMID:33748312

Haskell, H. (2020). Cumberlege review exposes stubborn and dangerous flaws in healthcare. *BMJ (Clinical Research Ed.)*, *370*, m3099. Advance online publication. doi:10.1136/bmj.m3099 PMID:32763955

Hossain, S. (2013). 5G wireless communication systems. *American Journal of Engineering Research*, *2*(10), 344–353.

Eze, Sadiku, & Musa. (2018). 5G Wireless Technology: A Primer. International *Journal of Science, Engineering and Technology*, *7*(July), 62–64.

Jagadeeswari, V., Subramaniyaswamy, V., Logesh, R., & Vijayakumar, V. (2018). A study on medical Internet of Things and Big Data in personalized healthcare system. *Health Information Science and Systems*, *6*(1), 14. Advance online publication. doi:10.100713755-018-0049-x PMID:30279984

Jain, R. (2016). Introduction to 5G. *Washington University in St. Louis*. Retrieved from https://www.cse.wustl.edu/~jain/cse574-16/

Jia, M., Gu, X., Guo, Q., Xiang, W., & Zhang, N. (2016). Broadband hybrid satellite-terrestrial communication systems based on cognitive radio toward 5G. *IEEE Wireless Communications*, *23*(6), 96–106. doi:10.1109/MWC.2016.1500108WC

Khan, R., Kumar, P., Jayakody, D. N. K., & Liyanage, M. (2020). A Survey on Security and Privacy of 5G Technologies: Potential Solutions, Recent Advancements, and Future Directions. *IEEE Communications Surveys and Tutorials*, *22*(1), 196–248. doi:10.1109/COMST.2019.2933899

Korhonen, J. (2003). Introduction to 3G Mobile Communications. Artech House.

Latif, S., Qadir, J., Farooq, S., & Imran, M. A. (2017). How 5G wireless (and Concomitant Technologies) will revolutionize healthcare? *Future Internet*, *9*(4), 93. Advance online publication. doi:10.3390/fi9040093

Lee, J., Tejedor, E., Ranta-Aho, K., Wang, H., Lee, K. T., Semaan, E., Mohyeldin, E., Song, J., Bergljung, C., & Jung, S. (2018). Spectrum for 5G: Global Status, Challenges, and Enabling Technologies. *IEEE Communications Magazine*, *56*(3), 12–18. doi:10.1109/MCOM.2018.1700818

Liu, E., Effiok, E., & Hitchcock, J. (2020). Survey on health care applications in 5G networks. *IET Communications*, *14*(7), 1073–1080. doi:10.1049/iet-com.2019.0813

M, F., G, G., K, B., N.d, P., & L, H. (2013). P4 medicine: How systems medicine will transform the healthcare sector and society. *Personalized Medicine*, 565–576.

Magsi, H., Sodhro, A. H., Chachar, F. A., Abro, S. A. K., Sodhro, G. H., & Pirbhulal, S. (2018). Evolution of 5G in Internet of medical things. *2018 International Conference on Computing, Mathematics and Engineering Technologies: Invent, Innovate and Integrate for Socioeconomic Development, ICoMET 2018 - Proceedings,* 1–7. 10.1109/ICOMET.2018.8346428

Marković, G. Z. (2017). Routing and spectrum allocation in elastic optical networks using bee colony optimization. *Photonic Network Communications*, *34*(3), 356–374. doi:10.100711107-017-0706-z

Mercola, J. (2020). *EMFD 5G, Wi-Fi Cell Phones Hidden Harms and How to Protect Yourself.* Hay House.

Naveen, S., & Kounte, M. R. (2020). In Search of the Future Technologies: Fusion of Machine Learning, Fog and Edge Computing in the Internet of Things. Lecture Notes on Data Engineering and Communications Technologies, 31, 278–285. doi:10.1007/978-3-030-24643-3_33

Nayak, S., & Patgiri, R. (2020). A Vision on Intelligent Medical Service for Emergency on 5G and 6G Communication Era. *EAI Endorsed Transactions on Internet of Things*, *6*(22), 166293. doi:10.4108/eai.17-8-2020.166293

Oleshchuk, V., & Fensli, R. (2011). Remote patient monitoring within a future 5G infrastructure. *Wireless Personal Communications*, *57*(3), 431–439. doi:10.100711277-010-0078-5

Palattella, M. R., Dohler, M., Grieco, A., Rizzo, G., Torsner, J., Engel, T., & Ladid, L. (2016). Internet of Things in the 5G Era: Enablers, Architecture, and Business Models. *IEEE Journal on Selected Areas in Communications*, *34*(3), 510–527. doi:10.1109/JSAC.2016.2525418

Paramita, S., Das Bebartta, H. N., & Pattanayak, P. (2021). IoT Based Healthcare Monitoring System Using 5G Communication and Machine Learning Models. *Studies in Computational Intelligence*, *932*, 159–182. doi:10.1007/978-981-15-9735-0_9

Peersman, G., Cvetkovic, S., Griffiths, P., & Spear, H. (2000). Global system for mobile communications short message service. *IEEE Personal Communications*, *7*(3), 15–23. doi:10.1109/98.847919

Qiao, J., Shen, X., Mark, J., Shen, Q., He, Y., & Lei, L. (2015). Enabling device-to-device communications in millimeter-wave 5G cellular networks. *IEEE Communications Magazine*, *53*(1), 209–215. doi:10.1109/MCOM.2015.7010536

Rajput, D. S., Basha, S. M., Xin, Q., Gadekallu, T. R., Kaluri, R., Lakshmanna, K., & Maddikunta, P. K. R. (2021). Providing diagnosis on diabetes using cloud computing environment to the people living in rural areas of India. *Journal of Ambient Intelligence and Humanized Computing*. Advance online publication. doi:10.100712652-021-03154-4

Ramaraju, A. (2020). Unlocking the Potential of 5G for Content Production. *Psychology and Education Journal, 57*(9), 5912-5917. Retrieved from https://www.bbc.co.uk/rd/blog/2019-03-5g-production-media-broadcasting

Ren, H., Shen, J., Tang, X., & Feng, T. (2020). 5G Healthcare Applications in COVID-19 Prevention and Control. *2020 ITU Kaleidoscope. Industry-Driven Digital Transformation, ITU K, 2020*, 1–4. Advance online publication. doi:10.23919/ITUK50268.2020.9303191

Sagner, M., McNeil, A., Puska, P., Auffray, C., Price, N. D., Hood, L., Lavie, C. J., Han, Z.-G., Chen, Z., Brahmachari, S. K., McEwen, B. S., Soares, M. B., Balling, R., Epel, E., & Arena, R. (2017). The P4 health spectrum–a predictive, preventive, personalized and participatory continuum for promoting healthspan. *Progress in Cardiovascular Diseases*, *59*(5), 506–521. doi:10.1016/j.pcad.2016.08.002 PMID:27546358

Schulz, P., Matthe, M., Klessig, H., Simsek, M., Fettweis, G., Ansari, J., Ashraf, S. A., Almeroth, B., Voigt, J., Riedel, I., Puschmann, A., Mitschele-Thiel, A., Muller, M., Elste, T., & Windisch, M. (2017). Latency Critical IoT Applications in 5G: Perspective on the Design of Radio Interface and Network Architecture. *IEEE Communications Magazine*, *55*(2), 70–78. doi:10.1109/MCOM.2017.1600435CM

Sharma, S. K., Bogale, T. E., Le, L. B., Chatzinotas, S., Wang, X., & Ottersten, B. (2018). Dynamic Spectrum Sharing in 5G Wireless Networks with Full-Duplex Technology: Recent Advances and Research Challenges. *IEEE Communications Surveys and Tutorials*, *20*(1), 674–707. doi:10.1109/COMST.2017.2773628

Sigwele, T., Hu, Y. F., Ali, M., Hou, J., Susanto, M., & Fitriawan, H. (2018). Intelligent and Energy Efficient Mobile Smartphone Gateway for Healthcare Smart Devices Based on 5G. *2018 IEEE Global Communications Conference, GLOBECOM 2018 - Proceedings*. 10.1109/GLOCOM.2018.8648031

Soldani, D., Fadini, F., Rasanen, H., Duran, J., Niemela, T., Chandramouli, D., ... Nanavaty, N. (2017). 5G Mobile Systems for Healthcare. *IEEE Vehicular Technology Conference*. 10.1109/VTC-Spring.2017.8108602

Trends, G. (2021). Retrieved May 17, 2021, from https://trends.google.com/trends/explore?date=now7-d&q=5G in healthcare

Ullah, H., Gopalakrishnan Nair, N., Moore, A., Nugent, C., Muschamp, P., & Cuevas, M. (2019). 5G communication: An overview of vehicle-to-everything, drones, and healthcare use-cases. *IEEE Access: Practical Innovations, Open Solutions*, *7*, 37251–37268. doi:10.1109/ACCESS.2019.2905347

Vatandsoost, M., & Litkouhi, S. (2019). The Future of Healthcare Facilities: How Technology and Medical Advances May Shape Hospitals of the Future. *Hospital Practices and Research*, *4*(1), 1–11. doi:10.15171/hpr.2019.01

Wang, Y., Li, J., Huang, L., Jing, Y., Georgakopoulos, A., & Demestichas, P. (2014). 5G mobile: Spectrum broadening to higher-frequency bands to support high data rates. *IEEE Vehicular Technology Magazine*, *9*(3), 39–46. doi:10.1109/MVT.2014.2333694

Yang, J. J., Li, J., Mulder, J., Wang, Y., Chen, S., Wu, H., Wang, Q., & Pan, H. (2015). Emerging information technologies for enhanced healthcare. *Computers in Industry*, *69*, 3–11. doi:10.1016/j.compind.2015.01.012

Yellowlees, P., Nakagawa, K., Pakyurek, M., Hanson, A., Elder, J., & Kales, H. C. (2020). Rapid conversion of an outpatient psychiatric clinic to a 100% virtual telepsychiatry clinic in response to COVID-19. *Psychiatric Services (Washington, D.C.)*, *71*(7), 749–752. doi:10.1176/appi.ps.202000230 PMID:32460683

Yu, K., Lin, L., Alazab, M., Tan, L., & Gu, B. (2020). Deep Learning-Based Traffic Safety Solution for a Mixture of Autonomous and Manual Vehicles in a 5G-Enabled Intelligent Transportation System. *IEEE Transactions on Intelligent Transportation Systems*. Advance online publication. doi:10.1109/TITS.2020.3042504

Zahoor, M. I., Dou, Z., Shah, S. B. H., Khan, I. U., Ayub, S., & Gadekallu, T. R. (2020). Pilot decontamination using asynchronous fractional pilot scheduling in massive MIMO systems. *Sensors (Switzerland)*, *20*(21), 1–21. doi:10.339020216213 PMID:33143363

Zaidi, A. A., Baldemair, R., Tullberg, H., Bjorkegren, H., Sundstrom, L., Medbo, J., Kilinc, C., & Da Silva, I. (2016). Waveform and Numerology to Support 5G Services and Requirements. *IEEE Communications Magazine*, *54*(11), 90–98. doi:10.1109/MCOM.2016.1600336CM

Zhou, J., Cao, Z., Dong, X., & Vasilakos, A. V. (2017). Security and Privacy for Cloud-Based IoT: Challenges. *IEEE Communications Magazine*, *55*(1), 26–33. doi:10.1109/MCOM.2017.1600363CM

Chapter 5
High-Speed Connectivity:
Potential Impact on the Quality of Life

Vijay Prakash
Thapar Institute of Engineering and Technology, India

Lalit Garg
https://orcid.org/0000-0002-3868-0481
University of Malta, Malta

Jack Azzopardi
University of Malta, Malta

Thomas Camilleri
University of Malta, Malta

ABSTRACT

Since the early 1990s, there has been a lot of enthusiasm for using high-speed connectivity to develop local community links through education, employment possibilities, fostering community events, and enhancing overall sociability within a local region. 5G is the 5th iteration of a broadband network operating on cellular systems. 5G is not only for mobile phones, but it is also the foundation for virtual reality (VR); the internet of things (IoT); and autonomous transport, immersive services, and public infrastructure; and connecting many electronic devices to the internet. In this chapter, first, the authors have discussed the evolution of 1G network to 6G networks by focussing on its potential impact on the quality of life. Further, 5G applications in IoT, autonomous transport, immersive services, and public infrastructure have been discussed. Then the chapter discusses popular advantages, limitations in the current technologies, implementations, and future perspective.

DOI: 10.4018/978-1-7998-6988-7.ch005

INTRODUCTION

Today's standard of high-speed connectivity is 5G, the 5th iteration of a broadband network operating on cellular systems (Hossain, 2013). Similar to what came before it, 5G works mainly in small geographical areas known as cells (Zhang, Dong, Cheng, Hossain, & Leung, 2016). The transmission medium throughout the various compartments is radio waves that travel through the air, used to send or receive Internet and telephony packets (Capozzi, Piro, Grieco, Boggia, & Camarda, 2013). These duplicate packets are sent (or received) via Antennas located in every cell, depending on their contents. Telephone data (made up of sounds and images) is transmitted as a stream of bits and converted using a converter found in Antennas (Steinmetz, 2012). Internet packets sent over high-bandwidth fibre optic cable other than wireless medium. Speeds of 5G promised to go up to 10GB/s, the bandwidth being shared amongst users of a cell. Realistically, the typical end-user will more or less benefit from a speed between 50 and 100MB/s on wireless devices. It is statistically proven, wired devices are bound to make better use of the total available bandwidth due to less resistance in their connection. This depends on the infrastructure as it will require specific cabling to be made available to homes (Wey & Zhang, 2019). When moving from one geographical area to another, a user's device is automatically handed over without any intervention needed (Lei, Zhong, Lin, & Shen, 2012). 5G can cater for up to 1 million devices/square km., as opposed to 4G, which can only support a tenth of this. Most devices nowadays operate on wireless networks (or Wi-Fi), and recently launched smartphones, for example, can already boast the availability of a 5G-capable network unit (Rommer et al., 2019).

To truly understand the background behind network connectivity, it is necessary to plot the evolution of wireless connectivity. It was starting from the first generation (1G) to the sixth generation (6G) (Anju & Gawas, 2015; Pereira & Sousa, 2004). The first generation (1G) was launched by Nippon Telegraph and Telephone (NTT) in Tokyo around 1979. Although it was a big hit, it suffered from various limitations such as low coverage and poor sound quality. The second-generation (2G) was launched in Finland in 1991 (Bhalla & Bhalla, 2010). This brought about improvements upon the previous generation, including encrypted and digital voice calls with less background static noise. This generation also allowed text messages and picture messages to be sent and received on their phone. Then came the third-generation (3G), also known as the 'Packet-Switching' generation (Chiussi, Khotimsky, & Krishnan, 2002). Launched by NTT DoCoMo in 2001, its primary goal was to allow users to access data from anywhere with a connection that made international roaming services possible. Through this, video conferencing and VoIP were introduced. After that, the fourth generation (4G) was launched in Sweden, Stockholm and Norway in 2009 as LTE (Cox, 2012; Shikhare, G., & Shaikh, 2014). This generation has also brought about the streaming era as it was introduced globally and made high-quality video streaming accessible for many people. The only problem with this is that mobile phones need to be designed to support 4G connectivity, whereas the switch from 2G to 3G only required a change of SIM card. Although 4G is currently the standard for most countries, some countries still experience network patchiness. This comes about due to the higher frequency required for faster speeds. Now we are entering into the Internet of Things Era, the fifth generation (5G) (Anju & Gawas, 2015; Eze, N. O. Sadiku, & M. Musa, 2018). This network has been years in the making; in 2008, NASA launched the technology required to support 5G speeds, and in the same year, South Korea started a 5G R&D program. Finally, carriers in South Korea have rolled out 5G services in December of 2019. With 4G experiencing limitations in terms of coverage due to the high frequency. Why is 5G being implemented already with even higher frequencies? The far superior speeds and latency offered by 5G can transform industries such as banking and healthcare(Latif, Qadir,

Farooq, & Imran, 2017). Although the implementation of the fifth generation globally is still in its initial phase, research of the sixth generation (6G) has already started taking place. It is expected to be implemented between 2027 and 2030 (Strinati, E. C., Barbarossa, S., Gonzalez-Jimenez, J. L., Ktenas, D., Cassiau, N., Maret, L., & Dehos, 2019).

Diabetes is one of the most common illnesses, impacting 400 million people globally. Regrettably, over half of them reside in rural regions and are ignorant of the disease's seriousness. Diabetes treatment is feasible, but it is difficult and costly. The authors (Rajput et al., 2021) create a reference model for rural diabetics. It assists rural Indians in detecting diabetes two sufferers at an early stage. In addition, this paradigm facilitates patient-physician contact and engagement. The purpose of their research study's investigation is to compile a list of existing risk variables and the relationships between them.

MIMO has become a promising technology for 5G systems because of its enormous spectrum capacity and low power consumption. On the other hand, Pilot contamination (PC) severely inhibits the performance of MIMO systems. As a result, two pilot scheduling strategies have been presented, such as Fractional Pilot Reuse (FPR) and Asynchronous Fractional Pilot Scheduling Scheme (AFPS). These scheduling strategies are beneficial to minimize significantly the number of PCs utilized by PCs in uplink time division duplex (TDD) gigantic MIMO systems. Customers are assigned to the central cell and edge cell depending on their signal-to-interference-plus-noise ratio (SINR) (Zahoor et al., 2020). 5G isn't just about speed; it has the potential to transform the whole healthcare system by overcoming all of the limits of today's wireless networking technologies while also providing a fantastic user experience (Dananjayan & Raj, 2020).

The significant contribution to this chapter is the evolution of 1G network to 6G networks by focussing on its potential impact on the quality of life. Further, 5G in IoT, 5G in autonomous transport, 5G in immersive services and 5G in public infrastructure have been discussed along with their popular advantages and limitations. Limitations in the current technologies and implementations also been considered along with its future perspective.

LITERATURE REVIEW

Network operators managing the technology are also entitled to make millimetre waves available (Bojic et al., 2013). These waves assist the rest of the microwaves by providing additional capacity and increasing throughput. Their range, however, is shorter than that of microwaves, meaning they can only be used with geographically smaller cells and find it challenging to go through walls (or any obstacles). In addition, the antennas use for microwaves are significantly larger and will require a decent-sized footprint. In contrast, millimetre wave antennas are smaller, making them more feasible for congested areas (Curwen & Whalley, 2014). Though it might take time for the whole infrastructure to get on par, the process of transitioning towards 5G speeds will be an iterative one. As the technology itself becomes more of a standard, products utilizing it will be normalized, and the accessibility of high-speed connectivity will steadily grow. Due to the process not being a simple overnight change, 4G can not be dumped as soon as its more modern replacement shows up. Instead, the older technologies will remain supported as a fall-back for failures and still give connectivity to those who are yet to switch. The new networks will also use their predecessors, as 4G will be useful for the initial connection with the cell.

The Whale optimization algorithm (WOA) (Pham, Mirjalili, Kumar, Alazab, & Hwang, 2020) has recently gotten a lot of attention from the academic community as a viable solution to various performance and optimization issues. It is a viable alternative to the current methods. The major goal is to see how successful WOA is in resolving resource allocation issues on wireless networks. First, we introduce the primary domains and the WOA binary version as well as introduce a penalty mechanism for dealing with performance and optimization difficulties. The energy-and-spectral tradeoff of power distribution includes wireless interference networks, power allocation for secure throughput maximization, and compot mobile edge loading. Finally, the authors discuss how WOA adoption may help with various problems, such as resource allocation for 5G and wireless networks. Because the application of WOA to wireless and communication networks is still unexplored, and the WOA-based algorithm can provide competitive performance compared to state-of-the-art algorithms. WOA can be used as a benchmark for performance comparison when someone proposes his methods for performance comparison any optimization problem.

Automotive communication augmented reality (AR), virtual reality (VR), Industry 4.0 and remote healthcare are among the vertical sectors and new services that the 5G communications infrastructure is expected to support. The authors (Abidi et al., 2021) have worked on the optimal 5G network slicing approach and proposed a glowworm swarm-based deer hunting optimization algorithm (GS-DHOA). The three main stages of the proposed method are data collection, optimal weighted feature extraction (OWFE) and different slicing classification for 5G networks. The proposed approach works effectively for network slicing. Although, the developed method need improvements to solve more complex problems.

5G in the Internet of Things

The Internet of things has already made its mark and is becoming increasingly popular as we speak. IoT can expect an even more significant rise in popularity which will be supported by 5G; after all, they are

Figure 1. 5G in the Internet of Things (Trends, 2021a)

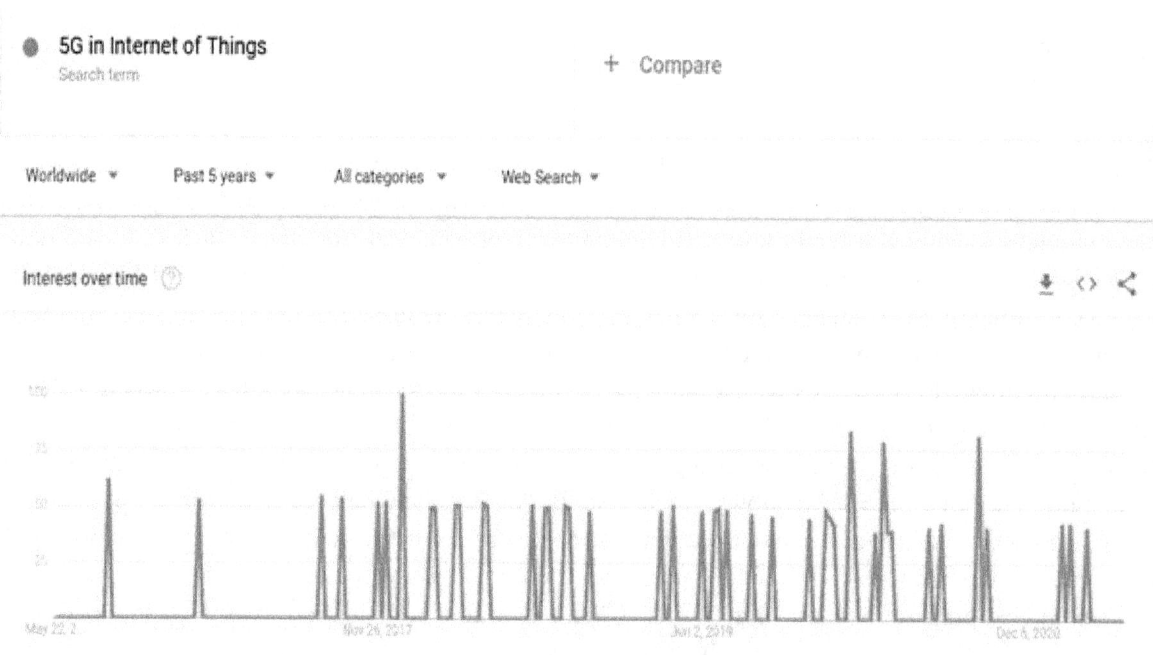

calling it the Internet of Things era for a reason (Saxena, Roy, Sahu, & Kim, 2017). Figure 1 depicts the past five years Google Trends for 5G in the Internet of Things worldwide.

The introduction of wearable devices and network-connected appliances/devices will generate vast amounts of data that 5G can support. The information generated will be stored in storage devices which are then later processed using big data technologies. IoT may be categorized into three main categories.

Figure 2. High-Speed Connectivity (Trends, 2021b)

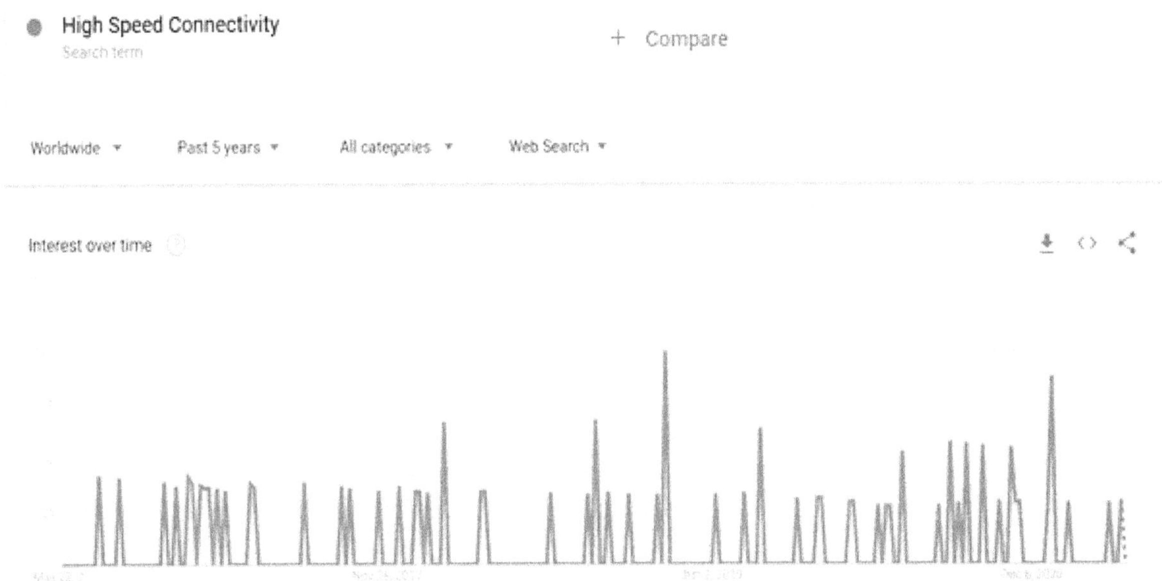

Firstly, smart personal networks move on to smart buildings and smart cities (Qadri, Nauman, Zikria, Vasilakos, & Kim, 2020). Smart private networks are made up of smartwatches, smart jackets (detect motion), etc. These smart personal networks are great as they can monitor and sometimes detect health issues that could save a life. Smart buildings are made up of various sensors ranging from light sensors to temperature sensors (Minoli, Sohraby, & Occhiogrosso, 2017). They also help regulate energy consumption through the use of the data being generated. Finally, smart cities will also become more popular using efficient traffic light control to communicate with vehicles on the road. IoT may also be used to conduct an efficient city infrastructure, develop highly efficient public transport routes, and reduce congestion on the streets. This is all possible with the introduction of 5G networks, as it increases overall bandwidth and allows massive IoT devices to connect (Arasteh et al., 2016). Figure 2 depicts the past five years of Google Trends for high-speed connectivity worldwide.

The authors (Bera et al., 2020) discussed the challenges and issues that apply to blockchain on the Internet of Drones (IoD) environment launched by the 5G-based Internet of Things (IoT). Furthermore, the authors proposed and explored the new blockchain-based secure framework for data management of IoD communication between entities. This plan can prevent many of the attacks required in an IoT-enabled IoD environment. Furthermore, the proposed scheme provides better protection, operational requirements, minimal communication and computational overhead than other related schemes.

5G in Autonomous Transport

Driving has grown to become a mainstay in the lives of recent generations. Other than its functionality, humans have found ways to conduct business or maintain their hobbies using vehicles. However, it comes with its risk. Be it due to the probability of human error or unpredictable circumstances, driving can be dangerous even as a necessity (Senders & Moray, 1991). Several innovations have become normalized in modern cars to help make driving as safe as possible, but certain aspects remain almost impossible to attain. Luckily, with the promising signs shown by 5G, these obstacles can be overcome to help make driving a better overall experience (X. Cheng, Chen, Zhang, & Yang, 2017). For example, reckless driving or lack of consideration often results in collisions between vehicles (or at best a near-miss) (Marchant & Lindor, 2012). With the help of 5G, sensors can provide warning signals in a fast enough time to avoid a collision and ensure that a minimum safe distance is always maintained between vehicles. In addition, traffic safety base stations operating through 5G can assist by foreseeing the ongoings of traffic and intervening whenever necessary (Polese et al., 2020).

As it has been labelled, vehicle platooning is another aspect of autonomous transport that will operate on 5G (Höyhtyä, M., Corici, M., Covaci, S., & Guta, 2021). Using the identical antennas as mentioned earlier, vehicles will follow a leading automotive towards a common destination. This helps to fight congested traffic while also increasing the efficiency of the commute itself. Autonomous vehicles constantly intercommunicate with others in their vicinity, allowing for decisions to be made according to the situation at hand (as opposed to a one size fits all) (Talebpour & Mahmassani, 2016). Such an approach can benefit the endless possibilities found in daily traffic at urbanized junctions, for example. Vehicle platooning also ensures that the best possible route is constantly being followed, considering current traffic conditions and distance travelled. The bandwidth required would mean that similar systems would not be possible today and can only be fully viable with 5G speeds (Shafi et al., 2017).

Lastly, autonomous transport will also relieve humans of a burden. To some, driving can be no different to a chore and an inducer of stress or anxiety. Self-driving cars operating on 5G will allow the humans inside the vehicle to be relieved of their duties (Yaqoob et al., 2020). Besides improving safety, this innovation can mean that public transport and product logistics will also grow to become more efficient than ever before. Once programmed with the appropriate regulations, self-driving vehicles will always follow speed limits, road signs and other traffic-related guidelines.

The authors (Yu, Lin, Alazab, Tan, & Gu, 2020) have studied Intelligent Transportation Systems (ITS). Autonomous vehicles face poor intention detection rates and real-time performance when predicting driving direction as ITS systems become more complicated. These concerns can have a significant impact on the safety and comfort of mixed-traffic systems. As a result, autonomous cars' capacity to forecast driving direction in real-time based on the surrounding traffic situation should be enhanced. In 5G-enabled ITS, the authors developed a deep learning-based traffic safety solution for a mix of autonomous and manual vehicles. The driving path datasets and the natural-driving datasets for long-term memory networks in 5G-enabled ITS will be employed as network inputs in the proposed system.

5G in Immersive Services

Many technologies will experience a performance enhancement with the introduction of 5G. The fast speeds and low latency can transform industries such as Immersive Services completely, allowing us to perform actions that are impossible for us to do today (Ghosh, Maeder, Baker, & Chandramouli, 2019).

Latency will be approximately ten times faster than that of 4G speeds. Lower latency allows for much greater immersive experiences in Virtual Reality and Augmented Reality (Baratè et al., 2019). This is because the lower latency better simulates real-life scenarios, and the devices being used will respond to the movement in real-time. The availability of Virtual Reality headsets has increased, and subsequently, the price has decreased, generating a more considerable demand. However, long periods of use on these devices currently cause side effects such as motion sickness and dizziness. It has been discovered that if the refresh rate is increased to at least 100Hz and the motion-to-photon latency is decreased, these side effects can be reduced (Lincoln et al., 2016). Although 5G cannot impact the screens refresh rate, it significantly impacts the motion-to-photon latency, which describes the time it takes for the user's motion to reflect on the net. The extremely low latency allows for this to happen. Furthermore, the increase in network speed will allow for better resolution to be used, further improving the user's immersive experience.

Additionally, in the immersive services sector, Massive Contents Streaming will also become possible (Qi, Hunukumbure, Nekovee, Lorca, & Sgardoni, 2016). The multimedia content demand has been rapidly increasing over the past decade, in forms such as broadcasting and movies. 5G will handle the much higher resolution and resource-demanding 4K and 8K streaming (Nightingale, Salva-Garcia, Calero, & Wang, 2018). Although HD video conferencing is currently available through our current technology, the experience is still nowhere near the face-to-face interaction experiences in real life. This is where 5G comes in; it is expected to bridge this gap and make this more of a reality through 5G massive content streaming services. The best way to bridge this gap is through the use of holographic technology. Holographic technology is required to generate and deliver a realistic hologram which requires several terabytes per second for a smooth experience. Unfortunately, this may not be achieved by 5G as of yet, but local hologram services with low data rate have the possibility of being achieved (Slinger, Cameron, & Stanley, 2005).

The authors (Numan et al., 2020) conducted a comprehensive review of available clone node identification algorithms in the literature. Furthermore, they have offered a theoretical and analytical overview of the existing centralized and distributed strategies for detecting clone nodes in static wireless sensor networks (WSNs), as well as their shortcomings and problems.

5G in Public Infrastructure

5G has proven to be of assistance in matters regarding public safety, contributing to the benefit of the population. Traditional systems involved a warning system in providing an alarm whenever torrential rain flowed through some infrastructure (Einfalt et al., 2004). This was occasionally met with issues, going off on days with no rain or taking too long to respond. 5G aims to iron out such nuances and improve on the same theory but with a more secure and stable approach. Using the inner workings of public safety networks, sensors operating close to mountains or seas closely monitored any signs of avalanches or tsunamis and sent a warning ahead of time. Preventing a natural disaster before its occurrence can mean the difference between thousands of people being dead or alive by the end of it all

(Bostrom, 2013). Thus, 5G, with its low latency and high stability infrastructure, can be the medium used for such systems to get their message across (Sachs et al., 2019). Minimal time is required for the message itself to arrive, compared to what it would take prior. Other innovations include the monitoring of forest or nuclear power plants in a forest fire or radioactive leak. In the future, 5G networks can also start to accommodate satellites forming part of their structure to keep long-distance networks alive even when disaster strikes. The essential requirements for such systems can only be fully viable when combined with hyper-reliability and hyper-energy efficiency to bring the risk rate down as much as physically possible (Alcaraz-Calero et al., 2018).

Another application of this innovation built upon a similar idea would be the need for networks throughout disaster situations. Critical data will have to be sent in real-time (other than the standard communication that goes through), as can be the case with patients requiring immediate treatment (Epstein, R. H., Dexter, F., & Patel, 2015). For example, governments had to set up institutions to treat public members who fell ill from the virus regarding the Coronavirus pandemic. Due to the lack of preparation for the large-scale task, temporary hospitals were also set up in places that had been previously used for other activities, such as sporting centres. In addition, connectivity in homes will have to be set up to monitor the area's safety and be done through infrastructure-less networks (Daneels et al., 2017). These are made up of mobile devices to ensure that minimal telecommunications devices are needed and the place is up and running as soon as possible. Should a disaster happen, 5G is also prepared to accustom to it with little-to-no extra hassle needed since its network can operate functionally on publicly accessible devices such as mobile phones or routers similar to those found in homes (García Moro, 2020).

After the 5G standards are finalized, everybody may expect a slew of improvements in their daily lives. There will also be substantial advancements in artificial intelligence, self-driving cars, IoT devices, and security. In many respects, the accomplishments discussed in this paper will make life easier and more comfortable. With efficient transportation, fewer accidents, less pollution, fewer criminal chances, and safer living, smart cities are the way of the future. However, there are potential health problems with 5G networks, which will hopefully be comprehensively addressed soon, allowing all of the benefits of fifth-generation mobile technology to be deployed with the fewest possible health hazards (Irving, 2006). The significant findings of the literature survey are summarized in Table 1.

ADVANTAGES OF 5G

The newer technology brings with it several benefits as 5G in comparison to its older counterparts. What interests end-users is the higher speeds, as they promise faster downloads, more streaming and a more stable Internet experience. Excluding the hype being built for marketing purposes, 5G will be a noteworthy improvement on what already exists. It will also contribute to making Internet use more stable (Oughton et al., 2021). By implementing cells, urbanized areas will be able to have their connectivity more centralized. Even with their limited strength against obstacles, millimetre antennas can be more accustomed to densely populated areas and targeting use within a neighbourhood or two. The end-users with compatible devices will see boosts in speeds and lower latency when connecting to other devices (Kar & Sanyal, 2018).

5G wireless network connectivity is set to transform nearly every industry; it will change how citizens learn, work, communicate and commute. It will make companies more productive, manufacturers more competitive, and healthcare better and accessible. In our case, we will look at how 5G is transforming

High-Speed Connectivity

Table 1. Significant attributes of high-speed connectivity.

Attribute	Description	Reference
5G in the Internet of Things	• Next-generation 5G wireless networks are expected to employ emerging technologies like mmWave, massive MIMO, and C-RAN to deliver the broad connectivity, resource pooling, and energy efficiency required for commercial IoT deployments. • On IoT-enabled devices, application traffic, is likely to range from static, intermittent, delay-tolerant tiny packets to mobile, frequent, delay-sensitive huge packets. • A wide range of IoT applications should be delay-tolerant. • The Internet of Things (IoT) has had a significant impact on the evolution of the healthcare industry, leading to the creation of Healthcare IoT (H-IoT) solutions. • The fog/edge concept brings processing power closer to the deployed network, reducing the number of issues. • Software Defined Networks (SDNs) provide the system more flexibility, while blockchains find new H-IoT systems applications. H-IoT applications are being driven by the Internet of Nano Things (IoNT) and Tactile Internet (TI). • Smart cities, smart grids, smart homes, physical security, e-health, asset management, and logistics are just a few of the areas where the Internet of Things (IoT) is being used. • With city-wide deployments of enhanced street lighting controls, infrastructure monitoring, public safety and surveillance, physical security, gunshot detection, meter reading, and transportation analysis and optimization systems, the concept of smart cities is gaining popularity across the world. • Support for IoT-enabled smart buildings is a relevant and cost-effective user-level IoT application.	(Saxena et al., 2017); (Qadri et al., 2020); (Minoli et al., 2017); (Arasteh et al., 2016); (Raddo et al., 2021); (Alli & Alam, 2020).
5G in Autonomous Transport	• Several businesses, like Google, Tesla Motors, and Baidu, have invested significant time and money in developing self-driving automobiles. • Self-driving vehicles have aroused unprecedented media attention, sparking conjecture about their impact and ramifications on societal concerns such as road safety, privacy, traffic flow, energy and environmental challenges, land usage, automobile industry profitability, and cybersecurity. • Using wireless backhaul lines to relay access traffic, Internet Access and Backhaul (IAB) is being studied to minimize the deployment costs of ultra-dense 5G mmWave networks. • The development and administration of a network topology that is optimal are critical for efficient IAB operations. Indeed, the whole network's end-to-end performance. • Intelligent activities like collision avoidance, lane departure warning and traffic sign detection relieve human drivers' responsibilities. • Autonomous cars help people in their everyday lives by providing dependable and safe transportation for the elderly and disabled, resolving parking issues, and reducing the number of accidents caused by human mistake.	(X. Cheng et al., 2017); (Polese et al., 2020); (Höyhtyä, M., Corici, M., Covaci, S., & Guta, 2021); (Shafi et al., 2017); (Yaqoob et al., 2020)
5G in Immersive Services	• Autonomous cars help people in their everyday lives by providing dependable and safe transportation for the elderly and disabled, resolving parking issues, and reducing the number of accidents caused by human mistake. • Intelligent activities like collision avoidance, lane departure warning and traffic sign detection relieve human drivers' responsibilities. • Augment Reality aids students with disabilities by integrating their learning through suitable visual, aural, and haptic interfaces. • Virtual Reality is ideal for lab activities that realistically require manipulating items and conducting instruction without the risk of learners making mistakes. • Users' accurate geolocation and orientation, as well as low-latency two-way interactions. These experiences may now be enjoyed on personal devices, distance, and mobility, but using 5G, opening up a slew of new educational possibilities. • Two main expectations for 5G networks are the capacity to handle ultra-high-definition (UHD) video streaming and the delivery of services that meet the demands of the end user's perceived quality by employing quality of experience (QoE) aware network management technology. • Video applications that need real-time and ultra-high quality transmission, such as mobile broadcasting, remote surgery, and augmented reality, are projected to dominate traffic on 5G mobile networks.	(Ghosh et al., 2019); (Baratè et al., 2019); (Qi et al., 2016); (Nightingale et al., 2018);
5G in Public Infrastructure	• Remote industrial applications in hazardous environments, such as remote-controlled driving in a fully automated intelligent transportation system, to remote surgery. Unique expert skills can be delivered to various locations worldwide, which are all examples of tactile internet use cases. • 5G provides the necessary capabilities for wirelessly linked operators or teleoperated systems to satisfy their high communication needs regarding reliability and reduced latency. • High data speeds, low end-to-end latency, massive connection, ultra-reliability, and ubiquitous support for extremely high mobility offer end-users an unparalleled experience. • Thousands of researchers and developers from throughout Europe are working on 5G architecture, a flexible RAN, new spectrum options, front haul and backhaul solutions, virtualized networks, and other issues. • Enhanced mobile broadband (EMBB), ultra-reliable and low latency communications (URLLC), and colossal machine type communications are among the 5G service classifications (MMTC).	(Sachs et al., 2019); (Alcaraz-Calero et al., 2018); (García Moro, 2020)
5G applications	• 5G can be applied in various healthcare sectors, including telemedicine, remote surgery and wearable notifications. • 5G network sends data to the based cloud server. • 5G improves collaboration between healthcare systems. • Connecting physicians/surgeons in different hospitals. • Telesurgery becoming more feasible. • Monitoring of irregular health signs • Better ambulances with a real-time connection with the hospital • 5G will bring around a revolution in healthcare.	Latif et al., 2017), (Ullah et al., 2019), (Sigwele et al., 2018); (Soldani et al., 2017); (Choi, P. J., Oskouian, R. J., & Tubbs, 2018); (Oleshchuk & Fensli, 2011) (Dananjayan & Raj, 2020)
5G Specifications	• MIMO technology • High carrier frequencies, many base stations and antennae • Better capacity, low latency, higher data rates and a higher number of linked devices • Low-band, mid-band, and high-band • 5G devices need to be effective, comfortable, personalized, sustainable and smart	(Bogale & Le, 2016); (Andrews, J. G., Buzzi, S., Choi, W., Hanly, S. V., Lozano, A., Soong, A. C., & Zhang, 2014); (Sharma et al., 2018); (Wang et al., 2014); (Yang et al., 2015).

the healthcare sector and the industry that comes with it (Latif et al., 2017). 5G could aid healthcare in significant ways by providing telehealth, remote surgery, real-time monitoring, and delivering treatment information and support to patients continually because of the reliable networking, speed, and scale (Darrell M. West, 2009).

5G will also contribute to the IoT devices around us, making communication between them more efficient. Given these technologies constantly send and receive data as streams, devices of all uses and sizes will be available to operate at improved speeds and a greater degree of reliability. Vehicles, for example, will be available to work autonomously and be safer since they can potentially put people's lives at risk in the event of a disconnection or a disruption of the network. Online gaming and real-time applications of the Internet will see improvement(Jiang et al., 2016). They will begin to operate on lower latencies and have a minor delay from the end users' perspective. Multisensory digital content such as VR, AR and 3D experiences will communicate their information between the network of devices also using 5G, making the experience all the more immersive. The newer infrastructure is responsible for all of this, as it allows for more simultaneous communication to occur with little-to-no interruptions. As can be depicted, the primary point to take away from 5G as a whole will be the applications making use of improved speeds, improved latency and a more stable connection (J. Cheng, Chen, Tao, & Lin, 2018).

LIMITATIONS AND PROBABLE SOLUTIONS

The fifth-generation (5G) is making the Internet of Things (IoT) reality possible as it can handle more devices simultaneously (Li, Xu, & Zhao, 2018). However, just like with any other new technology, this generation also brings about several limitations.

They are starting with its most prominent limitation, its inability to penetrate efficiently through physical objects. The range of 5G has been dramatically reduced compared to its predecessors. Therefore, obstructions may impact the connectivity significantly. Obstacles such as trees, walls, and buildings may block or even absorb the high-frequency signals leading to a non-stable connection. Secondly, the initial costs to deploy a 5G infrastructure are very high. Existing cellular infrastructure will need to be adapted, which will require a lot of funding. But the cost does not stop there, as the price will keep increasing by the continuous need for maintenance to guarantee the high-speed connectivity it promises to deliver. Thirdly, 5G connectivity is mainly focused on benefitting people in urban areas. People living in rural areas will find it challenging to benefit from 5G unless they install the required technology close by. Carriers will target big cities first to cater to most people and deploy 5G technology in rural areas, but this will not happen anytime soon (Wigren, Colombi, Thors, & Berg, 2016).

Additionally, most people already have a problem with their current battery capacity on their mobile devices, as it does not last the whole day in most cases. The introduction of 5G will cause additional strain on battery usage and causes the battery to drain much faster (Ng, Peng, Faegh, & Mustain, 2020). Along with this, users have been reporting their devices getting increasingly hot whilst operating on 5G. Thus, advancement in battery technology is required to be able to handle 5G. Furthermore, the download speeds of 5G are breakneck and may reach up to almost 2 Gbps. However, upload speeds are not as impressive, rarely reaching speeds over 100Mbps, which is about 5% the speed of the download speed. However, this being said, upload speeds are still higher than what is experienced on 4G networks. Finally, from a visual perspective, the increase in the number of cellphone towers will be quite an eye-sore and will not be welcomed easily by many communities. With the rise in development, the older generation will certainly not appreciate the increase in speed with the cost of depleting the visual scene (Narayanan et al., 2020).

5G is projected to bring around a revolution in communication. This will enable healthcare systems to provide a more dynamically and enhanced service. However, 5G cannot do everything as, like all other technologies, it has its limitations. Apart from this, some theories noted in academic articles suggested that 5G creates health risks due to 5G's higher radiofrequency (Dananjayan & Raj, 2020). The scientific evidence, however, does not support any of these claims. This is agreed upon by worldwide known NGOs such as the National Cancer Society, WHO, FCC and others.

The first drawback 5G has is the small range compared to previous generations since 5G uses higher frequency bands. Furthermore, the connection is easily disrupted when there is an obstacle between the device and the antennae. This limitation will make it easier to lose access to 5G and switch back to other networks (Boccardi, Heath, R. W., Marzetta, Popovski, & Lozano Solsona, 2013).

In some places where 5G antennas have already been installed, some residents are resisting these installations. This is because these antennae are being installed near homes, presenting two main concerns for residents. Some complained about the aesthetics of their antennas, while some are worried about harmful electromagnetic waves being emitted near where they live. The reason for installing them near homes is due to the short effective range of this technology. These cellular radios are installed as close as hundreds of feet apart (Mercola, 2020).

Another significant limitation is the costs of developing 5G and other related infrastructure. That is just the upfront cost. There are also increased upholding costs which is a big issue. New strategies have to be adopted to offset the vast costs as only some of these costs will be put on the customers. Standard measures such as cost-saving efforts may not be enough to bear the price, so that some alternative approaches may be needed. These may include network sharing between the service providers and finding new revenue models (Mercola, 2020).

One of the most significant issues for 5G is rural access. Some of the population may not even have a cellular connection, let alone providing them with 5G access. As many people live in cities rather than scattered evenly around the country, it might be easier to offer 5G access. However, a percentage of the population may and will be left behind in advancing this technology (Mercola, 2020). Rural areas lack access to the best healthcare; eliminating this problem would be great, but it is impractical from an economic point of view. It is proposed to get rural areas a form of 5G called "low-band", which has less capacity, it still has lower latency which is very important for healthcare with 5G. Speeds, however, may not even be equal to the average of 4G speeds we are used to today (Akpakwu, Silva, Hancke, & Abu-Mahfouz, 2017).

APPLICATIONS AND CURRENT IMPLEMENTATIONS

As previously explained briefly, 5G will bring about several applications to make use of the improvements being made. The IoT devices that consist of smartphones, smart devices, wearable technologies, and other micro-devices will all use 5G (Miraz, Ali, Excell, & Picking, 2018). Their requirement for a connection to send and receive a stream of data continuously and constantly can be fully utilized. Communication using the devices mentioned above can happen faster, allowing for analysis of such data in analytics to be done and processed more quickly. 5G technology will assist those working with big data to receive the data quicker and always expect more (Memon et al., 2019).

Another application can be seen in medical treatment for a similar reason. Doctors can make their services available across long distances and overseas. Remote parts of the world can be given dedicated assistance in a situation similar to the one we are in now (regarding the pandemic). Of course, sensitive information would require a degree of technological security to be transmitted over the Internet. Still, it can say that once such hurdles are accustomed to, this solution can become a very much viable one. To a similar extent, education and businesses can use 5G to benefit for the same reasons. What might have disrupted online lecturing and distanced learning previously, such as delayed responses and connections being cut off, will now be a thing of the past thanks to 5G. On the other hand, it means the greater spread of helpful information occurs in an ideal world and attracts those to harm. Spam and fake news, amongst other things, will increase correspondingly, and malicious content ensues anyway, unfortunately (SJ, J., & RD, 2015).

Immersive experiences that are based on Virtual and Augmented Reality will have their features enhanced. Lower delays between the interconnected devices, especially those wireless, will be next to nothing due to the improved technology. The previously mentioned medical treatment can occur through a merger of two technologies - VR and 5G. With enhanced speeds and reduced latency times, operations taking place through Virtual Reality can occur with more accuracy and less risk involved, something very much crucial with such things (Orlosky, Kiyokawa, & Takemura, 2017). Minds will be put to rest (if not totally, at least more than previously) when medical interventions start to occur in this way and become more normalized. Furthermore, 5G will allow the technology to be marketed as a safe option, regardless if it is a bit abstract to the average person, further making it easier to accept for the general public.

CONCLUSION AND FUTURE PERSPECTIVES

With the 5G infrastructure being deployed currently, the research on 6G connectivity has already begun. 6G is the next generation of telecommunication which will be the successor to 5G. With a new generation being implemented once approximately every ten years, it is expected for 6G rollouts to begin before 2030. The rapid development of various emerging technologies created a need for the sixth generation (6G). Artificial Intelligence, Virtual Reality, 3D media and the Internet of Things are a few examples of such rapid technologies. The rapid increase in data creation may be seen from the difference between 2010 and the present. Whereas in 2010, the global mobile traffic was at 7.462 EB/month, it is expected to reach 5016 EB/month by 2030, which 5G networks will not support. Additionally, 6G is expected to be extremely fast, reaching speeds up to 100 times faster than 4G and five times faster than speeds reached by 5G.

6G mobile communication will be implemented on technology that has already been established for 5G communication, but it will require further development. 6G makes use of millimetre-wave technologies. The use of these higher frequencies in the spectrum allows much larger bandwidth, saying it might enter the Terahertz region. Furthermore, dense networks will become a must because of the short range of the higher frequencies. Reducing the size of cells will allow more efficient use of the 6G communication; these will be deployed in the forms of femtocells, which can easily be installed indoors. Finally, massive MIMO will be required; the use of microwave frequencies opens up using multiple antennas on one device. To handle the problematic issue of UAVs and ground infrastructure is adopting different communication technologies, the potential solutions by analyzing several wireless technologies such as cellular (3G, 4G LTE), WiMAX, and Wi-Fi. The drones establish a FANET to achieve effective UAV networking in which they communicate ad hoc with one another. They show device-to-device or drone-to-drone conversations, both of which are viable options for this form of communication.

There are too many applications of 6G to name them all, but to name a few, these would include; Split Computing, which is considered one of the most sought after features coming from 6G. This application will allow computing power to be split between mobile devices and cloud servers, which will lower equipment costs and increase the processing power of mobile devices, all performed through the cloud. Another application could be the super-smart society. The speeds that 6G benefits from will accelerate the building of super-smart communities, leading to improved quality of life and further introducing automation. Finally, with 5G not having the capability to support real-time holographic technology, 6G will be the key to making this a possibility. With its breakneck speeds and large bandwidth, not to mention the extremely low latency, real-time holographic technology will no longer be considered the future.

REFERENCES

Abidi, M. H., Alkhalefah, H., Moiduddin, K., Alazab, M., Mohammed, M. K., Ameen, W., & Gadekallu, T. R. (2021). Optimal 5G network slicing using machine learning and deep learning concepts. *Computer Standards & Interfaces*, *76*, 103518. Advance online publication. doi:10.1016/j.csi.2021.103518

Akpakwu, G. A., Silva, B. J., Hancke, G. P., & Abu-Mahfouz, A. M. (2017). A Survey on 5G Networks for the Internet of Things: Communication Technologies and Challenges. *IEEE Access: Practical Innovations, Open Solutions*, *6*, 3619–3647. doi:10.1109/ACCESS.2017.2779844

Alcaraz-Calero, J., Belikaidis, I. P., Cano, C. J. B., Bisson, P., Bourse, D., Bredel, M., ... Wang, Q. (2018). Leading innovations towards 5G: Europe's perspective in 5G Infrastructure Public-Private Partnership (5G-PPP). *IEEE International Symposium on Personal, Indoor and Mobile Radio Communications, PIMRC,* 1–5. 10.1109/PIMRC.2017.8292654

Alli, A. A., & Alam, M. M. (2020). The fog cloud of things: A survey on concepts, architecture, standards, tools, and applications. *Internet of Things*, *9*, 100177. doi:10.1016/j.iot.2020.100177

Andrews, J. G., Buzzi, S., Choi, W., Hanly, S. V., Lozano, A., Soong, A. C., & Zhang, J. C. (2014). What Will 5G Be? *IEEE Journal on Selected Areas in Communications*, *32*(6), 1065–1082. doi:10.1109/JSAC.2014.2328098

Anju, M., & Gawas, U. (2015). An Overview on Evolution of Mobile Wireless Communication Networks : 1G-6G. *Ijritcc. Org*, *3*(5), 3130–3133.

Arasteh, H., Hosseinnezhad, V., Loia, V., Tommasetti, A., Troisi, O., Shafie-Khah, M., & Siano, P. (2016). Iot-based smart cities: A survey. *EEEIC 2016 - International Conference on Environment and Electrical Engineering*. 10.1109/EEEIC.2016.7555867

Baratè, A., Haus, G., Ludovico, L. A., Pagani, E., & Scarabottolo, N. (2019). 5G Technology for Augmented and Virtual Reality in Education. *Education and New Developments*, *2019*(1), 512–516. doi:10.36315/2019v1end116

Bera, B., Saha, S., Das, A. K., Kumar, N., Lorenz, P., & Alazab, M. (2020). Blockchain-Envisioned Secure Data Delivery and Collection Scheme for 5G-Based IoT-Enabled Internet of Drones Environment. *IEEE Transactions on Vehicular Technology*, *69*(8), 9097–9111. doi:10.1109/TVT.2020.3000576

Bhalla, M. R., & Bhalla, A. V. (2010). Generations of Mobile Wireless Technology: A Survey. *International Journal of Computers and Applications*, *5*(4), 26–32. doi:10.5120/905-1282

Boccardi, F., & Heath, R. W., Marzetta, T. L., Popovski, P., & Lozano Solsona, A. (2013). Five Disruptive Technology Directions for 5G. *IEEE Communications Magazine*, (February), 74–80.

Bogale, T. E., & Le, L. B. (2016). Massive MIMO and mmWave for 5G Wireless HetNet: Potential Benefits and Challenges. *IEEE Vehicular Technology Magazine*, *11*(1), 64–75. doi:10.1109/MVT.2015.2496240

Bojic, D., Sasaki, E., Cvijetic, N., Ting Wang, T., Kuno, J., Lessmann, J., Schmid, S., Ishii, H., & Nakamura, S. (2013). Advanced wireless and optical technologies for small-cell mobile backhaul with dynamic software-defined management. *IEEE Communications Magazine*, *51*(9), 86–93. doi:10.1109/MCOM.2013.6588655

Bostrom, N. (2013). Existential risk prevention as global priority. *Global Policy*, *4*(1), 15–31. doi:10.1111/1758-5899.12002

Capozzi, F., Piro, G., Grieco, L. A., Boggia, G., & Camarda, P. (2013). Downlink packet scheduling in LTE cellular networks: Key design issues and a survey. *IEEE Communications Surveys and Tutorials*, *15*(2), 678–700. doi:10.1109/SURV.2012.060912.00100

Cheng, J., Chen, W., Tao, F., & Lin, C. L. (2018). Industrial IoT in 5G environment towards smart manufacturing. *Journal of Industrial Information Integration*, *10*, 10–19. doi:10.1016/j.jii.2018.04.001

Cheng, X., Chen, C., Zhang, W., & Yang, Y. (2017). 5G-enabled cooperative intelligent vehicular (5GenCIV) Framework: When Benz Meets Marconi. *IEEE Intelligent Systems*, *32*(3), 53–59. doi:10.1109/MIS.2017.53

Chiussi, F. M., Khotimsky, D. A., & Krishnan, S. (2002). Mobility management in third-generation all-IP networks. *IEEE Communications Magazine*, *40*(9), 124–135. doi:10.1109/MCOM.2002.1031839

Choi, P. J., Oskouian, R. J., & Tubbs, R. S. (2018). Telesurgery: Past, Present, and Future. *Cureus*, *10*(5). https://www.ncbi.nlm.nih.gov/pmc/articles/PMC6067812/ PMID:30079282

Cox, C. (2012). An Introduction to LTE: LTE, LTE-Advanced, SAE and 4G Mobile Communications. doi:10.1002/9781119942825

Curwen, P., & Whalley, J. (2014). Mobile Telecommunications Networks: Restructuring as a Response to a Challenging Environment. Academic Press.

Dananjayan, S., & Raj, G. M. (2020). 5G in healthcare: How fast will be the transformation? *Irish Journal of Medical Science*. Advance online publication. doi:10.100711845-020-02329-w PMID:32737688

Daneels, G., Municio, E., Spaey, K., Vandewiele, G., Dejonghe, A., Ongenae, F., ... Famaey, J. (2017). Real-Time data dissemination and analytics platform for challenging IoT environments. *2017 Global Information Infrastructure and Networking Symposium, GIIS 2017*, 23–30. 10.1109/GIIS.2017.8169799

Darrell, M. W. (2009). How 5G technology enables the health Internet of Things. *Cyber Resilience of Systems and Networks*, (July), 1–150. Retrieved from https://link.springer.com/10.1007/978-3-319-77492-3_16

Einfalt, T., Arnbjerg-Nielsen, K., Golz, C., Jensen, N.-E., Quirmbach, M., Vaes, G., & Vieux, B. (2004). Towards a roadmap for use of radar rainfall data in urban drainage. *Journal of Hydrology (Amsterdam)*, *299*(3–4), 186–202. doi:10.1016/S0022-1694(04)00365-8

Epstein, R. H., Dexter, F., & Patel, N. (2015). Influencing Anesthesia Provider Behavior Using Anesthesia Information Management System Data for Near Real-Time Alerts and Post Hoc Reports. *Anesthesia and Analgesia*, *121*(5), 1404. doi:10.1213/ANE.0000000000001038 PMID:26262500

García Moro, F. (2020). The Death and Life of Hong Kong's Illegal Façades. *ARENA Journal of Architectural Research*, *5*(1), 2. Advance online publication. doi:10.5334/ajar.231

Ghosh, A., Maeder, A., Baker, M., & Chandramouli, D. (2019). 5G Evolution: A View on 5G Cellular Technology beyond 3GPP Release 15. *IEEE Access: Practical Innovations, Open Solutions*, *7*, 127639–127651. doi:10.1109/ACCESS.2019.2939938

Hossain, S. (2013). 5G wireless communication systems. *American Journal of Engineering Research*, *2*(10), 344–353.

Höyhtyä, M., Corici, M., Covaci, S., & Guta, M. (2021). 5G and beyond for new space: vision and research challenges. *Advances in Communications Satellite Systems: Proceedings of the 37th International Communications Satellite Systems Conference (ICSSC-2019)*, 387–402. 10.1049/PBTE095E_ch30

Eze, Sadiku, & Musa. (2018). 5G Wireless Technology: A Primer. International *Journal of Science, Engineering and Technology*, *7*(July), 62–64.

Irving, K. E. (2006). The impact of technology on the 21st century. *Teaching Science in the 21st Century*, (March), 3–19. Retrieved from http://books.google.com/books?id=g5NflcuxkJcC&pgis=1

Jiang, J., Das, R., Ananthanarayanan, G., Chou, P. A., Padmanabhan, V. N., Sekar, V., . . . Zhang, H. (2016). VIA: Improving internet telephony call quality using predictive relay selection. *SIGCOMM 2016 - Proceedings of the 2016 ACM Conference on Special Interest Group on Data Communication*, 286–299. 10.1145/2934872.2934907

Kar, U. N., & Sanyal, D. K. (2018). An overview of device-to-device communication in cellular networks. *ICT Express*, *4*(4), 203–208. doi:10.1016/j.icte.2017.08.002

Latif, S., Qadir, J., Farooq, S., & Imran, M. A. (2017). How 5G wireless (and Concomitant Technologies) will revolutionize healthcare? *Future Internet*, *9*(4), 93. Advance online publication. doi:10.3390/fi9040093

Lei, L., Zhong, Z., Lin, C., & Shen, X. (2012). Operator controlled device-to-device communications in LTE-advanced networks. *IEEE Wireless Communications*, *19*(3), 96–104. doi:10.1109/MWC.2012.6231164

Li, S., Da Xu, L., & Zhao, S. (2018). 5G Internet of Things: A survey. *Journal of Industrial Information Integration*, *10*, 1–9. doi:10.1016/j.jii.2018.01.005

Lincoln, P., Blate, A., Singh, M., Whitted, T., State, A., Lastra, A., & Fuchs, H. (2016). From Motion to Photons in 80 Microseconds: Towards Minimal Latency for Virtual and Augmented Reality. *IEEE Transactions on Visualization and Computer Graphics*, *22*(4), 1367–1376. doi:10.1109/TVCG.2016.2518038 PMID:26780797

Marchant, G. E., & Lindor, R. A. (2012). The Coming Collision Between Autonomous Vehicles and the Liability System. *Santa Clara Law Review*, *52*(4), 1321.

Memon, I., Fazal, H., Ahmed Shaikh, R., Muhammad, G., Arain, Q. A., & Khatri, T. K. (2019). *Big Data, Cloud, 5G Networks Create Smart and Intelligent World: A Survey*. Academic Press.

Mercola, J. (2020). *EMFD 5G, Wi-Fi Cell Phones Hidden Harms and How to Protect Yourself*. Hay House.

Minoli, D., Sohraby, K., & Occhiogrosso, B. (2017). IoT Considerations, Requirements, and Architectures for Smart Buildings-Energy Optimization and Next-Generation Building Management Systems. *IEEE Internet of Things Journal*, *4*(1), 269–283. doi:10.1109/JIOT.2017.2647881

Miraz, M. H., Ali, M., Excell, P. S., & Picking, R. (2018). *Internet of nano-things, things and everything: Future growth trends*. ArXiv.

Narayanan, A., Ramadan, E., Carpenter, J., Liu, Q., Liu, Y., Qian, F., & Zhang, Z. L. (2020). A First Look at Commercial 5G Performance on Smartphones. *The Web Conference 2020 - Proceedings of the World Wide Web Conference, WWW 2020*, 894–905. 10.1145/3366423.3380169

Ng, B., Peng, X., Faegh, E., & Mustain, W. E. (2020). Using nanoconfinement to inhibit the degradation pathways of conversion-metal oxide anodes for highly stable fast-charging Li-ion batteries. *Journal of Materials Chemistry. A, Materials for Energy and Sustainability*, *8*(5), 2712–2727. doi:10.1039/C9TA11708C

Nightingale, J., Salva-Garcia, P., Calero, J. M. A., & Wang, Q. (2018). 5G-QoE: QoE modelling for ultra-HD video streaming in 5G networks. *IEEE Transactions on Broadcasting*, *64*(2), 621–634. doi:10.1109/TBC.2018.2816786

Numan, M., Subhan, F., Khan, W. Z., Hakak, S., Haider, S., Reddy, G. T., Jolfaei, A., & Alazab, M. (2020). A Systematic Review on Clone Node Detection in Static Wireless Sensor Networks. *IEEE Access: Practical Innovations, Open Solutions*, *8*, 65450–65461. doi:10.1109/ACCESS.2020.2983091

Oleshchuk, V., & Fensli, R. (2011). Remote patient monitoring within a future 5G infrastructure. *Wireless Personal Communications*, *57*(3), 431–439. doi:10.100711277-010-0078-5

Orlosky, J., Kiyokawa, K., & Takemura, H. (2017). Virtual and augmented reality on the 5G highway. *Journal of Information Processing*, *25*(0), 133–141. doi:10.2197/ipsjjip.25.133

Oughton, E. J., Lehr, W., Katsaros, K., Selinis, I., Bubley, D., & Kusuma, J. (2021). Revisiting Wireless Internet Connectivity: 5G vs Wi-Fi 6. *Telecommunications Policy*, *45*(5), 102127. Advance online publication. doi:10.1016/j.telpol.2021.102127

Pereira, V., & Sousa, T. (2004). *Evolution of Mobile Communications: from 1G to 4G*. Academic Press.

Pham, Q. V., Mirjalili, S., Kumar, N., Alazab, M., & Hwang, W. J. (2020). Whale Optimization Algorithm with Applications to Resource Allocation in Wireless Networks. *IEEE Transactions on Vehicular Technology*, *69*(4), 4285–4297. doi:10.1109/TVT.2020.2973294

Polese, M., Giordani, M., Zugno, T., Roy, A., Goyal, S., Castor, D., & Zorzi, M. (2020). Integrated Access and Backhaul in 5G mmWave Networks: Potential and Challenges. *IEEE Communications Magazine*, *58*(3), 62–68. doi:10.1109/MCOM.001.1900346

Qadri, Y. A., Nauman, A., Zikria, Y., Vasilakos, A. V., & Kim, S. W. (2020). The Future of Healthcare Internet of Things: A Survey of Emerging Technologies. *IEEE Communications Surveys and Tutorials*, *22*(2), 1121–1167. doi:10.1109/COMST.2020.2973314

Qi, Y., Hunukumbure, M., Nekovee, M., Lorca, J., & Sgardoni, V. (2016). Quantifying data rate and bandwidth requirements for immersive 5G experience. *2016 IEEE International Conference on Communications Workshops, ICC 2016*, 455–461. 10.1109/ICCW.2016.7503829

Raddo, T. R., Rommel, S., Cimoli, B., Vagionas, C., Perez-Galacho, D., Pikasis, E., … Tafur Monroy, I. (2021). Transition technologies towards 6G networks. *Eurasip Journal on Wireless Communications and Networking, 2021*(1). doi:10.1186/s13638-021-01973-9

Rajput, D. S., Basha, S. M., Xin, Q., Gadekallu, T. R., Kaluri, R., Lakshmanna, K., & Maddikunta, P. K. R. (2021). Providing diagnosis on diabetes using cloud computing environment to the people living in rural areas of India. *Journal of Ambient Intelligence and Humanized Computing*. Advance online publication. doi:10.100712652-021-03154-4

Rommer, S., Hedman, P., Olsson, M., Frid, L., Sultana, S., & Mulligan, C. (2019). *5G core networks: Powering digitalization. In 5G Core Networks*. Powering Digitalization. doi:10.1016/C2018-0-01335-3

Sachs, J., Andersson, L. A. A., Araujo, J., Curescu, C., Lundsjo, J., Rune, G., Steinbach, E., & Wikstrom, G. (2019). Adaptive 5G low-latency communication for tactile internet services. *Proceedings of the IEEE, 107*(2), 325–349. doi:10.1109/JPROC.2018.2864587

Saxena, N., Roy, A., Sahu, B. J. R., & Kim, H. (2017). Efficient IoT Gateway over 5G Wireless: A New Design with Prototype and Implementation Results. *IEEE Communications Magazine*, *55*(2), 97–105. doi:10.1109/MCOM.2017.1600437CM

Senders, J., & Moray, N. (1991). *Human error: Cause, prediction and reduction*. Academic Press.

Shafi, M., Fellow, L., Molisch, A. F., Smith, P. J., Haustein, T., Zhu, P., ... Member, S. (2017). 5G: A Tutorial Overview of Standards, Trials, Challenges, Deployment, and Practice. *IEEE Journal on Selected Areas in Communications*, *35*(6), 1201–1221. doi:10.1109/JSAC.2017.2692307

Sharma, S. K., Bogale, T. E., Le, L. B., Chatzinotas, S., Wang, X., & Ottersten, B. (2018). Dynamic Spectrum Sharing in 5G Wireless Networks with Full-Duplex Technology: Recent Advances and Research Challenges. *IEEE Communications Surveys and Tutorials*, *20*(1), 674–707. doi:10.1109/COMST.2017.2773628

Shikhare, G., & Shaikh, A. (2014). 4G LTE Technology. *International Journal of Networking and Parallel Computing*, *2*(03), 110–117.

Sigwele, T., Hu, Y. F., Ali, M., Hou, J., Susanto, M., & Fitriawan, H. (2018). Intelligent and Energy Efficient Mobile Smartphone Gateway for Healthcare Smart Devices Based on 5G. *2018 IEEE Global Communications Conference, GLOBECOM 2018 - Proceedings*. 10.1109/GLOCOM.2018.8648031

SJ. C., J., K., & RD, G. (2015). Teledermatology: From historical perspective to emerging techniques of the modern era: Part I: History, rationale, and current practice. *Journal of the American Academy of Dermatology, 72*(4), 563–574. Retrieved from http://www.embase.com/search/results?subaction=viewrecord&from=export&id=L604736181%255Cnhttp://dx.doi.org/10.1016/j.jaad.2014.07.061%255Cnhttp://elvis.ubvu.vu.nl:9003/vulink?sid=EMBASE&issn=10976787&id=doi:10.1016%252Fj.jaad.2014.07.061&atitle=Teledermatology%25

Slinger, C., Cameron, C., & Stanley, M. (2005). Computer-generated holography as a generic display technology. *Computer*, *38*(8), 46–53. doi:10.1109/MC.2005.260

Soldani, D., Fadini, F., Rasanen, H., Duran, J., Niemela, T., Chandramouli, D., ... Nanavaty, N. (2017). 5G Mobile Systems for Healthcare. *IEEE Vehicular Technology Conference*. 10.1109/VTC-Spring.2017.8108602

Steinmetz, R. (2012). *Multimedia: Computing communications & applications*. Academic Press.

Strinati, E. C., Barbarossa, S., Gonzalez-Jimenez, J. L., Ktenas, D., Cassiau, N., Maret, L., & Dehos, C. (2019). sixth generation (6G) has already started taking place, and it is expected to be implemented between 2027 and 2030. *IEEE Vehicular Technology Magazine*, *14*(3), 42–50.

Talebpour, A., & Mahmassani, H. S. (2016). Influence of connected and autonomous vehicles on traffic flow stability and throughput. *Transportation Research Part C, Emerging Technologies*, *71*, 143–163. doi:10.1016/j.trc.2016.07.007

Trends, G. (2021a). *5G in Internet of Things*. Retrieved May 15, 2021, from 2021 website: https://trends.google.com/trends/explore?date=today5-y&q=5GinInternetofThings

Trends, G. (2021b). *High Speed Connectivity*. Retrieved May 15, 2021, from 2021 website: https://trends.google.com/trends/explore?date=today5-y&q=HighSpeedConnectivity

Ullah, H., Gopalakrishnan Nair, N., Moore, A., Nugent, C., Muschamp, P., & Cuevas, M. (2019). 5G communication: An overview of vehicle-to-everything, drones, and healthcare use-cases. *IEEE Access: Practical Innovations, Open Solutions*, *7*, 37251–37268. doi:10.1109/ACCESS.2019.2905347

Wang, Y., Li, J., Huang, L., Jing, Y., Georgakopoulos, A., & Demestichas, P. (2014). 5G mobile: Spectrum broadening to higher-frequency bands to support high data rates. *IEEE Vehicular Technology Magazine*, *9*(3), 39–46. doi:10.1109/MVT.2014.2333694

Wey, J. S., & Zhang, J. (2019). Passive Optical Networks for 5G Transport: Technology and Standards. *Journal of Lightwave Technology*, *37*(12), 2830–2837. doi:10.1109/JLT.2018.2856828

Wigren, T., Colombi, D., Thors, B., & Berg, J. E. (2016). Implication of RF-EMF exposure limitations on 5g data rates above 6 GHz. *2015 IEEE 82nd Vehicular Technology Conference, VTC Fall 2015 - Proceedings*. 10.1109/VTCFall.2015.7390974

Yang, J. J., Li, J., Mulder, J., Wang, Y., Chen, S., Wu, H., Wang, Q., & Pan, H. (2015). Emerging information technologies for enhanced healthcare. *Computers in Industry*, *69*, 3–11. doi:10.1016/j.compind.2015.01.012

Yaqoob, I., Khan, L. U., Kazmi, S. M. A., Imran, M., Guizani, N., & Hong, C. S. (2020). Autonomous Driving Cars in Smart Cities: Recent Advances, Requirements, and Challenges. *IEEE Network*, *34*(1), 174–181. doi:10.1109/MNET.2019.1900120

Yu, K., Lin, L., Alazab, M., Tan, L., & Gu, B. (2020). Deep Learning-Based Traffic Safety Solution for a Mixture of Autonomous and Manual Vehicles in a 5G-Enabled Intelligent Transportation System. *IEEE Transactions on Intelligent Transportation Systems*. Advance online publication. doi:10.1109/TITS.2020.3042504

Zahoor, M. I., Dou, Z., Shah, S. B. H., Khan, I. U., Ayub, S., & Gadekallu, T. R. (2020). Pilot decontamination using asynchronous fractional pilot scheduling in massive MIMO systems. *Sensors (Switzerland)*, *20*(21), 1–21. doi:10.339020216213 PMID:33143363

Zhang, H., Dong, Y., Cheng, J., Hossain, M. J., & Leung, V. C. M. (2016). Fronthauling for 5G LTE-U ultra-dense cloud small cell networks. *IEEE Wireless Communications*, *23*(6), 48–53. doi:10.1109/MWC.2016.1600066WC

Chapter 6
A Rabin Cryptosystem-Based Lightweight Authentication Protocol and Session Key-Generation Scheme for IoT Deployment:
Authentication in IoT

Priyanka Ahlawat
National Institute of Technology, Kurukshetra, India

Ankit Attkan
National Institute of Technology, Kurukshetra, India

ABSTRACT

Handling unpredictable attack vulnerabilities in self-proclaiming secure algorithms in WSNs is an issue. Vulnerabilities provide loop holes for adversary to barge in the privacy of the network. Attacks performed by the attacker can be active or passive. Adversary may listen to the sensitive information and exploit its confidentiality which is passive, or adversary may modify sensitive information being transferred over a WSN in case of active attacks. As Internet of things has basically three layers, middle-ware layer, Application layer, perceptron layer, most of the attacks are observed to happen at the perceptron layer in case of both wireless sensor network and RFID Tag implication Layer. Both are a major part of the perceptron layer that consist a small part of the IoT. Some of the major attack vulnerabilities are exploited

DOI: 10.4018/978-1-7998-6988-7.ch006

by executing the attacks through certain flaws in the protocol that are difficult to identify and almost complex to identify in complicated bigger protocols. As most of the sensors are resource constrained in terms of memory, battery power, processing power, bandwidth and due to which implementation of complex cryptosystem to keep the data being transferred secure is a challenging phase. The three main objectives studied in this scenario are setting up the system, registering user and the sensors via multiple gateways. Generating a common key which can be used for a particular interaction session among user, gateway and the sensor network. In this paper, we address one or more of these objectives for some of the fundamental problems in authentication and mutual authentication phase of the WSN in IoT deployment. We prevent the leakage of sensitive information using the rabin cryptosystem to avoid attacks like Man-in-the-middle attack, sensor session key leakage, all session hi-jacking attack and sniffing attacks in which data is analyzed maliciously by the adversary. We also compare and prove the security of our protocol using proverif protocol verifier tool.

1. INTRODUCTION

Authentication is a procedure of assuring the validity, integrity and trust-worthiness of information. Most basic form of authentication technique is approving the identity/ID of a communicating peer or node, and this ID is provided by the node which has a valid evidence that proves with strong validity that the identity being claimed is correct. The trust among the peers and other communicating pairs of nodes is established by known individuals with their respective verifiable digital IDs that are validated using digital signatures or digital finger-printing.For example, one kind of authentication mechanism is exhibited using the properties and primary attributes to identify digital objects and entities uniquely. In cybersecurity, a human being on a computer node terminal can be denoted as User node which has the privileges only after that individual successfully logs into the computer. According to the level of access provided, the user node has access to resources and data to a certain level. This is where authorization is marked upto a level and the a particular user node has authorized access to only allocated resources and data files access. Root server node is the hub to which network administrator has full access for manipulating, change, deleting or even adding newer data. Large scaled number of IoT edge devices in WSN are not supported by the IPv4, so IPv6 is required which has a wide range of IP addresses. Ipv6 needs a heavy load of battery support and hence making lightweight protocols like ZigBee[1] or 6LowPAN and hash approach based authentication schemes is preferable. Some of the most frequently occurring sensor node attacks in WSNs are node capture attacks[2], smart-card stealth and manipulative forgery attack[3], replay attacks, DOS attacks, session key leakage, user terminal node forgery attack, gateway node (foreign or home does not matter) forgery attack[4], MITM attack etc. Major cause of there adversarial attacks on the wireless sensor networks were an inefficient vulnerable protocol for communication which is unable to authenticate the component nodes of WSN or in simpler terms their cryptographic key generation and maintenance mechanism was not secure enough. By authentication we want to convey the following: i)It is a property that makes sure that an exchange of information is received exactly from the source it claims to be, ii) It associates the proof of identity of a smart IoT edge device connected to a network, iii) IoT edge devices communicating with each other directly or indirectly should be able to verify and validate their identities on the regards of some metric(s) to maintain authenticity. WSNs are a part of the IoT architecture which has three major layers that are application layer, middle-ware layer, perceptron layer, sensor node contains memory, battery, communicating components like transmitter, receiver, transducer etc. [5] Perceptron layer deals with the components that are sensors, tags, readers, actuators etc. These sensors can be deployed in two ways namely regular deployment and the other as random deployment. In regular deployment, constant distance among sensors is maintained for as to cover all the area whereas in random deployment, random placing of sensors is exhibited over the region which may or may not cover all the area. Middle-ware collects the continuous stream of data being generated and sends it to the application layer through Bluetooth or wi-fi techniques. Application layer analyzes the information received from the perceptron layer [6]. To have secured transmission of data over the network, proper authentication is mandatory. To have proper authentication, we need a secure access control mechanism corresponding to a secure communication channel. The major challenges while building a secure WSN involves:

1) **Heterogeneity in WSN:** Various devices run on different technologies and platforms using multiple communicational ways, which brings the challenge of connecting, managing and maintaining security for such devices.
2) **Scalability:** Handling large sized acaled WSN network while minimizing the exploitable vulnerabilities is a challenge.
3) **Data tranfer via wireless communication:** It involves issues and challenges associated with the wireless heterogeneous technologies like availability, network delays, congestion etc.
4) **Optimized Efficiency:** All the energy which is consumed for data communication will be high. The challenge is to minimize the use of power requirements for communication between different devices of a wireless sensor network considering minimum wastage of power. [7]
5) **IoT Edge Node Tracking Ability:** Tagged IOT edge devices and sensor nodes must be recognizable, identified and tracking of them is a challenge while maintaining range of communication.

Handling unpredictable attack vulnerabilities in self-proclaiming secure algorithms is an issue. Vulnerabilities provide loop holes for adversary to barge in the privacy of the network. Attacks performed by the attacker can be active or passive. Adversary may listen to the sensitive information and exploit its confidentiality which is passive, or adversary may modify sensitive information being transferred over a wireless sensor network in case of active attacks. As IoT (Internet of Things) has basically three layers, Middle-ware layer, Application layer, Perceptron layer, most of the attacks are observed to happen at the perceptron layer in case of both wireless sensor network and RFID Tag implication Layer[8] . Both are a major part of the perceptron layer that consists a small part of the IoT. Some of the major attack vulnerabilities are exploited by executing the attacks through certain flaws in the protocol that are difficult to identify and almost complex to identify in complicated bigger protocols. For Wireless sensor networks, Major attacks possible are given in Figure.1.

It is a well known fact that WSN is being adapted widely in numerous areas of human activities and day to day life routines, especially in the remote and environmentally suffering harsh areas where human reach is lethal, there WSNs are used to gather information and analyze the data deduce a suitable solution. Therefore preventing loss of human lives is depending on the way we include technological

Figure 1. Classification of attacks in WSN in IoT

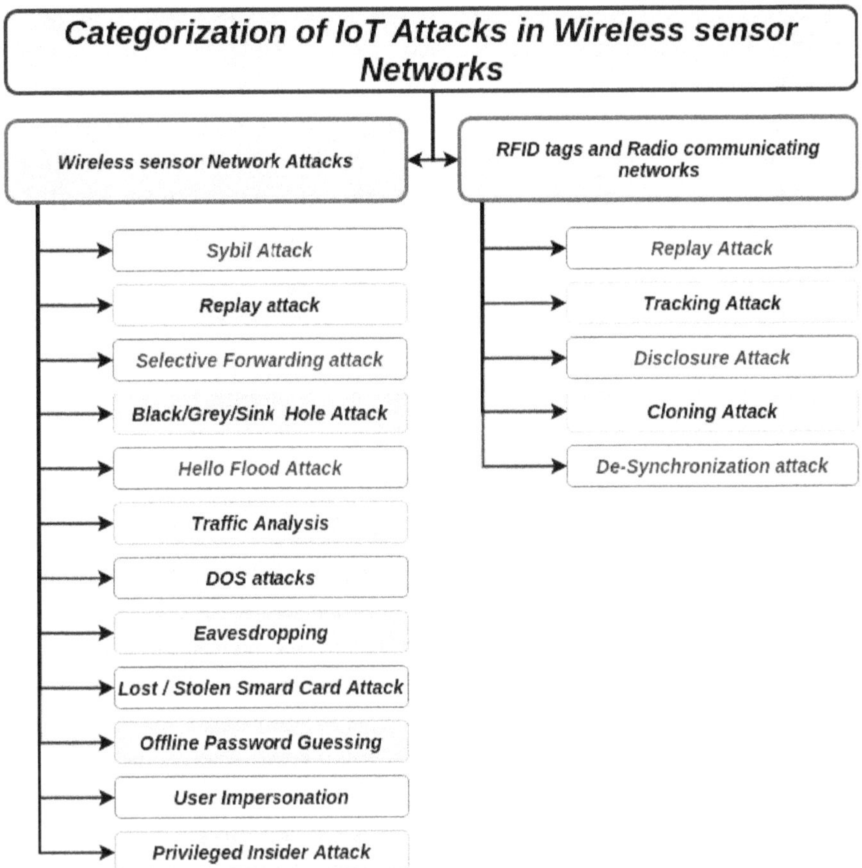

aspects in day to day routines. If so many delicate and sensitive responsibilities are being handled by the WSNs like in hospitals, land mines, flood prone areas, heavy rain detection or even military areas, the information gathered, transferred and stored by WSNs should be protected, monitored and prevented from adversarial changes and attacks. There are a number attacks that work on both WSNs and RFID layers to favor the adversary by exploiting vulnerabilities of the WSN itself [9]. Some of the major attack vulnerabilities are exploited by executing the attacks through certain flaws in the protocol that are difficult to identify and almost complex to identify in complicated bigger protocols. In WSN, major attacks possible are Sybil attack, Replay attack, Selective forwarding of data packets, Hello flood attack, Black hole attack, Man-in-the-middle attack, DoS attack, Sensor capture attack, Gateway Forgery attack, Stolen verifier attack, Stolen smart-card attack [10][11][12][13][14]. Radio frequency identification layer has its own share of attacks that it is vulnerable like Replay attack, Tracking of User attack, Packet Cloning attack, De-synchronization and impair sequence attack, Disclosure attack, Manual tampering of RFID Tag (Hardware privilege attack) etc [15][16][17]. To prevent above attacks and maintain CIA-Triad properties, that are, confidentiality, integrity and availability of information, there are various protocols available that provide or not so claim to provide proper security features. Protocols providing actual security are too heavy weight for resource constrained sensors and very few protocols are available that are light weight as well as provide security to the expectations. We are motivated with the work proposed in three major contributions by A.K Das's three factor scheme [18], Jiang and Ma's Crypto secured three factor bio hashing based scheme & Amin and Biswas's hashing based on common session key generation which is lightweight. We propose an improved two factor rabin cryptosystem and hashing based protocol which includes not all but certain features from those protocols mentioned earlier. It provides enhanced security features with secure common session key generation as well as address some of the previous unaddressed attacks .We used Proverif cryptographic protocol verifier to prove the security features of our protocol. The paper is organized as follows: In Section 2, discusses related work so far, Revisiting Biswas's protocol for IoT security for multifactor schemes are discussed in Section 3, whereas in Section 4, we discuss the proposed protocol for smart card based authentication protocol for IoT security. Section 5 gives implementation details of our protocol. Section 6 discusses result analysis of key secrecy, security and immunity towards certain attack while providing the comparative improvements in our protocol over the previous protocol and Section 7 concludes this paper with possible future directions.

2. RELATED WORK

In the mechanism proposed by Das in 2009, an RSA based crypto-authentication protocol was devised using a 2-factor approach. Watro [19] provided a diffie hellman based common shared key protocol that authenticates its peers using large sized keys that are complex to be constructed or guessed by the attacker. In 2010, Khan [20] proposed a similar mutual authentication mechanism but later it was proven to be incapable of providing proper mutual authentication. Wong [21] deduced a scheme that exploits theperls of hash based approach but it is still found weak against stolen verifier adversary attacks. A.K.Das [22] introduced a login credential based scheme that is meant for WSN in IoT but later it was detected with certain vulnerabilities which made it weaker against DOS attacks. In a lightweight scheme based on shared key proposed by Xue [23] in year 2013, there was a shared key used which found to be vulnerable towards simple brute force attacks. None of the schemes for authentication discussed so far provides any proper mutual authentication. Wang [24] and Ashok [25] provided an authentication protocol that was focused on credentials which were temporary and randomly assigned. Turkanovic [26] in 2014 and Farash [27] in 2016 devised a simplex authentication scheme that was meant for sensor networks but the implementation was found to be inefficient putting a greater toll on the power source of sensors, leading to quicker running out of battery power. Jiang [28] used a credential login password based authentication scheme and also discussed how Xue's [29] is exploitable in terms of user tracking exploitable attacks and node impersonation via SQL injection and brute force credential reveal technique. Jiang's protocol had an efficient user node registration phase that was meant for only non-dynamic sensors. Wu [31] gave a scheme whose security was supported via provable verification techniques like BAN logic and Pi-calculus. It was not able to provide proper authentication among peer-to-peer authentication modes. Huang [32] also focussed on multi-factor authentication protoccols that utilized the smart card technology, bio-fingerprint and login credential hybrid. Wang [33] proved that the mechanism provided by Huang is highly vulnerable to credential stealth from stolen smart and forgery attacks. Ma [34] proved yet again that a 2-factor based approach for authentication is updatable to a more resilient 3-factor scheme but only if hashing[35] based parameter passing is exploited for data transmission. Biswas and Amin [36] used the protocol that uses a similar set of rules that follow hash based parameter passing. This protocol was deduced in year 2016. Improper mutual authentication and numerous vulnerabilities called for improvements in Amin and Biswas's mechanism. These updates and higher security were observed in an hybrid mutli-factor authentication scheme provided by Saru, Wazid and Wu [37]. In June 2017, Ahlawat et al.[38] proposed a scheme that asses the vulnerability of network towards adversarial attacks by building an attack matrix. It is shown that this matrix can be effectively hold in countering attacks. Whether it is symmetric key cryptosystem or asymmetric key cryptosystem, Even if a key is generated, sharing the key secretly among components of a network without the adversary knowing it is an issue of security. So, Confidentiality, Integrity and Availability of services provided by a wireless sensor network are to be efficiently authenticated because a sensor running out of battery in critical situations is the worst one can ask for. Keys are highly used in WSNs, it maybe secret key or public key, depending on the mechanism of cryptosystem, whether it is symmetric key cryptosystem or asymmetric key cryptosystem. Even if a key is generated, sharing the key secretly among components of a network without the adversary knowing it is an issue of security. Also maintaining the security features without delegating the complex nature of protocol is quite difficult.

3. EXISTING TECHNIQUES

Amin and Biswas(2016) proposed a protocol that claimed it was secure against a lot of major attacks and sufficiently provided mutual authentication. Eventually it was proven that it was not secure to sensor capture attack and didn't provided proper mutual authentication. Their solution has seven major phases of which five phases are responsible for secure key generation for communication of a user with the wireless sensor network via monitoring gateway nodes. Gateway Nodes may be HGWN-Home gateway Node or FGWN-Foreign Gateway Node depending upon the location of User and the corresponding sensor node being accessed by the User. It has initialization/system setup, registration of User terminal, node registration, login credentials parameters and key agreement phases.

1. **System setup and initialization phase:** The system's administrator, initiator and handler, *system's Network Administrator(SNA)* chooses $SensorID_j$ for the S_j, selects a random nonce r_{sr} and calculates $xj = h(Send\_ID_j \| r_{sr})$.*system's Network Administrator(SNA)* stores *(SensorIDj, x_j, rsr)* into S_j. Here r_{sr} is known to all home and foreign Gateway Nodes and confidentially stored.
2. **Sensor Node registration phase:** S_j calculates $A_j = x_j \oplus r_{sr}$ and sends $\{Send\_ID_j, A_j\}$ to gateway node through a public wireless communication channel. Gateway node calculates $x_j = A_j \oplus r_{sr}$ *and saves* $(Send\_ID_j, x_j)$ in DBMS and maintains it regularly. Then gateway node sends a grant of request to S_j. S_j then erases r_{sr}.
3. **User registration phase:** In this phase of protocol, user is registered and issued a smart card that has authenticating information stored within its memory. Smart card is issued as follows:

Step 1: User U_i chooses UID_i, UPW_i and a random nonce r_0, calculates, $GID_i = h(UID_i \| r_0)$ *and* $Pwd_i = h(UPW_i \| r_0)$, *& sends* $\{GID_i, Pwd_i\}$ to Gateway node via a secure channel.

Step 2: Home gateway node (HGWN) produces a time bounded non-permanent id $Temp\_ID_i$ for terminal node user U_i *then, computed* B_1 *and* B_2 *as,* $B_1 = h(GIDi \| Pwdi)$ and $B_2 = h(GIDi \| Temp\_IDi \| Xgwn) \oplus h(GIDi \oplus Pwdi)$.

Step 3: Gateway saves $(Temp\_ID_i, GID_i)$ in database, maintains data regularly and issues a smart-card containing parameters $(B1, B2, ID_{hgn}, Temp\_ID_i)$ to user U_i through a guarded secure channel. Ultimately, User U_i stores r_0 in the memory of the smartcard.

4. **Login phase for ith User:** In this phase, user feeds its credentials in the user terminal node after inserting and verification of the smart card. Credentials are username and password and if the credentials computed and credentials on the smartcard match, then only further access is granted. Login phase has following steps of execution to log the user in the WSN:

Step 1: User U_i enters the credentials of his/her smartcard on the terminal interactive node and provides UID_i and UPW_i. The smartcard calculates $GID_i = h(UID_i \| r_0)$ and $Pwd_i = h(UPW_i \| r_0)$, and checks to verify whether $B_1 ?= h(GID_i \| Pwd_i)$. If yes, then the next step will be executed.

Step 2: A sending ID, $Send\_ID_j$, is chosen for the smart card and a random nonce r_u, a time defining timestamp as T_1, calculates the parameter $D_0 = [B_2 \oplus h(GID_i \oplus Pwd_i)]$, $D_1 = h(ID_{hgn} \| D_0 \| r_u \| T_1)$ and $D_2 = D_0 \oplus r_u$, and sends the $MSG_1 = \{ID_{hgn}, Temp\_ID_t, Send\_ID_j, D_1, D_2, T_1\}$ to the required gateway which is interacting with the sensor node.

5. **Authentication and key agreement phase**: In this phase, user and gateway node authenticate each other, gateway and sensor network authenticate each other mutually and user and sensors authenticate each other via sharing of secret parameters over the communication channel and a successful communicational common key of generated if and only if all three, sensor, gateway nodes and user terminal have authenticated each other mutually. Following are the steps for this phase of key agreement and authentication.

Step 1: Gateway Node picks up a timestamp T_2 and evaluates if $|T_2 - T_1| \leq \Delta T$ maintains delay limit. Then it obtains GID_i from *DBMS* according to $Temp\_ID_t$, computes the $D_0 = h(GID_i \| Temp\_ID_i \| Xgwn)$ and $r_u = D_2 \oplus D_0$, and verifies if $D_1 ?= h(ID_{hgn} \| D_0 \| r_u \| T_1)$. The session will be discarded if either checking fails to obtain correctness.

Step 2: Gateway node produces a nonce r_{hgn}, calculates, $D_3 = h(ID_{hgn} \| GID_i \| x_j \| r_{hgn} \| T_2)$, $D_4 = x_j \oplus r_{hgn}$, $D_5 = r_u \oplus h(r_{hgn})$ and $D_6 = GID_i \oplus h(ID_{hgn} \| r_{hgn})$, and then forwards the message $MSG_2 = \{D_3, D_4, D_5, D_6, T_2\}$ to S_j.

Step 3: S_j chooses T_3 according to clock and verifies if $|T_3 - T_2| \leq \Delta T$. Then it calculates $r_{hgn} = D_4 \oplus x_j$, $r_u = D_5 \oplus h(r_{hgn})$ and $GID_i = D_6 \oplus h(ID_{hgn} \| r_{hgn})$ and checks whether $D_3 ?= h(ID_{hgn} \| GID_i \| x_j \| r_{hgn} \| T_2)$. The session will be discarded if any of the checking is unfruitful that is not successful.

Step 4: S_j produces a nonce r_s, computes $D_7 = h(D_3 \| GID_i \| r_s \| T_3)$ and $D_8 = r_{hgn} \oplus r_s$, and also sends the message $MSG_3 = \{D_7, D_8, T_3\}$ to the gateway node.

Step 5: Gateway node selects T_4, and checks if $|T_4 - T_3| \leq \Delta T$. Then it calculates $r_s = D_8 \oplus r_{hgn}$ and checks whether $D_7 ?= h(D_3 \| GID_i \| r_s \| T_3)$. The complete session will be discarded if either of the checking is failed. After the checking, gateway node calculates $D_9 = h(D_3 \| GID_i \| r_s \| r_{hgn} \| T_4)$ and $D_{10} = r_s \oplus r_u$, and also sends the message $MSG_4 = \{D_3, D_8, D_9, D_{10}, T_4\}$ to User U_i (the ith User).

Step 6: User U_i picks up T_5 and checks if $|T_5 - T_4| \leq \Delta T$. Then user U_i computes $r_s = D_{10} \oplus r_u$ and $r_{hgn} = D_8 \oplus r_s$, and checks to verify if $D_9 ?= h(D_3 \| GID_i \| r_s \| r_{hgn} \| T_4)$. The session will be discarded if any of the checking is failed. *Ui*, *Sj* and gateway node generate a common secret key, $SK_u = SK_s = SK_{hgn} = h(GID_i \| r_u \| r_s \| r_{hgn})$.

However, the above scheme is found to be vulnerable and weak from security point of view as gateway impersonation and sensor node impersonation and sensor node capture attacks are possible.

4. PROPOSED PROTOCOL

The proposed protocol for security is based on the rabin cryptosystem that provides cryptographic security. It also uses random salt values to enhance the authentication mechanism and provide higher security. Rabin cryptosystem is a public key based crptosystem that relies on the prime factorization of integer(tougher to do for prime numbers). During the synthesis of secret key, a,b prime numbers are chosen to compute $N = a * b$. Encryption process is $Ciphertext, Ciper = (m * m) modN$ and (a,b) is denoted as the the private key pair. Whereas N is the desired shared public key. The decryption of ciphertext is done by under rooting the cipher in first step and then performing modulus by N with the resultant.Two cases arise in our protocol, Case 1, copes with the circumstances where the user and the sensor(s) being accessed are in the same cluster under a common home gateway node. In Case 2, our protocol deals with situations where the sensor(s) are under a different gateway node known as Foreign gateway node than the home gateway node of the user itself. Both Case 1 and Case 2 have nine phases, however, the phases with no changes are not discussed here and only 5 major phases are discussed step by step below:

Case 4.1: Data Accessing Within Home Gateway Node Cluster

In this case the sensors are accessed are in the same cluster under a common home gateway node (HGWN). It has four major phases that are discussed below:

i) **System setup and initialization Phase:** In this phase the system startup and initializations settings are configured and necessary parameters are passed and stored for further proceedings of the protocol. System setup and initialization phase has following steps of execution:
   Step1: *System and Network Administrator(SNA)* generates two primes, *"a"* and *"b"* to calculate *N=a*b,* and preserves (a,b) as private key and "N" as the public key. *System and Network Administrator(SNA)* also selects X_{gn}, which is the private key for communication session for the gateway node and an integer "l" as the parameter for the verifier which relies on fuzzy logic.
   Step2: *System and Network Administrator(SNA) selects* a UID_j for the jth User and calculates the key $X_j = h(UID_j \| X_{gn})$ for every S_j $(1<j<m)$.
   Step3: *System and Network Administrator(SNA)* randomly picks a nonce/salt "*R*" which is shared among gateway node(GWN) and S_j. Finally, S_j saves < UID_j, X_j, R > in its memory.

ii) **User Registration Phase: When user register**, ith User chooses UID_i and credentials are send to System and Network Administrator(SNA). At *System and Network Administrator(SNA)*, necessary parameters α_i and β_i are calculate and are saved into the database. A smartcard is then to the user when necessary input of vector *[* UID_i, UPW_i, *URN]* is fed. The contents stored in the smartcard has values C_i, f_i, g_i, *SCID, l, n,* r_i, *FF() and h(),* while $\boldsymbol{\alpha}_i$ and $\boldsymbol{\beta}_i$ are erased from the memory to enhance system security.

iii) **Login Phase** (User logins successfully only if smartcard and credentials are available): In this phase, user feeds its credentials in the user terminal node after inserting and verification of the smart card. Credentials are username and password and if the credentials computed and credentials on the smartcard match, then only further access is granted. Login phase has following steps of execution to log the user in the WSN for information sharing and listening for further requests.

Step 1: U_i inserts the smartcard and inputs the identity UID_i', password UPW_i', and nonce F_i. Then, the smartcard calculates $F_i^{*}= FF\ (r_i, URN)$ and $C_i^{*} = h(h\ (UID_i^{*} \| UPW_i^{*} \| F_i^{*})\ mod\ l)$. The smartcard aborts *User's* login request if C_i^{*} *which is required not to be equal to* C_i.

Step 2: The smartcard system manager creates a random salt "K_i" and a time tracking time-stamp T_1, which are further used to compute $\alpha_i^{*}= f_i \oplus h\ (UID_i' \| UPW_i' \| F_i')$, $\beta_i^{*} = g_i \oplus h\ (UID_i^{*} \oplus UPW_i^{*} \oplus F_i^{*}$, $M_1 = (UID_i \| SCID \| K_i)^2 mod(n)$, $MSG_2 = h\ (\alpha_i^{*} \| \beta_i^{*} \| K_i \| T_1)$.

Step 3: $U_{i\,(jth\ User)}$ *selects* the identity UID_j of the sensor node that the user wants to access, control or manipulate and then the smart card computes $CUID_j = UID_j \oplus h\ (UID_i \| K_i \| T_1)$*, then finally sending*$MSG_1 = < MSG_1, MSG_2, T_1, CUID_j >$ to the Gateway node.

iv)Authentication and Session key generation phase: The **user and the gateway node mutually authenticate each other, user and sensors authenticate each other via sharing of secret parameters over the communication channel and a successful communicational common key of generated if and only if all three, sensor, gateway nodes and user** terminal have authenticated each other mutually. Following are the steps for this phase of key agreement and authentication.

Step 1: On receiving MSG_1 from U_i, Gateway node decrypts MSG_1 using *"a"* and *"b"* which were calculated earlier to obtain UID_i', $SCID'$, K_i', get x_i as per the UID_i'. It checks whether $SCID'$ matches the value in the record. If there is no matching, then Gateway node discards or destroys the request and then terminates the process. Gateway node computes $\beta' = h(SCID' \| X_{gn})$, $\alpha'= h\ (UID_i' \| X_{gn} \| x_i)$, $K_i'= M_2 \oplus h\ (\alpha' \| T_1)$*, and* $M_2'= h(\alpha' \| \beta' \| K_i' \| T_1)$. Gateway node, F- or H- GWN terminates the present ongoing session if MSG_2' *not equal to* MSG_2 ; Oherwise, Gateway node calculates $UID_j' = CUID_j \oplus h\ (UID_i \| K_i \| T_1)$, $X_j' = h\ (UID_j' \| X_{gn})$, $MSG_3 = h\ (UID_i' \| UID_j' \| ID_{gn} \| X_j' \| K_i' \| T_2)$, $MSG_4 = UID_i' \oplus h(ID_{gn} \| X_j' \| T_2)$, $MSG_5 = K_i \oplus h\ (ID_i' \| ID_j' \| X_j' \| T_2)$ *and then sends* $MSG_2 = < ID_{gn}, MSG_3, MSG_4, MSG_5, T_2 >$ *to* S_j .

Step 2: S_j verifies if $| T_3 - T_2 | \leq \Delta T$ condition is satisfied or not, where T_3 is the present timestamp. If found invalid, S_j immediately terminates the session; but if the condition holds "True", it computes $UID_i''= MSG_4 \oplus h(ID_{gn} \| X_j \| T_2)$, $K_i'' = MSG_5 \oplus h(UID_i'' \| UID_j \| X_j \| T_2)$, $MSG_3'' = h(UID_i'' \| UID_j \| ID_{gn} \| X_j \| K_i'' \| T_2)$. S_j ends the connection if MSG_3'' *is not equal to* MSG_3; else, it accepts that U_i and Gateway nodes are authentic. Next is, S_j calculates $SECK_i = h(UID_i'' \| UID_j \| K_i'' \| K_j)$, $MSG_6 = h\ (SECK_j \| X_j \| K_j \| T_3)$*, and* $MSG_7 = K_i'' \oplus K_j$, *where* K_j is the randomly generated numeric value by S_j. Ultimately, S_j sends the $MSG_3 = < MSG_6, MSG_7, T_3 >$ to Gateway node.

Step 3: Gateway node verifies if the $|T_4 - T_3 | \leq \Delta T$ condition holds true, where "T_4" is the present timestamp. If it is computed to be negative, Gateway node (Foreign or Home gateway) terminates the session [53]; else it computes $K_j' = MSG_7 \oplus K_i'$, $SECK_{gn} = h(ID_{gn}' \| UID_j \| K_i' \| K_j')$*, and* $MSG_6' = h\ (SECK_{gn} \| X_j' \| K_i' \| T_3)$. Gateway node terminates the present communication session if $MSG_6' \neq MSG_6$; Otherwise, it calculates $MSG_8 = h\ (SECK_{gn} \| UID_i' \| \alpha' \| K_j')$. Ultimately, the gateway node sends$MSG_4 = < MSG_7, MSG_8 >$ *to* U_i.

Step 4: U_i(jth user) calculates $D_7 = MSG_7 \oplus K_i$, $SECK_i = h(UID_i \| UID_j \| K_i \| K_j')$, also $MSG_8' = h\ (SECK_i \| UID_i \| \alpha \| K_j')$. U_i rejects the current ongoing communication session if $MSG_8' != MSG_8$; else, U_i approves the privileged access of that gateway node and S_j request is considered as legitimate. In the end, a common session key for secure communication, $SECK_i = SECK_{gn} = SECK_j$ is constructed among User(U_i), Sensor(S_j) and Gateway Node (GWN).

Figure 2. User Registration phase before session key agreement in our proposed protocol

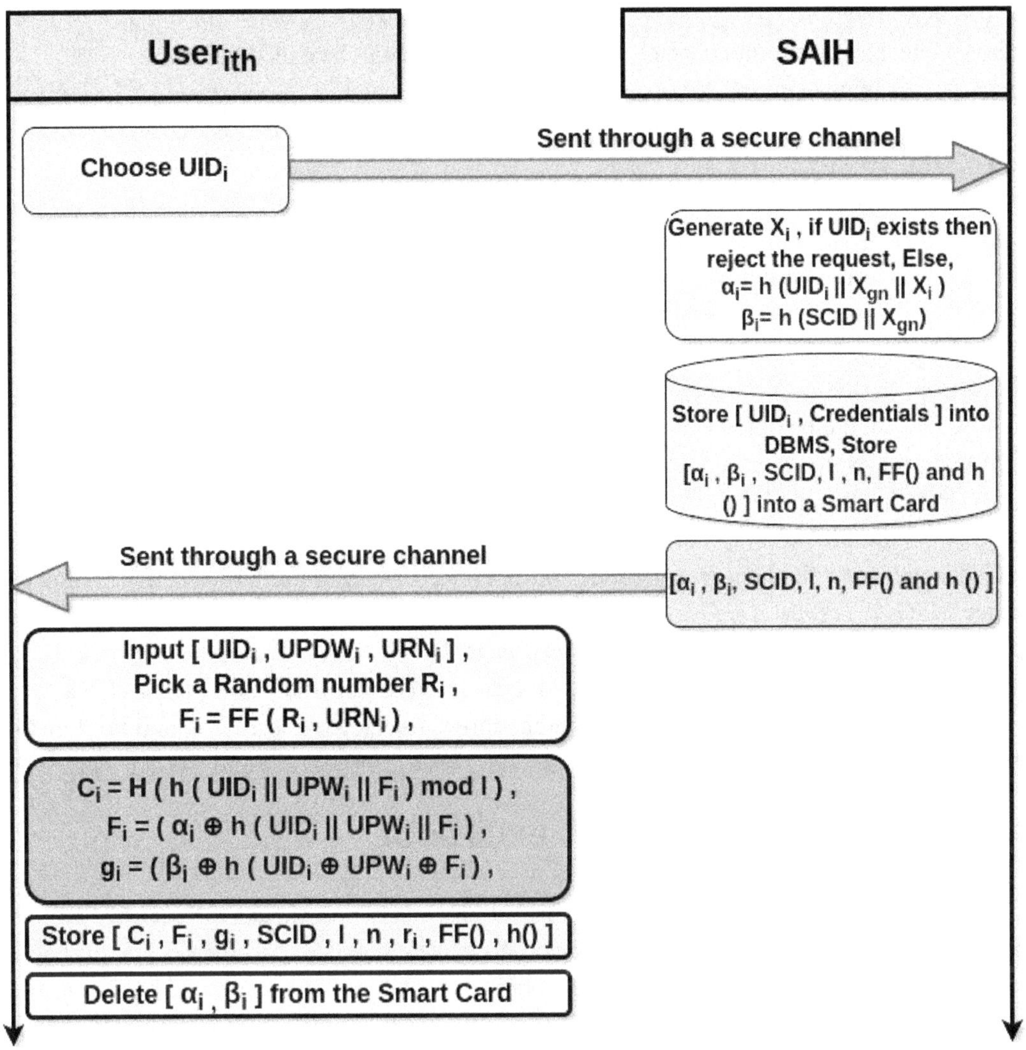

Case 4.2: Data Accessing Outside HGWN With Foreign Gateway Node

Initialization and setup phase, Sensor registration phase, User registration phase and Login phase for case 2 are same as in case 1, variations are found in the session key generation phase as discussed below in stepwise manner:

Step 1: If one foreign gateway node computes *Sensor ID, Send_ID*$_j$ from its database, it finds x_j according to *Send_ID*$_j$, and computes $A_1 = h(Temp\_ID_i \| x_{fgn})$ and $A_2 = A_1 \oplus r_{sr}$, and sends $\{A_2, ID_{fgn}\}$ to Home Gateway node. Home Gateway node computes D_0, $A_1 = A_2 \oplus r_{sr}$ and $A_3 = D_0$? A_1, and sends $\{A_3, ID_{fgn}\}$ to U_i.

Step 2: U_i extracts $A_1 = A_3 \oplus D_0$, selects timestamp T_6 and a random *nonce* r_{u2}, computes $D_{11} = h(Temp\_ID_i \| A_1 \| r_{u2} \| T_6)$ and $D_{12} = A_1 \oplus r_{u2}$, and sends the message $MSG_5 = \{Temp\_ID_i, D_{11}, D_{12}, T_6\}$ to Foreign Gateway Node.

Step 3: Foreign Gateway Node chooses timestamp T_7, and evaluates if $|T_7 - T_6| =? T$. Then it calculates $A_1 = h(Temp\_ID_i \,||\, x_{fgn})$ and $r_{u2} = D_{12} \oplus A_1$, and checks $D_{11} ?= h(Temp\_ID_i \,||\, A_1 \,||\, r_{u2} \,||\, T_6)$. If both terms of checking are found correct, the succeeding step can be executed.

Step 4: Foreign Gateway Node computes a random nonce r_{fgn}, calculates $D_{13} = h(Temp\_ID_i \,||\, A_1 \,||\, r_{fgn} \,||\, x_j \,||\, T_7 \,||\, r_{u2})$, $D_{14} = r_{fgn} \oplus x_j$ and $D_{15} = h(x_j) \oplus (A_1)$, and transfers the message $MSG_6 = \{Temp\_ID_i, D_{12}, D_{13}, D_{14}, D_{15}, T_7\}$ to S_j.

Step 5: S_j selects T_8 and checks if $|T_8 - T_7| ?= T$. It further calculates $r_{fgn} = D_{14} \oplus x_j$, $A_1 = h(x_j) \oplus D_{15}$ and $r_{u2} = D_{12} \oplus A_1$, and computes if $D_{13} ?= h(Temp\_ID_i \,||\, A_1 \,||\, r_{fgn} \,||\, x_j \,||\, T_7 \,||\, r_{u2})$. If both computational checks performed are found right, further steps will be continued.

Step 6: S_j generates r_{s2}, computes $B_1 = h(Send\_ID_j \,||\, x_j)$, $D_{16} = h(Temp\_ID_i \,||\, r_{s2} \,||\, B_1 \,||\, T_8)$, $D_{17} = r_{s2} \oplus B_1$ and $D_{18} = A_1 \oplus B_1$, and transfers the message $MSG_7 = \{Temp\_ID_i, D_{16}, D_{17}, D_{18}, T_8\}$ to Foreign Gateway Node.

One thing that should be noted is that D_{17} is missing in Amin and Biswas's protocol released in 2016. If Foreign Gateway Node can't get D_{17}, Foreign Gateway Node and U_i will not haver to generate the session key. Hence, we added the changes to overcome previous flaw.

Step 7: Foreign Gateway Node chooses timestamp T_9 and verifies if $|T_9 - T_8| ?= T$. If true, Foreign Gateway Node calculates $B_1 = h(Send\_ID_j \,||\, x_j)$, $r_{s2} = D_{17} \oplus B_1$ and $D_{19} = r_{s2} ?= r_{fgn}$ and transfers the message $MSG_8 = \{Temp\_ID_i, D_{16}, D_{18}, D_{19}, T_8, T_9\}$ to U_i.

Step 8: U_i chooses timestamp T_{10} and verifies if $|T_{10} - T_9| ?= T$. Later, the smart card calculates $B_1 = D_{18} \oplus A_1$, $r_{s2} = D_{17} \oplus B_1$ and $r_{fgn} = D_{19} \oplus r_{s2}$ and checks whether if $D_{16} ?= h(Temp\_ID_i \,||\, r_{s2} \,||\, B_1 \,||\, T_8)$. Even if one of the two checking fails, complete procedure is rejected. At last U_i, S_j and Foreign Gateway Node share the common session key $SK_u = SK_s = SK_{fgn} = h(Temp\_ID_i \,||\, Send\_ID_j \,||\, r_{u2} \,||\, r_{s2} \,||\, r_{fgn})$.

5. IMPLEMENTATION OF PROPOSED PROTOCOL

Proverif [32] is a protocol verifier tool based on provable security that checks the security promises of a protocol via breaching and checking whether protocol's private sensitive parameters are reachable or not and decides the security on the basis of degree of breach possible. It as developed by Bruno Blanchet, Vincent Cheval and Marc Sylvestre in OCaml. It is a tool for automatically testing and checking the security and hardness of cryptographic protocols. It supports cryptographic base protocols including symmetric & asymmetric encryption and decryption of data packets, digitally signed data via digital signatures, hash functions, or even bit-commitment operators and non-interactive types of zero knowledge proofs for complicated protocols. ProVerif is self sufficient of proving reachability of parameters

Figure 3. Case 1 Output: Successfully avoiding leakage of sensitive data parameters to the adversary during the Interaction of user with the wireless sensor networks sensor nodes via Home Gateway node(HGWN)

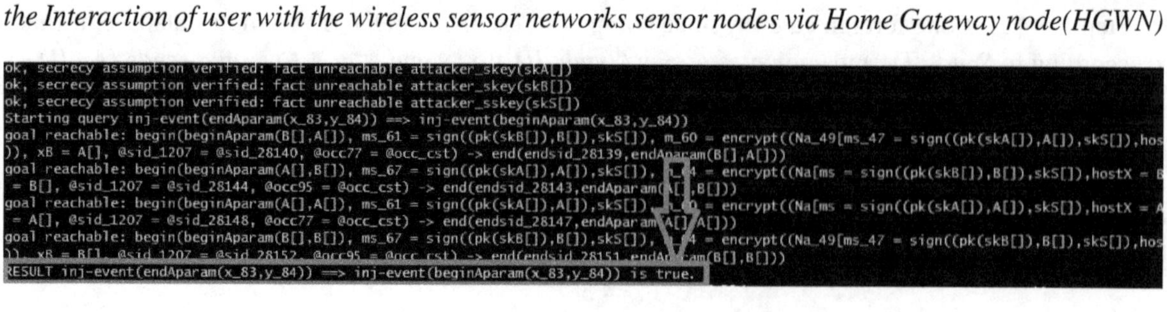

from the attackers point of view, correspondence and its assertions, and observational equivalence are some highlight key points of the same. These capabilities are specifically beneficial to the computer security areas as they allow the analysis of secrecy of protocols as well as verification of authentication properties of the protocols that are fed to it. We implemented our protocol in proverif cryptographic protocol verifier and verified that our protocol successfully avoids certain major attacks discussed earlier providing better security features and light weight session key agreement procedure which is practically feasible and secure to a great extent.

Figure 4. Case 2 Output: Successfully avoiding leakage of sensitive data parameters to the adversary during the Interaction of user with the wireless sensor networks sensor nodes via Indirect communication through Foreign Gateway node (FGWN)

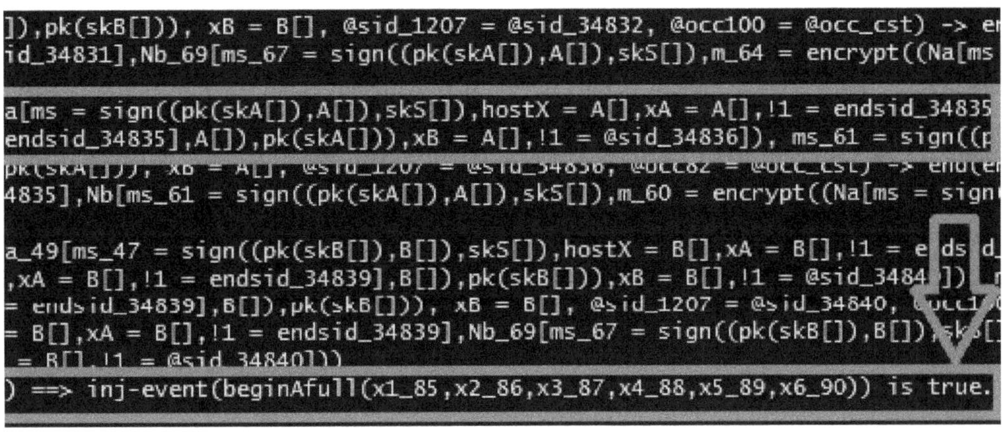

The outcome after executing the processes in ProVerif 2.0 is given below, which demonstrates that our protocol achieves session key secrecy and mutual authentication.

RESULT event(serverDoneAccept (user[]))==>event (AdminscAccept (user[])) is **true**.
RESULT inj-event(sensorGen (user[], server[]))==>inj-event (serverDoneAccept (user[])) is **true**.
RESULT inj-event (serverGen (sensor[]))==> inj-event (sensorGen (user[], server[])) is **true**.
RESULT inj-event(userGen(server[],sensor[]))==> inj-event (serverGen(sensor[])) is **true**.
RESULT not attacker (SECKj[]) is **true**. RESULT not attacker (SECKGWN[]) is **true**.
RESULT not attacker (SECKi[]) is **true**.
"True"- means the sensitive parameters are NOT reachable by the adversary, showing positive results regarding security of the protocol towards certain attacks tested during verification.

6. SECURITY ANALYSIS OF PROPOSED PROTOCOL

This section discusses the attack and security analysis of the previous existing protocols with the newly proposed protocol and compares the run time cost thatis, efficiency of the protocols in terms of time cost parameters. Some of the major attacks that were prevented in our protocol are explained below:

6.1 Session key Hi-jacking Attack Avoidance

Sensors are physical devices that can be personally tampered by the attacker via using hardware tools. If a sensor is compromised and adversary has physical access to the sensor node, the information like node's ID, its secret key, configuration meta-data can be disclosed. A shared key is the same as for gateway's and the user node, complete communication session can be overtaken. Using separate random nonces to create separate keys for every communicating channel between a user node which is interacting with the sensor. During sensor capture attack, that one particular session, between physically captured sensor node, the gateway node and user node remains slave of the attacker serving limited information to the attacker. Limited because only one session is compromised here.Rebooting the communication session can solve the spreading of the infected sessions. Discarding that sensor is another choice. Figure 5 shows how sensor capture attack from an adversary in Amin & Biswas's Algorithm can be executed to in order to take over the maximum possible sessions for data packet exploitation. For example, the adversary physically captures a node N_1 and the obtains the key creating parameters like private key of the node, X_i, and its ID, SID. Now it is to be noted that attacker is simultaneously sniffing the wireless traffic and tries to capture every data packet (a passive copy of data packet). This captured copy is used to obtain the

Figure 5. Sensor capture attack in Amin and Biswas's authentication protocol

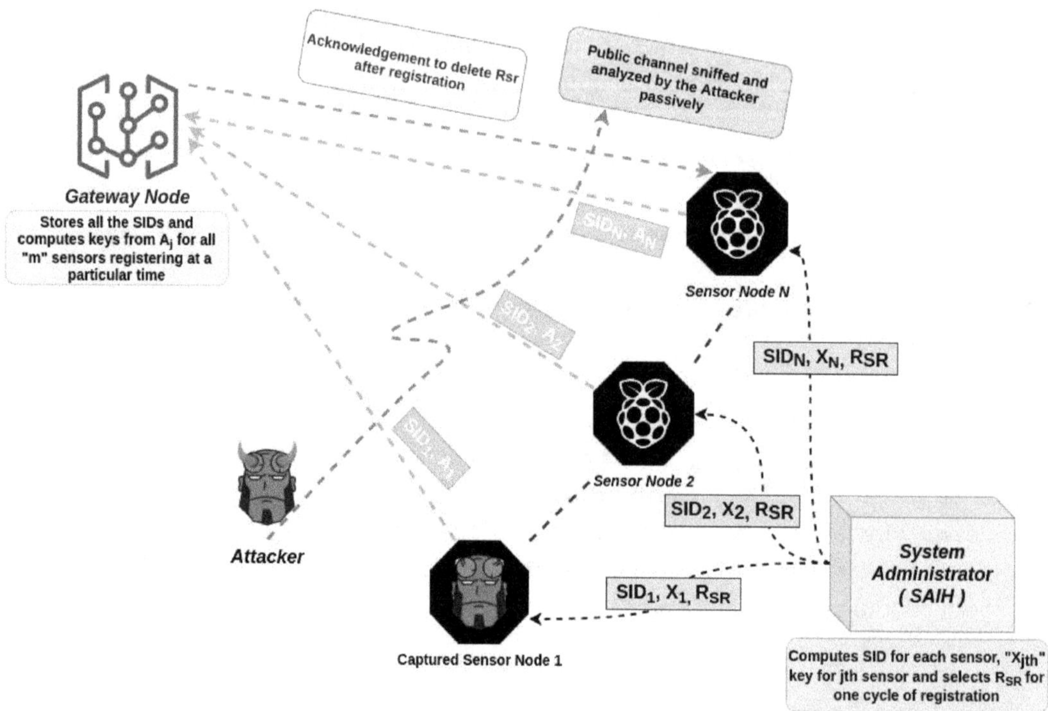

private keys of other session cryptographic communicational keys. A_N is computed by exclusively OR-ing the private key with the R_{SR}. The weakness of OR function is that it is reversible if matched with the right number. Attacker captures the sensor node, obtains the common R_{SR} number/random nonce, which is common for all the sensor node. This number is used by the attacker to reverse the A_N values, which in return reveal the private keys of the respective sensor nodes, whose data packet was passively copied at the attackers end. This is a primitive example of how an attacker actually breaches into the network and compromises the security.Our proposed protocol avoids this sensor node attack as for every group of sensors trying to register at a particular period of time, the system's Network Administrator(SNA) generates a unique randomly picked nonce for every sensor node instead of using a common shared nonce R_{sr}. It leads to the feature that even if a sensor is captured, the complete Hijacking of the session of the group of sensor nodes trying to register at a particular having a common R_{sr} earlier is not possible. Only the session of the captured sensor node gets compromised in this case.

6.2 Curbing the *Gateway and Sensor Node Impersonation Attack*

In our method, the attacker is not capable of obtaining the MSG_3 to impersonate as gateway node to either U_i or S_j. To impersonate or steal identity as a gateway node to S_j, It is necessary to compute the message, MSG_3= h (UID_i || UID_j || ID_{gn} || X_i || K_j || T_2). However, without this value X_j'= h(UID_j' || X_{gn}), it is not feasible for attacker to calculate MSG_3. As we used the hashing derived algorithm and time keeping, timestamps, attacker is unable to obtain any exploitable meta form of information from the messages captured through network sniffing. Attacker needs to calculate a protected value MSG_8 = h ($SECK_{gn}$ || UID_i || α_i || K_j). Attacker needs to have information of "K_i" to find the value "$SECK_{gn}$" = h (UID_i || UID_j || K_i || K_j). To obtain K_i, Attacker has to know the secret key "a" and "b" of gateway node. It is not possible because the secret key, $SECK_{gn}$ is protected by the SNA-system network administrator. Attacker tries to impersonate the identity as S_j (sensor node) after copying the data packets in passive mode and the messages transferred during the previous communicated authentication protocol phase sessions. Adversary has to compute MSG_3 = <MSG_6, MSG_7,T_3 > to impersonate as S_j, where $SECK_j$ = h (UID_i || UID_j || K_i || K_j), MSG_6 = h($SECK_j$ || X_j || K_j || T_3), & MSG_7= $K_i''\oplus K_j$. Attacker must know "K_i" in aim to calculate MSG_6 = h ($SECK_j$ || X_j || K_j || T_3). So, Adversary is incapable of obtaining K_i. Adversary cannot execute the SN impersonation attack.

Table 1. Comparison Analysis between previous and proposed protocol

S.No.	Immunity Towards Attack	Biswas (2016)	Proposed Protocol
1.	SSLA(Type 2 Attacks)	Strong	Strong
2.	Gateway Forgery	Weak	Strong
3.	Sensor Node Impersonation	Weak	Strong
4.	IoT edge device capture Attack	Weak	Strong

Now, Efficiency comparison of existing and proposed protocol We evaluate the time efficiency of our proposed protocol with the pre-existing authentication protocol in terms of time delay observed at sensor node(s), gateways and user terminal. The overall time complexity in terms of delay cost is given in Table 2. Since, the sensor nodes are constrained in terms of resources and some critical resources are memory, computational strength of processor and energy, special analysis and care should be taken to

Table 2. Efficiency comparison of existing and Proposed protocol (Time Units standardized upto micro-seconds)

Script running on Protocol	U_i terminal ith User	Gateway node GWN	S_j jth Sensor node	Total Time Cost (Standardized upto microseconds)
Das's Protocol	$10T_{hash}$	$10\ T_{hash}$	$5\ T_{hash}$	$25\ T_{hash}$
Li et al's Protocol	$6\ T_{hash}+2\ T_{SED}$	$8\ T_{hash}+ 5\ T_{SED}$	$6\ T_{hash} +1\ T_{SED}$	$18\ T_{hash} +10T_{SED}$
Amin and Biswas's Protocol	$13\ T_{hash}$	$14\ T_{hash}$	$5\ T_{hash}$	$32\ T_{hash}$
Our Proposed Protocol	$7\ T_{hash}+2T_{MOD}$	$11\ T_{hash}+2\ T_{SQRT}$	$5\ T_{hash}$	$25\ T_{hash}+ T_{MOD+}\ T_{SQRT}$

the computational delay cost of authentication protocols for WSN in IoT deployment.

While observing and comparing the time cost, we mainly focus on login and authentication phase and neglect the bit-XOR operation as it requires negligible time computation cost. Whereas T_{hash}, T_{SED}, T_{MOD}, $T_{SQRT,}$ T_{ECC} are used to denote the cost of hash, symmetric encryption/decryption cryptosystem, modular squaring, square root modulo N and ECC-point computation respectively. It is good to know here that modular squaring is as efficient as the hashing operation of the protocol. Also the calculation of the square root modulo N is way more similar to the modular encryption procedure. The protocol we proposed is almost as efficient as the protocols proposed earlier but still successfully mitigates more attacks than the previous ones. Even though the computational cost at gateway is higher in our proposed protocol but enhanced security features are a support pillar to our proposal.

7. CONCLUSION AND FUTURE SCOPE

WSNs have become a highly proactive research field because of their capabilities of providing variant services to a wide range of applications. The tremendous growth of applications for the present and emerging user services introduces a number of issues in WSNs. Growing technologies and application services have posed various challenges to both academic and industrial users, especially for the development of efficiency, scalability, and reliability on WSNs. These challenges motivate us to design and develop new protocols, architectures, and services for future. WSN sensor nodes are actually unevenly distributed as well as sensors themselves are not homogenous in hardware specifications in real situations. However there are still certain vulnerabilities persisting in WSNs and with introduction of new technologies, new vulnerabilities are bound to appear as no algorithm or protocol is perfect, there's always a possibility to counter these vulnerabilities. For resource constrained limitations, it is proposed that three-step systematic method to calculate the battery's utilization under different temperature of a mobile node's battery. Just because an algorithm is self sufficient doesn't impose that its cent per cent immune to attacks. Hardware limitations also affect the security like in case of sensor capture attacks. All the sensors, gateways or participant nodes in a WSNs are physical devices that can be attacked manually, individually or a certain cluster in WSNs being targeted at any point of time i.e; via hardware interception or impersonation. Creating devices that are immune to such interceptions is still a challenge. Situations are dynamic in actual practice and developing dynamic authentication schemes that are efficient as well attack tolerant is an issue to persist even in the upcoming years. Mutual authentication is just a step towards security, with Introduction of new technologies and rapid improvements in the battery capacities of the sensors along with increasing computational capability, the term "Resource constrained" is bound to fade in upcoming years. One of the main aim of WSN is to transmit the sensitive information over a network in a secure and efficient manner. To achieve this, a strong authentication protocol is the basic requirement. WSN are placed in hostile environment thus making them vulnerable to many attacks. Thus, to make the defensive mechanism effective, we should consider the possible attacks on it. The proposed protocol makes use of rabin cryptosystem and random nonces and fuzzy logic to secure channel, prevent session hijacking and add probabilistic randomness respectively. We require a an efficient tolerant authentication protocol, If the sensor nodes are subjected to geological conditions or even hostile environment. This makes the sensor nodes exploitable in various terms to numerous passive and active attacks. This article proposed a rabin crytposystem based protocol for IoT security interms of authentication and key generation. It is capable of being resilient towards session key exploit and session hi-jacking attacks. Utilizing probabilistic randomness using random nonces adds up to security and increases the difficulties for the attacker to hack such crypto-secured sessions. We used protocol verifier tool, Proverif [40] that provides proof that the proposed protocol is secure against gateway impersonation, sensor node impersonation, communication channel overhearing attack and sensor capture attack. In future, we try to integrate sound based commands and optical signal and speech recognition functionality to further improve authentication schemes. We will also work in the future direction to use the same protocol in order to mutually authenticate network devices among one another. It is aimed at enhancing security features of the overall system.

REFERENCES

Adat, V., & Gupta, B. (2018). Security in internet of things: Issues, challenges, taxonomy, and architecture. *Telecommunication Systems*, *67*(3), 423–441. doi:10.100711235-017-0345-9

Ahlawat, & Dave, M. (2017, June). A Hybrid Approach for Path Vulnerability Matrix on Random Key Predistribution for Wireless Sensor Networks. *Wireless Personal Communications*, *94*(4), 3327–3353. doi:10.100711277-016-3779-6

Amin, R., & Biswas, G. (2016). A secure Light weight scheme for user authentication & key agreement in Multi gateway based wireless sensor networks. *Ad Hoc Networks*, *36*, 58–80. doi:10.1016/j.adhoc.2015.05.020

Ammar, M., Russello, G., & Crispo, B. (2018). Internet of things: A survey on the security of iot frameworks. *Journal of Information Security and Applications*, *38*, 8–27. doi:10.1016/j.jisa.2017.11.002

Ammara, M., & Giovanni, R. B. C. (2018). Internet of Things: A survey on the security of IoT frameworks. *Journal of Information Security and Applications*, *38*(February), 8–27. doi:10.1016/j.jisa.2017.11.002

Chi, Q., Yan, H., Zhang, C., Pang, Z., & Da Xu, L. (2014). A reconfigurable smart sensor interface for industrial wsn in iot environment. *IEEE Transactions on Industrial Informatics*, *10*(2), 1417–1425. doi:10.1109/TII.2014.2306798

Das. (2012). A random key establishment scheme for multi-phase deployment in large-scale distributed sensor networks. *International Journal of Information Security, 11*(3), 189-211.

Das, A. K. (2012). A random key establishment scheme for multi-phase deployment in large-scale distributed sensor networks. *International Journal of Information Security*, *11*(3), 189–211. doi:10.100710207-012-0162-9

Das, A. K., Zeadally, S., & He, D. (2018). Taxonomy and analysis of security protocols for internet of things. *Future Generation Computer Systems*, *89*, 110–125. doi:10.1016/j.future.2018.06.027

Das, M. L. (2009). Two-factor user authentication in Wireless sensor network. *IEEE Transactions on Wireless Communications*, *8*(3), 1086–1090. doi:10.1109/TWC.2008.080128

Farash, M. S., Turkanovic, M., Kumari, S., & Holbl, M. (2016). An efficient user authentication and key agreement scheme for heterogeneous wireless sensor network tailored for the internet of things environment. *Ad Hoc Networks*, *36*, 152–176. doi:10.1016/j.adhoc.2015.05.014

Huang, X., Xiang, Y., Bertino, E., Zhou, J., & Xu, L. (2014). Robust Multi-factor authentication for fragile communications. *IEEE Transactions on Dependable and Secure Computing*, *11*(6), 568–581. doi:10.1109/TDSC.2013.2297110

Jiang, Q., Ma, J., Lu, X., & Tian, Y. (2015). An efficient two-factor user authentication scheme with unlinkability for wireless sensor networks. *Peer-to-Peer Networking and Applications*, *8*(6), 1070–1081. doi:10.100712083-014-0285-z

Khader, R., & Eleyan, D. (2021). Survey of DoS/DDoS attacks in IoT. *Sustainable Engineering and Innovation*, *3*(1), 23–28. doi:10.37868ei.v3i1.124

Khan, M. K., & Alghathbar, K. (2010). Cryptanalysis and security improvements of two-factor user authentication in wireless sensor networks. *Sensors (Basel)*, *10*(3), 2450–2459. doi:10.3390100302450 PMID:22294935

Kumari, S., & Om, H. (2016). Authentication protocol for wireless sensor networks applications like safety monitoring in coal mines. *Computer Networks*, *104*, 137–154. doi:10.1016/j.comnet.2016.05.007

Li, X., Ma, J., Wang, W., Xiong, Y., & Zhang, J. (2013). A novel smart card and dynamic ID based remote user authentication scheme for multi-server environments. *Mathematical and Computer Modelling*, *58*(1), 85–95. doi:10.1016/j.mcm.2012.06.033

Lounis, K., & Zulkernine, M. (2020). Attacks and defenses in short-range wireless technologies for IoT. *IEEE Access: Practical Innovations, Open Solutions*, *8*, 88892–88932. doi:10.1109/ACCESS.2020.2993553

Mathew, A., & Terence, J. S. (2017, April). A survey on various detection techniques of sinkhole attacks in WSN. In *2017 International Conference on Communication and Signal Processing (ICCSP)* (pp. 1115-1119). IEEE. 10.1109/ICCSP.2017.8286550

Oua & Phan. (2008). Traceable privacy of recent provably-secure rfid protocols. In *International conference on applied cryptography and network security*. Springer.

Phan, R. C.-W. (2009). Cryptanalysis of a new ultra lightweight rfid authentication protocols as IEEE Transactions on Dependable and secure. *Computing*, *6*(4), 316–320.

Riaz, M. N., Buriro, A., & Mahboob, A. (2018). Classification of attacks on wireless sensor networks: A survey. *International Journal of Wireless and Microwave Technologies*, *8*(6), 15–39. doi:10.5815/ijwmt.2018.06.02

Sengupta, J., Ruj, S., & Bit, S. D. (2020). A comprehensive survey on attacks, security issues and blockchain solutions for IoT and IIoT. *Journal of Network and Computer Applications*, *149*, 102481. doi:10.1016/j.jnca.2019.102481

Shafiq, M., Ashraf, H., Ullah, A., & Tahira, S. (2020). Systematic Literature Review on Energy Efficient Routing Schemes in WSN–A Survey. *Mobile Networks and Applications*, *25*(3), 1–14. doi:10.100711036-020-01523-5

Sharma, M., Tandon, A., Narayan, S., & Bhushan, B. (2017, September). Classification and analysis of security attacks in WSNs and IEEE 802.15. 4 standards: A survey. In *2017 3rd International Conference on Advances in Computing, Communication & Automation (ICACCA)* (pp. 1-5). IEEE.

Sinha, P., Jha, V. K., Rai, A. K., & Bhushan, B. (2017, July). Security vulnerabilities, attacks and countermeasures in wireless sensor networks at various layers of OSI reference model: A survey. In *2017 International Conference on Signal Processing and Communication (ICSPC)* (pp. 288-293). IEEE. 10.1109/CSPC.2017.8305855

Srivastava, K., Awasthi, A. K., Kaul, S. D., & Mittal, R. (2015). A hash based mutual rfid tag authentication protocol in telecare medicine information system. *Journal of Medical Systems*, *39*(1), 153. doi:10.100710916-014-0153-7 PMID:25491577

ToolP. (n.d.). https://prosecco.gforge.inria.fr/personal/bblanche/proverif/

Turkanovic, M., Brumen, B., & Holbl, M. (2014). A novel user authentication and key agreement scheme for heterogeneous ad hoc wireless sensor networks, based on the internet of things notion. *Ad Hoc Networks*, *20*, 96–112. doi:10.1016/j.adhoc.2014.03.009

Wang, D., & Wang, P. (2014). On the anonymity of two-factor authentication schemes for wireless sensor networks: Attacks, principle and solutions. *Computer Networks*, *73*, 41–57. doi:10.1016/j.comnet.2014.07.010

Watro, R., & Kong, D. (n.d.). Securing sensor networks with public key technology. In *Proceedings of the 2nd ACM Workshop On Security of Ad hoc and Sensor Networks*. ACM.

Wong, K. H. M., Zheng, Y., Cao, J., & Wang, S. (2006). A dynamic user authentication scheme for wireless sensor networks. *IEEE International Conference on Sensor-Networks, Ubiquitous & Trust-worthy Computing*.

Wu, F. (2017). An Efficient Authentication & Key agreement protocol for Multi-Gateway wireless sensor Networks. *Journal of Network and Computer Applications*, *89*, 72–85. doi:10.1016/j.jnca.2016.12.008

Wu, F., Xue, L., Kumari, S., & Li, X. (2015). A novel and provably secure biometrics-based three-factor remote authentication scheme for mobile client-server networks. *Computers & Electrical Engineering*, *45*, 274–285. doi:10.1016/j.compeleceng.2015.02.015

Xue, K., Ma, C., Hong, P., & Ding, R. (2013). A temporal-credential-based mutual authentication and key agreement scheme for wireless sensor networks. *Journal of Network and Computer Applications*, *36*(1), 316–323. doi:10.1016/j.jnca.2012.05.010

Yu, J., Wang, G., Mu, Y., & Gao, W. (2014). An efficient generic framework for 3-factor authentication with provably secure instantiation. *IEEE Transactions on Information Forensics and Security*, *9*(12), 2302–2313. doi:10.1109/TIFS.2014.2362979

Chapter 7
Multi-Keyword Searchable Encryption for E-Health System With Multiple Data Writers and Readers

Dhruti P. Sharma
Sarvajanik College of Engineering and Technology, India

Devesh C. Jinwala
https://orcid.org/0000-0003-4830-1702
S. V. National Institute of Technology, India

ABSTRACT

E-health is a cloud-based system to store and share medical data with the stakeholders. From a security perspective, the stored data are in encrypted form that could further be searched by the stakeholders through searchable encryption (SE). Practically, an e-health system with support of multiple stakeholders (that may work as either data owner [writer] or user [reader]) along with the provision of multi-keyword search is desirable. However, the existing SE schemes either support multi-keyword search in multi-reader setting or offer multi-writer, multi-reader mechanism along with single-keyword search only. This chapter proposes a multi-keyword SE for an e-health system in multi-writer multi-reader setting. With this scheme, any registered writer could share data with any registered reader with optimal storage-computational overhead on writer. The proposed scheme offers conjunctive search with optimal search complexity at server. It also ensures security to medical records and privacy of keywords. The theoretical and empirical analysis demonstrates the effectiveness of the proposed work.

DOI: 10.4018/978-1-7998-6988-7.ch007

INTRODUCTION

Since the last decade, several countries are moving towards digitization of medical records to improve data availability, data accessibility, data interoperability and data exchange (Akinyele et al., 2011; Löhr et al., 2010). Such digitized medical data would be effectively used in several applications concerning maintenance of health records in terms of EHR (electronic health record)(Rau et al., 2010; Schabetsberger et al., 2006), accounting and billing (Macdonald, 1986), medical research (Sunyaev et al., 2009). In practice, to offer ubiquitous access of data in cost effective manner, the exiting E-Health systems store medical data onto third party cloud server. Since such storage outsourcing may introduce risks of data leakage and security breach, most e-health systems offload encrypted data onto cloud server and subsequently use Searchable Encryption(SE) to search across the stored encrypted data. SE offers two significant features besides data privacy - (1) The data can be shared by the data owners (writers) to data readers and the reader has capability to query the shared data, (2) Query keywords and search operation would be secured in such a way that the service provider will be unable to access the unauthorized medical data stored over it(R. Zhang et al., 2017). There exist several different types of searchable encryptions based on - the cryptographic key(s) used for construction of ciphertext and search token, the structure of the search index used to compute ciphertexts, the number of keywords used to query data, the number of data writers and readers existed in system. Different E-Health systems require different searchable encryption schemes. Considering the number of writers/readers, the authors identify 4 different types of E-Health systems and suggest their suitable SE schemes - (1) When the outsourced data is created and accessed by the same user, then a Symmetric Searchable Encryption (SSE) could be used. For example, a hospital wants to maintain staff payroll, then an authorized accountant could store data onto could server and then would be able to search data from any location, (2) When a single data writer shares data with a single data reader, then any Public Key Searchable Encryption (PKSE) (Baek et al., 2008a; Boneh et al., 2004; Boneh & Waters, 2007) can utilize. Example, a patient shares his medical history with a doctor, (3) When a single data writer shares data with multiple data readers, then any multi-user searchable encryption scheme (Bao et al., 2008; Huang et al., 2016; Y. H. Hwang & Lee, 2007a; Kiayias et al., 2016; Lai et al., 2013; Wang et al., 2016; Ye et al., 2016; Y. Zhang et al., 2016)could be used. For example, a hospital wants to share the information about doctors currently working in that hospital with all registered patients, (4) When multiple data writers want to share data with multiple data readers, then either writer-managed multi-user SE(Bao et al., 2008; Huang et al., 2016; Y. H. Hwang & Lee, 2007b; J. Li & Chen, 2013) or trusted authority based multi-user SE could be used (M.-S. Hwang et al., 2014; Jingzhang et al., 2018; Kiayias et al., 2016; J. Li & Chen, 2013; Lv et al., 2014; Wang et al., 2016; Xu et al., 2019; Ye et al., 2016; Y. Zhang et al., 2016). An example of such E-Health system will be discussed in the Section **Problem Definition**.

Additionally, the existing SE schemes either offer search for a single keyword (Baek et al., 2008b; Boneh et al., 2004) or for multiple keywords (Ballard et al., 2005; Boneh & Waters, 2007; Byun et al., 2006; Z. Chen et al., 2012; Ding et al., 2012; M.-S. Hwang et al., 2014; Y. H. Hwang & Lee, 2007a; B. Zhang & Zhang, 2011)based on number of keywords allowed in search query. In practice, an E-Health system offering multi-keyword search by the stakeholders(viz. hospitals, pharmacy, insurance company etc.) is more desirable.

Figure 1. Scenario of an E-Health System

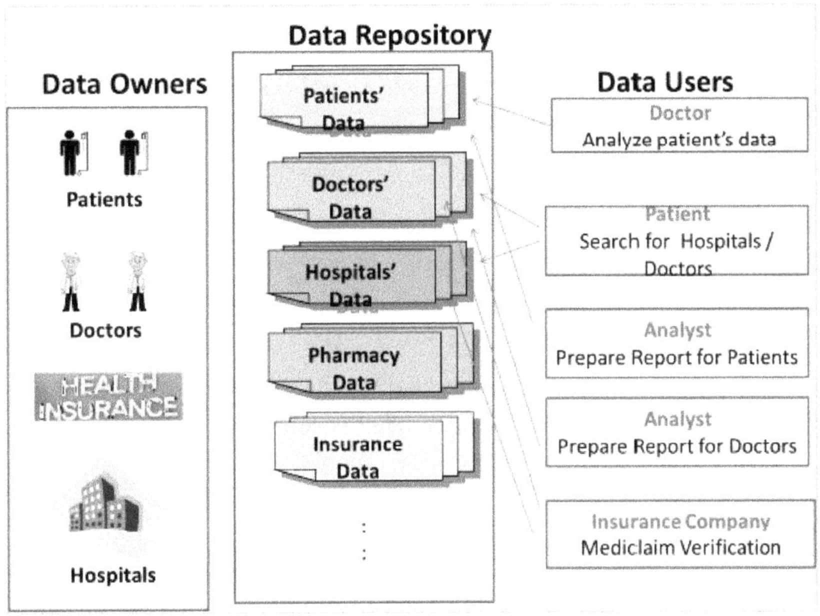

Problem Definition

Let us consider a scenario of an E-Health system depicted in Figure 1. The system has multiple data owners viz. Doctors, Patients, Hospitals etc. Concerning security, the data collected from data owners are in encrypted form and stored at the centralized data repository. Assume that the collected data are analyzed by several data users. For example, a doctor located at remote place accesses patients' current record to treat him remotely, a data analyst sitting at analysis centre analyzes several medical records to generate report on health analysis report. In addition, a patient wants to perform search for cancer specialist or an insurance company wants to investigate hospital data for mediclaim disbursement. To perform all such tasks, extraction of the desired data from central data collection is indeed essential.Such extraction requires *'search over encrypted data'* since stored data are in encrypted form. In general, one could say that for an E-Health system that includes several potential data generators and data users, any public key searchable scheme with multi-reader, multi-writer capability could be used for effective data analysis.

Table 1. Potential Search Queries

Data Users	Query
Patient	1. Find Cancer Specialists in Delhi. ***Query**= 'Cancer Specialist' AND 'Delhi'* 2. Find Cancer hospitals in Delhi. ***Query**='Cancer' AND 'Hospital' AND 'Delhi'*
Doctor	1. List blood Pressure data of patient P1 on 26th April 2018. ***Query**='Patient' AND 'P1' AND 'Blood-Pressure'*
Data Analyst	1. Find female patients suffering with breast cancer in Chennai. ***Query**='Patient' AND 'Female' AND 'Brest-Cancer' AND'Chennai'*

Additionally, let us consider some search queries generated by data users (Table 1). Taking the quoted texts in each query as keywords, it is determined that the search based on multiple keywords (in terms of conjunction (AND)) is required to be performed.

From the above discussion, the authors infer that a public key searchable encryption supporting multiple writers/readers along with multi-keyword search would be more effective in design of such E-Health system.

RELATED WORK

In typical public key searchable encryption (PKSE), since the writer generates searchable ciphertext by employing the public key of reader, his computational overhead would be O(D · W · R) where D=total number of documents required to share, W=total number of keywords to search and R=total number of readers in system. Aiming to reduce the computational complexity of a writer, several multi-user searchable encryption schemes (Bao et al., 2008; Huang et al., 2016; Y. H. Hwang & Lee, 2007a; Kiayias et al., 2016; J. Li & Chen, 2013; Wang et al., 2016; Ye et al., 2016; Y. Zhang et al., 2016)in public key setting have been proposed. However, the schemes (Bao et al., 2008; Huang et al., 2016; Y. H. Hwang & Lee, 2007a; J. Li & Chen, 2013)are writer-managed schemes where a writer manages multiple readers existed in system and so a prior communication amongst the writer and readers is utmost important. Consequently, a writer could share data only with theknown readers. Such schemes can be extended to multi-writer settings where each writer can share data with any randomly chosen readers. On the other hand, the schemes (M.-S. Hwang et al., 2014; Jingzhang et al., 2018; Kiayias et al., 2016; J. Li & Chen, 2013; Lv et al., 2014; Wang et al., 2016; Xu et al., 2019; Ye et al., 2016; Y. Zhang et al., 2016)employs Trusted Authority(TA) to manage multiple readers. In such schemes, TA first computes a master public key. He also prepares the secret keys for each registered readers. Afterwards, the writer uses master key to generate ciphertexts and the reader uses his secret key to compute a search token. Though each ciphertext in these schemes serves to multiple tokens (issued by different readers), they place optimal overhead (O(D ·W)) on the writer in supporting multiple readers in system. Furthermore, any writer in these schemes could share data with any unknown reader. However, support to a single keyword search is the major limitation of these schemes. On the other hand, the recent TA-based schemes (Sharma & Jinwala, 2017; Xu et al., 2019)support multiple writers/readers along with multi-keyword search. However, in these schemes, the TA is responsible to compute a search token for each reader based on the query keywords issued by the reader. Thus, the leakage of query keywords to TA is a major issue in these schemes. To overcome the issue, Sharma et al. have proposed a TA-based multi-writer, multi-reader conjunctive keyword searchable scheme (Sharma & Jinwala, n.d.)where each reader computes a search token from the search query (of keywords) using his private key issued by TA. However, all such TA-based schemes (Sharma & Jinwala, n.d., 2017; Xu et al., 2019)suffer from the lack of control on writers since any writer knowing master public key can compute searchable ciphertexts.

In the context of E-Health System, there exist several works offering security to medical records exchanged between entities viz. doctors, patients, hospitals etc. One of such work has been offered by Yu et al. (Yu et al., 2012)where a forward secure digital signature is employed to sign EMR (electronic medical record) shared between entities. With such signature, the authors ensure the validity of the old EMR record in case of signature key update. To exchange electronic patient record across different hospitals (potentially having heterogeneous record management system), Chen et al.(T.-L. Chen et al.,

2012) have given a mobile agent based secure scheme for EMR. In this scheme, the authors employs Lagrange interpolation based key management scheme to propose a specialized program (mobile agent) that provides an efficient as well as secure access control for medical records in cross-hospital system. Furthermore, the schemes (Akinyele et al., 2011; Liu et al., 2015)offer secure sharing of medical records among data provider and multiple data readers. Both of these schemes employ CPABE based access control mechanism (Goyal et al., 2006)that enables data readers possessing appropriate attributes to access medical records. A scheme (Eom et al., 2016) discusses an attribute based encryption where a patient being data provider can control accessing of his data to health service provider (i.e. doctor, nurse etc). However, none of these schemes (Akinyele et al., 2011; Eom et al., 2016; Liu et al., 2015)provide search across encrypted medical data. To offer solution, researchers have used searchable encryption to encrypt electronic health records. One of such solutions is given in(Wu et al., 2016) where authors have proposed a searchable encryptionhaving secure channel architecture. Additionally, this scheme offers support to more than one writers and readers in system. However, the computational complexity for a writer is linear to the total number of readers. Besides this, the scheme(Wu et al., 2016)provides a single keyword search only. Furthermore, the recent schemes (Jingzhang et al., 2018; H. Li et al., 2017; Xu et al., 2019)also offer SE for encrypted medical data in multi-reader setting. All these schemes employ the notion of attribute based encryption to control shared data access by multiple readers. In addition, the scheme (Jingzhang et al., 2018)provides attribute update capability whereas (H. Li et al., 2017; Xu et al., 2019)offer search across dynamic database using inverted index search structure. However, in scheme (Xu et al., 2019), database update (insertion/deletion of record) is performed by the centralized TA and hence introduces additional communication overhead between data provider and TA. Such overhead degrades the effectiveness of (Xu et al., 2019)as compared to (H. Li et al., 2017)where data provider himself could update database.

The more recent mechanism for secure sharing of sensitive data collected via smart devices viz. smart pacer, smart band,smart pulse rate monitor, smart glucometer among patients and doctors is proposed in [Swarna2020]. A prime objective of this mechanism is to define Deep Neural Network based Intrusion Detection System to prevent different cyberattacks potentially been performed by attackers on medical data before it reaches to the intended user (doctor). Furthermore, the workpresented in [] is targeted for the emerging field of preventive medicine. The authors in [] proposed two new approaches - variance ranking (attribute selection technique) and ranked order similarity (similarity measurement technique). With this approaches, they work over imbalanced medical dataset and compute minority target class which is especially used for predicting processes in preventive medicine.

Exploring the current literature, it is observed thatthere doesn't exists any PKSE scheme for encrypted medical records with inclusion of multi keyword search capability for the controlled multi-writer multi-reader settings with optimal computational complexity on the data owners (writers).

CONTRIBUTIONS

In this work, the authors propose a PKSE scheme for E-Health system - **Multi-Keyword Searchable Encryption in Multi-Writer Multi-Reader Setting (MKSE-MWMRS)** that securely shares medical records amongst multiple data writers and data readers. The scheme puts optimal ciphertext computational overhead (that is independent of the number of readers) on writer and optimal ciphertext storage overhead on server. Furthermore, it enables readers to search across the shared data for the chosen query of keywords. The following are the major contributions:

1. **Controlled multi-writer multi-reader settings**With the proposed scheme, registration is mandatory for each stakeholder (viz. Patient, Doctor, Hospital, Pharmacy etc.) of E-Health system. A stakeholder wishing to share data is registered as Writer whereas a stakeholder wishing to access data is registered as Reader. With registration, each writer (resp. reader) gets a write (resp. read) secret key from the trusted authority. The scheme offers a controlled environment where only the registered writer (possessing a write secret key) can upload data and registered reader (possessing a read secret key) can search data.
2. **Multi-keyword search** The proposed scheme offers a computationally efficient search algorithm that takes a search token as input. Such token is computed from the chosen query of multiple keywords in conjunctive relation.
3. **Search Keyword Privacy** In the proposed scheme each searchable ciphertext includes an encrypted payload message and the set of encrypted keywords. The keywords are encrypted using the proposed randomized encryption algorithm that uses write secret key and a random element. Thus, no two ciphertexts are same even if they computed from the same set of keywords. Hence, the storage server possessing a search token would not be able to find the association of keywords with payload unless having write secret key. Such keyword privacy offers security against chosen keyword attack potentially be performed by the server.
4. **Medical Record Privacy**In the proposed scheme, to offer privacy, a symmetric cipher is used to encrypt each medical record. The symmetric key used to encrypt such record is chosen by each writer separately. Furthermore, to enable a reader to decrypt the record, the writer sends the encrypted form of the chosen symmetric key along with searchable ciphertext. The proposed encryption algorithm defines the way to encrypt the chosen symmetric key besides the way to encrypt keywords. Apparently, no adversary without having symmetric key would be able to decrypt the medical record and hence the proposed scheme offers record privacy.

PRELIMINARIES

Bilinear Map

Let G_1 and $G_2$are groups of the prime order p. For G_1, assume P is any arbitrary generator. Assume the DLP (Discrete Logarithm Problem) i.e. the standard harness assumption (Pohlig & Hellman, 1978) is hard in G_1 as well as in G_2. A mapping e: $G_1 \times G_1 \rightarrow G_2$ satisfying the following properties is called **Cryptographic Bilinear Map**(Dutta et al., 2004).

1. **Bilinearity**$\forall P,Q \in G_1$ and $(a,b) \in Z_p$, $e(aP, bQ) = e(P, Q)^{ab}$.
2. **Non-degeneracy**$e(P, P) \neq 1$. i.e. if P is a generator of G_1, then e(P, P) is a generator of G_2 with order p.
3. **Computability**An efficient algorithm which could compute a pairing e(P, Q) is always existed for all $P, Q \in G_1$.

Assumption

Decisional Diffie-Hellman (DDH) Assumption Assume Gis a finite cyclic subgroups with order p for an elliptic curve. Assume P is a generator of G. Then DDH problem for the given input tuple (P, aP, bP, cP, abP) is to differentiate the tuple (P, aP, bP, abP) from (P, aP, bP,cP). Here a,b,c $\in Z_p$ are random elements. The advantage for an **A**can be defined as

$$Adv_{\mathbf{A}}^{DDH}(\lambda) = | \Pr[\mathbf{A}(aP, bP, abP)=1] - \Pr[\mathbf{A}(aP, bP, cP)=1] |$$

Here, $\lambda$is a security parameter. The DDH assumption holds if the advantage $Adv_{\mathbf{A}}^{DDH}(\lambda)$ is negligible.

FORMAL DEFINITION & SECURITY MODEL

This section describes asystem model, proposed algorithms and a security model.

System Model

A system model for the proposed MKSE-MWMRS (Figure 2) includes four entities:

1. **Trusted Authority (TA)** An entity that computes a master public-private key pair as well as system's global parameters. It also performs registration for writers and readers. As a part of registration, it issues a write secret key to the writer and a read secret key to the reader. Additionally, it is responsible to maintain the lists of registered writers and readers onto storage server.
2. **Writer (WR)** An entity that outsources encrypted medical record(s) to cloud server for providing convenient and reliable data access by the registered readers of the E-Health system. To enable search across encrypted data, it associates a set of encrypted keywords with each medical record. The keywords are encrypted using a write secret key issued by TA.
3. **Reader (RD)** An entity that generates a search token from the chosen query of keywords in conjunctive form. A token includes a set of query keywords encrypted using read secret key. By issuing a search token, the reader enables the server to search across available medical record(s).
4. **Cloud Server(CS)** An entity that offers storage for encrypted medical records along and their associated set of keywords. On receiving a search token from the reader, it performs search across the stored sets of keywords and forwards the encrypted medical record to the reader if search is successful. On unsuccessful search, it returns null to the reader.

Proposed Algorithms

The proposed MKSE-MWMRS provides the following polynomial time algorithms. The notations used are as in Table 2.

1. **Sys_Init(λ,n)** Executed by TA

Figure 2. System Model

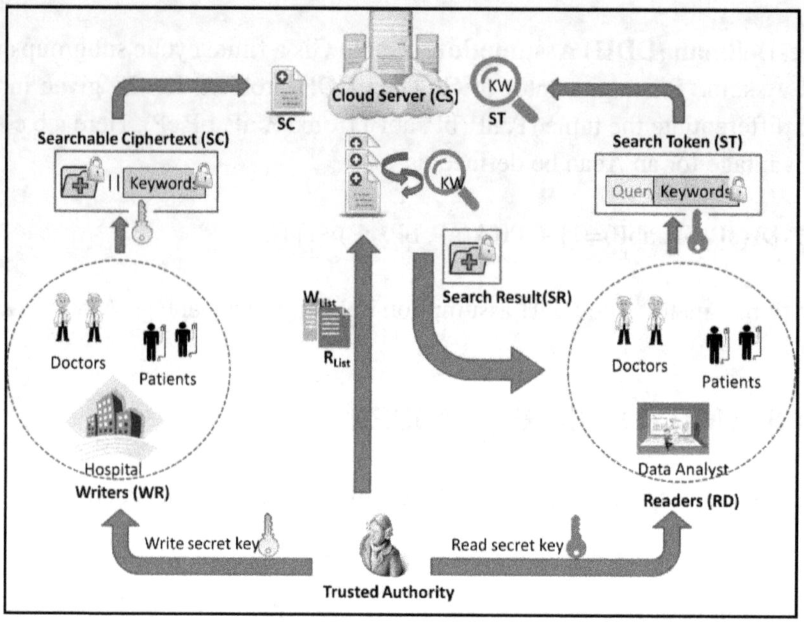

From security parameter λ, the ***System Initialization*** algorithm firstly defines a keyword space K for n keyword fields. It also considers any standard symmetric cipher for encryption of original medical

Table 2. List of Notations

Symbol	Description
K	Keyword Space
GP	Global parameter
(MPK, MSK)	Master Public Key, Master Secret Key
ID_r, ID_w	Unique Identifierfor a reader and a writer
W_{sc}, W_{sr}	Write secret key, Write server key
R_{sc}, R_{sr}	Read secret key, Read server key
W_{List}, R_{List}	Listsshowing registered Writers and Readers
M	Medical record in plaintext
ℓ	Total Keywords in query
SC	Searchable ciphertext
n	Total Keywords in ciphertext
K	$K=\{k_1,k_2,\ldots, k_n\}$, a set of keywords in ciphertext
K'	$K'=\{k_1,k_2,\ldots,k_\ell\}$, a set of query keywords where $\ell<=n$
Q	A conjunctive query Q comprises of two sets {K',P'} where K' is as mentioned above and $P'=\{p_1,p_2,\ldots,p_\ell\}$ is a set of positions of query keywords in ciphertext.
ST_Q	Search Token computed from query Q
SR_Q	Search Result for query Q

records. TA then generates the global parameter GP and a master key pair (MPK, MSK). It also generates the lists of registered writers and readers as W_{List} and R_{List} respectively. TA sends these lists with the initial empty value to CS.

2. **WR_Reg(ID_w,GP, MSK)** Executed by TA

The TA uses ***Writer Registration*** algorithm to register a new data writer possessing ID_w. The algorithm uses MSK to compute a write secret key W_{sc} for the registered writer and the associated server side key W_{sr}. For each new registered writer, TA updates W_{List} at CS.

3. **RD_Reg(ID_r,GP,MSK)** Executed by TA

The TA uses***Reader Registration*** algorithm toregister a new reader possessing ID_r. The algorithm uses MSK to compute a read secret key R_{sc} for the registered reader and theassociated server side key R_{sr}. For each new registered reader, TA updates R_{List} at CS.

4. **Encryption(ID_w,GP, MPK, W_{sc},K, M)** Executed by WR possessing ID_w

The ***Encryption*** algorithm first computes encrypted input medical record M'=Enc_{key}(M). Here, 'Enc' is any predefined symmetric cipher with key. The algorithm then encrypts 'key' using W_{sc} and generates encrypted key 'key''. It also computes a ciphertextCby encrypting keywords in list K={k_1,k_2,...,k_n} using (GP,MPK,W_{sc}). Finally, it outputs a searchable ciphertext SC=(C,M',key',ID_w).

5. **SC_Store(SC,W_{List})** Executed by CS

The ***Ciphertext Store*** algorithm first checks the authenticity of the received SC using available W_{List}. The algorithm stores SC only if it comes from the authenticated writer.

6. **TokGen(ID_r, GP, R_{sc}, Q)** Executed by RD possessing ID_r

The ***Token Generation*** algorithm generates a search token ST_Q for input query Q=(K',P') by encrypting keywords in query using (GP,R_{sc}). Here, K'={k'_1,k'_2,...,k'_ℓ} is a set of query keywords and P'={p_1, p_2,..., p_ℓ} shows their positions in ciphertext.

7. **Search(GP,SC, ST_Q,R_{sr},R_{List})** Executed by CS

The ***Search*** algorithm first checks the authenticity of the received ST_Q using available R_{List}. If token is from genuine reader, then the algorithm performs conjunctive search by applying ST_Q on ciphertext C of SC using R_{sr}. For the successful search, the algorithm returns SR_Q=(M',key'). Otherwise, it returns $SR_Q=\perp$.

8. **WR_Dereg(ID_w,W_{List})** Executed by TA

The ***Writer Deregistration*** algorithm revokes the writer WR (having ID_w) by removing its entry from W_{List}. It then updates W_{List} at CS.

9. **RD_Dereg(ID_r, R_{List})** Executed by TA

The ***Reader Deregistration*** algorithm revokes the reader RD (having ID_r) by removing its entry from R_{List}. It then updates R_{List} at CS.

10. **Decryption(SR_Q,R_{sc})** Executed by RD

The ***Decryption***algorithm first decrypts encrypted key key' using R_{sc}. It then applies the predefined symmetric cipher with key onto M' to generate original medical record M.

Assumptions: (i) There exist secure communication channels amongst entities, (ii) The CS is a semi-honest server which is curious to learn plaintext keywords from the set of encrypted keywords and token, (iii) TA assigns a unique identifier$ID_r$to each reader and ID_w to each data writer after physical document verification.

Security Model

The assumption for the proposed cloud storage based E-Health System is that the trusted authority as well as the registered writers and readers are honest entities and they follow the proposed algorithms. Besides this, assume that the server is semi-honest who honestly follows the algorithms but curious to acquire the underlined meaning of the stored medical records. Consequently the server could attempt to get information available within an encrypted record by decrypting it. The server could also try to identify the keywords from the available searchable ciphertexts and tokens that include list of encrypted keywords. Against this adversary, the authors proposes semantic security (aka. indistinguishability of ciphertext against chosen keyword attack (IND-CKA)) following the security game ICLR (Indistinguishability of Ciphertext from Limited Random) (Golle et al., 2004; M.-S. Hwang et al., 2014; Lee et al., 2013).

Game ICLR

Assume **A**is a polynomial bounded adversary and **B**is a challenger algorithm. In ICLR game, when **A**issues a set K of 'n' keywords and a subset $T \subseteq \{1,2,...,n\}$, **B**responses with two encrypted keyword sets that are associated with T in the way that**A**can'tdifferentiate the encrypted keyword sets created withT. Hence **A**can't deduce keywords from the other keyword sets. The phases defined for the proposed game are as follows:

1. **Setup Phase**Initially,**B** runs the algorithm **Sys_Init()** and generates the system's global public parameter GP as well as a master key pair (MPK, MSK). It also registers writers and readers by running the appropriate **WR_Reg()** or **RD_Reg()**. Accordingly, each registered writer has given a write secret key W_{sc} and each reader has given a read secret key R_{sc}. The corresponding server side keys i.e. $W_{sr}$and R_{sr} are also given to**A**.
2. **Query Phase** After system setup, an adversary **A** adaptively requests **B** for ciphertext and token for the chosen sets of keywords k_i and the chosen query Q_i. In response to the ciphertext

query for K keywords, **B**executes SC←Encryption(ID_w,GP,MPK,W_{sc},K,M) and sends SC to **A**. Furthermore, in response to search token query for the conjunctive query keywords Q, **B**executes ST_Q←TokGen(ID_r,GP,R_{sc},Q) and sends ST_Q to **A**. In addition, **A** may ask for ciphertextsand tokens for the chosen sets of keywords in this phase repeatedly.

3. **Challenge Phase** In this phase, **A**takesthe keyword set K, a subset T⊆{1,2,...,n} and σ ∈T in the way that no token given in **Query Phase** are distinguishing the sets generated byRand(K,T) and Rand(K,T−{σ}). Here, Rand(K,T) generates a set Kwhere keywords indexed by T (i.e. the set $\{k_i \mid i \in T\}$) are replaced by random values. **A** then sends (K,T,σ) to challenger **B**. In response,**B**generates two keyword sets K_0=Rand(K,T−{σ}) and K_1=Rand(K,T). Further, **B**randomly selectsb∈{0,1} and returns a challenge ciphertextSC←Encryption(ID_w,GP,MPK,W_{sc},K_b,M) to **A.**
4. **Repeat Query Phase** Here, **A** can again query for ciphertexts and tokens for the chosen sets of keywords. However, the restriction is that **A** can't make request for the token that is distinguishing for $K_0$and K_1.
5. **Guess Phase** From the available tokens, and challenge ciphertext, **A** tries to guess a bit b'∈{0,1}. Apparently, he could win ICLR game for b'=b.

The following is the advantage (ϵ) for adversary **A** in winning the ICLR

$Adv_A(\lambda) = |Pr[b' = b] - ½| \geq \epsilon$

For the negligible advantage ϵ, the proposed MKSE-MWMRS is IND-CKA secure.

DETAILS OF MKSE-MWMRS

This section presents the formal construction of MKSE-MWMRS along with the security proof for medical record as well as for encrypted keywords.

Construction

MKSE-MWMRS works in 5 stages: *System Initialization, Registration and Deregistration, Record Storage, Record Search, Record Retrieval.*

System Initialization

The Trusted Authority (TA) sets up a system by running the following system initialization algorithm.

Sys_Init(λ,n) Following this algorithm, TA takes security parameter λ to generate two bilinear group G_1 and G_2 of the same prime order p with P be the generator of G_1. Then TA takes a bilinear map $e:G_1\times G_1\rightarrow G_2$ and a hash function $H:\{0,1\}^*\rightarrow Z_p^*$. It defines a keyword spaceK of n keyword fields where position of each field is prefixed. TA also considers a standard symmetric cipher SYMM(Set(),Enc(),Dec(),key) where Set(),Enc()and Dec() are the algorithms as SGP← Set(), M'←Enc_{key}(M) and M←Dec_{key}(M'). Here SGP is the global parameter for SYMM.

Further, TA selects random elements α, β, $\gamma \in Z_p^*$ and computes $PK_1 = \alpha P$, $PK_2 = \beta P$ and $PK_3 = \gamma P$. It then sets the global parameter as $GP=\{G_1,G_2,e,P,H,n,K,SGP\}$, a master public key as $MPK=\{PK_1, PK_2\}$, a master secret key as $MSK=\{\alpha,\beta\}$ and an encryption element $E=\{PK_3\}$.

Additionally, TA prepares W_{List} and R_{List} with two fields i.e. $W_{List}=\{ID_w,W_{sr}\}$ and $R_{List}=\{ID_r,R_{sr}\}$. It then stores both lists with initial empty fields i.e. $W_{List}=\{\perp,\perp\}$, $R_{List}=\{\perp,\perp\}$ onto CS.

Registration and Deregistration

The TA manages all writers and readers by registering them on demand. TA can also revoke the registered writers and readers by deregistering them. The algorithms for registration and deregistration are as follows:

WR_Reg(ID_w,GP,MSK,E) To register a new writer possessing ID_w, TA first selects $w \in Z_p^*$ at random. Using input (GP, MSK,E), he sets a write secret key as $W_{sc}=\{w,E\}$ and the corresponding server side key as $W_{sr}=\{wP\}$. TA then issues W_{sc} to writer and updates W_{List} on CS by setting $W_{List} = W_{List} \cup \{ID_w,W_{sr}\}$.

RD_Reg(ID_r,GP,MSK,E) To register a new reader possessing ID_r, TA first selects $u \in Z_p^*$ at random. Using input (GP,MSK), he sets a read secret key $R_{sc}=\{u_1=k,u_2=(\beta-u)P,E\}$ and the corresponding server side key $R_{sr}=\{u\}$. TA then issues R_{sc} to reader and updates R_{List} on CS by setting $R_{List}=R_{List} \cup \{ID_r,R_{sr}\}$.

WR_Dereg(ID_w,W_{List}) To revoke a registered writer possessing ID_w, TA updates a local W_{List} as $W_{List}=W_{List}-\{ID_w,W_{sr}\}$. He then replaces the old W_{List} at CS with this new W_{List}.

RD_Dereg(ID_r,R_{List}) To revoke a registered reader possessing ID_r, TA updates a local R_{List} as $R_{List}=R_{List}-\{ID_r,R_{sr}\}$. He then replaces the old R_{List} at CS with this new R_{List}.

Record Storage

In this stage, a registered writer WR constructs a searchable ciphertext SC and forwards it to storage server CS. Further, CS checks the authenticity of the received ciphertext and stores it if valid. The algorithm for ciphertext construction i.e.*Encryption()* and for ciphertext storage i.e. *SC_Store()* are as follows.

Encryption (ID_w,GP,MPK,W_{sc},K,M) A writer possessing ID_w first selects a symmetric **key** for the cipher SYMM and computes encrypted medical record $M'=E_{key}(M)$. He then encrypts **key** using an encryption element $E \in W_{sc}$ as $key'=key \oplus E$. Further, from an input list of keywords $K= \{k_1,k_2,\ldots,k_n\}$, the writer WR computes a ciphertext $C=\{C_{1i},C_2\}$. Here, $C_{1i}=r(H(k_i)PK_1 + PK_2) + wP$ and $C_2=rP$ for a random $r \in Z_p^*$ and $1 \le i \le n$. Finally, WR offloads the searchable ciphertext as $SC=\{C, M', key', ID_w\}$ onto CS.

SC_Store(SC,W_{List}) Following this algorithm, CS first checks the authenticity of each input SC by verifying the entry of $ID_w \in SC$ in the available list W_{List}. If $ID_w \in W_{List}$, then CS updates SC by replacing $C_{1i}=C_{1i}-W_{sr}$ for $1 \le i \le n$. He then stores this updated SC onto storage space. On the other hand, if $ID_w \notin W_{List}$, then CS rejects the input SC.

Record Search

To search a medical record, a registered reader first computes a search token and issues it to CS. On receiving a token, CS first checks the authenticity of token. If token is sent by the registered reader, then CS uses this token to perform conjunctive search on the available SC. The algorithm for computation of a search token i.e. *TokGen()* and its application on SC to perform search i.e. *Search()* are as follows.

TokGen(ID_r,GP,R_{sc},Q) Any registered reader possessing ID_r and the corresponding R_{sc} can compute a search token from the chosen conjunctive query Q=(K',P'). Here, K'={k'_1,k'_2,...,k'_ℓ}is a set of query keywords and P'={p_1,p_2,...,p_ℓ} shows their positions in ciphertext. Such a search token is ST_Q={ST_1,ST_2,P',ID_r} where $ST_1=t(\sum(H(k'_j)u_1P + u_2))$ for $p_1 \leq j \leq p_\ell$ and ST_2 = tP for a random t $\in Z_p^*$. The reader RD then sends ST_Q to CS to perform search across available SCs.

Search(GP,SC,ST_Q,R_{sr},R_{List}) For the input search token ST_Q, the CS first checks its authenticity by verifying the entry of $ID_r \in ST_Q$ in the available list R_{List}. If $ID_r \in R_{List}$, then CS computes $C'=\sum C_{1i}$ and $T'=ST_1-(\ell R_{sr}ST_2)$ where ℓ=|P'|. He then perform search by checking the equality

$$e(C',ST_2) = e(T',C_2) \tag{1}$$

If the above Eq. (1) holds, the CS returns the search result SR_Q=(M',key'), else it returns $SR_Q=\perp$.

Record Retrieval

Once the receiver RD gets the search result SR_Q=(M',key') from CS, he retrieves the original medical record by using *Decryption()* algorithm as follows

Decryption(SR_Q,R_{sc}) The RD first decrypts key' using $E \in R_{sc}$ as key=key'$\oplus$E. He then generates the plaintext medical record as $M=Dec_{key}(M')$.

Correctness Analysis

The authors demonstrates that the correctly generated search token can search across correctly generated ciphertext. The proof for the correctness of the equality Eq. (1) is as follows.

For a valid ST_Q, the proposed *Search()* algorithm computes

$C' = \sum C_{1i}$ for $p_1 \leq i \leq p_\ell$

$= r(H(k_{p1})\alpha P + \beta P) + r(H(k_{p2})\alpha P + \beta P) + \ldots + r(H(k_{p\ell})\alpha P + \beta P)$

$= rP(\sum(H(k_i)\alpha) + (\ell\beta))$ for $p_1 \leq i \leq p_\ell$

$T' = ST_1 - (\ell R_{sr}T_2)$

$= t(\sum(H(k'_j)u_1P + u_2)) - (\ell R_{sr}T_2)$ for $p_1 \leq j \leq p_\ell$

$= t(\sum(H(k'_j)\alpha P + (\beta\text{-}u)P)) - (\ell utP)$

$= t(\sum(H(k'_j)\alpha P)) + (\ell t\beta P) - (\ell tuP) - (\ell utP)$

$= t(\sum(H(k'_j)\alpha P) + (\ell\beta P))$

$= tP(\sum(H(k'_j)\alpha) + (\ell\beta))$

The equality is then checked i.e. $e(C',ST_2) = e(T',C_2)$ where

$e(C',ST_2) = e(rP(\sum(H(k_i)\alpha) + (\ell\beta)),tP)$ for $p_1 \leq i \leq p_\ell$

$= e(P,P)^{(rt(\sum(H(ki)\alpha) + (\ell\beta))}$ (2)

$e(T',C_2)= e(tP((H(k'_j)\alpha) + (\ell\beta)),rP)$ $p_1 \leq j \leq p_\ell$

$= e(P,P)^{(rt(\sum(H(kj)\alpha) + (\ell\beta))}$ (3)

From Eq. (2) and (3), L.H.S. = R.H.S if ($k_i=k'_j$ and i=j).

Security Analysis

The authors here analyze security of MKSE-MWMRS. The security proof for the medical records is given in **Theorem 1** whereas security for thecorresponding keywordsisdescribed**Theorem 2.** With Theorem 1, the authors prove that the adversary including storage server would not be able to get plaintext from the encrypted medical records and thus security of medical records is proved. With Theorem 2, the authors prove that the cloud server with available search tokens and challenge ciphertext would not be able to learn keywords in plaintext under DDH assumption and so the security forciphertexts against chosen keyword attack (IND-CKA) has been proved.

Theorem 1: A medical record in the proposed MKSE-MWMRS is secure.

Proof: In the proposed scheme, the TA during system initialization phase chooses any standard symmetric cipher SYMM=(*Set(), Enc(), Dec()*) to encrypt/decrypt a medical record M. In addition, it publishes the global parameter of such a symmteric cipher SGPwith system's global parameter GP. As a result, SGP is available to each registered writer and reader. As discussed in the proposed *Encryption()*, a writer first selects a secret key '**key**' and computes an encrypted medical record as M' $\leftarrow Enc_{key}(M)$. Subsequently, he secures the selected **'key'** by encrypting it as key'=key$\oplus$ E where E is the encryption element issued by TA only to the registered writers and readers. The writer then sends (key',M') along with encrypted keywords as a searchable ciphertext SC to the cloud server CS for further storage.

For any adversary **A** (including CS) in such a scheme, to learn a plaintext record M from the available SC, it is indeed necessary to get decryption (symmetric) key 'key'. However, 'key' available within SC is already encrypted by writer using E. Such an encryption component E is actually been generated by TA using secret element γ. Thus, the probability of**A** to compute record M is equivalent to computing E. Since, E is only available to the registered reader and writer, no adversary would be able to learn M from the available (M',key') with negligible probability. Thus, it can be claimed that a medical record shared with the proposed MKSE-MWMRS is indeed secure.

Theorem 2: The MKSE-MWMRS is IND-CKA secure if DDH assumption holds.

Proof: Consider a polynomial time adversary **A** who makes maximum q(<p) token queries with a non-negligible advantage ϵ in solving DDH problem in G_1. Assume that there exist two bilinear groups G_1 and G_2 having same prime order p with P be the generator of G_1. For such a setup, there exist an algorithm**B**(simulator) as a challenger having an advantageϵ'to simulate the security game. Here $\epsilon' = \epsilon/(nqe^n)$ where eis a base of natural logarithm.

Consider a tuple (aP, bP, cP) as DDH challenge in $G_1$for **B** where a,b,c $\in Z_p^*$ are the randomly selected elements. The aim of **B** is to differentiate cP=abP from the random element of G_1. To do so, **B** uniformly selects a position z that is independent of position σ selected by**A** during *Challenge Phase* of the game ICLR. The simulation for ICLR can be as follows:

1. **Setup Phase** In this phase, TA selects random elements α, β, $\gamma \in Z_p^*$ and computes $PK_1 = \alpha P$, $PK_2 = \beta P$ and $PK_3 = \gamma P$. It then sets the Global Parameter GP={G_1,G_2,e,P,H,n,K},Master Public Key MSK={PK_1, PK_2}, Master Secret Key as MSK={α,β} and Encryption element E={PK_3}. Afterwards, TA publishes (GP, MPK). Thus,**A**and **B** have (GP,MPK). TA also provides W_{sc}={w,E} and $R_{sc}=\{u_1=k, u_2=(\beta - u)P, E\}$ to **B** for random element w,u $\in Z_p^*$. In addition, TA sends the server side keys W_{sr}={wP} and R_{sr}={u} to **A**.
2. **Query Phase** Here, **A** adaptively requests **B** for the ciphertexts for the chosen sets of keywords K_i and tokens for queries Q_i where $1 \leq i \leq q$.

As a ciphertext query, **A** sends a keyword set $K=\{k_1,k_2,...,k_n\}$ to**B**. In response, **B** executes *Encryption(ID_w,GP,MPK,W_{sc},K,M)* as follows:

For every keyword $k_i \in K$ where $1 \leq i \leq n$, **B** randomly selects $\gamma_i \in Z_p^*$, $r \in Z_p^*$ and computes $C=\{C_{1i},C_2\}_{\{1 \leq i \leq n\}}$. Here, $C_{11}=r(\gamma_1 PK_1 + PK_2)+(wP)$, $C_{12}=r(\gamma_2 PK_1 + PK_2)+(wP)$, $C_{1z}=br(\gamma_z PK_1 + PK_2 + (wP)$, ..., $C_{1n}=r(\gamma_n PK_1 + PK_2)+(wP)$, and $C_2=rP$. Finally,**B** sends C to**A**. Furthermore,**A** collects multiple such ciphertexts C_i of the chosen K_i where $1 \leq i \leq q$. For each available ciphertext, **A** checks authenticity and stores only valid ciphertexts by running *SC_Store()* algorithm. Note that the given theorem concerns only keywords security, and thus the other parameters of searchable ciphertext are ignored here.

To get a search token,**A** sends a token query $Q=(K',P')$ where $K'=\{k'_{p1},k'_{p2},...,k'_{p\ell}\}$ and $P'=\{p_1,p_2,...,p_\ell\}$ to**B**. In response **B** executes *TokGen(ID_r,*GP,*R_{sc},Q)* as follows:

B randomly selects $t \in Z_p^*$ and computes $ST_1=(t \cdot (\sum(H(k'_j)u_1P + u_2))$ for $p_1 \leq j \leq p_\ell$, $ST_2=tP$. **B** then sends ST_Q to **A**. Furthermore, **A**collects multiple such tokens ST_{Qi} for different queries Q_i for $1 \leq i \leq q$.

3. **Challenge Phase** In this phase, **A**issues a tuple (K,T, σ) tosimulator **B**withσ ∈T and T⊆{1,n}.

In response**B**checks the selected z. If $z \neq \sigma$, **B** issues a random guess as a response of DDH challenge. This implies that **B** sends challenge ciphertextC_0 with random values for all keywords in K_i.

If $z=\sigma$, **B**sends C_1 with encrypted keywords computed as follows

Bfirst sets $C_{1\sigma}^*=c(\gamma_\sigma PK_1+PK_2)$. Then for $i \neq \sigma$, $i \in T$, it sets $C_{1i}^*=\ell_i'$ where $\ell_i' \in Z_p^*$. Furthermore, for $i \neq \sigma$, $i \notin T$, **B** sets$C_{1i}^*=a(\gamma_i PK_1 + PK_2)$. Besides these, it computes $C_2^*=aP$.

Finally, **B** sends **A** the challenge ciphertext (C_{1i}^*,C_2^*) where $1 \leq i \leq n$. **A**then wins the security game if $z=\sigma$. Note that the received ciphertext is encryption of keywords k_i for every $i \notin T$. On the other hand, the received ciphertext is an encrypted form of w_σ for position σ where c=ab. The ciphertexts for the elements at the other positions are random values.

4. **Repeat Query Phase** In this phase, **A** asks for ciphertexts for various keyword sets. He also asks for tokens for different chosen queries. **B** responses in the similar way as there in *Query Phase*. The constraint here is **A** can't issue the above queries for a location σ.
5. **Guess Phase**Finally, from the available tokens, and challenge ciphertext, **A** guesses a bit b'∈{0,1}. If b' =1, **B** outputs 'Yes', and (aP,bP,cP) is considered as a DDH tuple. So, for z=σ, a proof showing (aP,bP,cP) as DDH tuple is as follows.

As know, *Search()*includes the equality check

$$e(C',ST_2) = e(T',C_2) \tag{4}$$

where

$$e(C',ST_2) = e(br(\gamma_z PK_1+PK_2), tP)$$

$$= e(br(\gamma_z \alpha P+\beta P), tP)$$

$$= e(P,P)^{(brt(\gamma z)\alpha+\beta)} \tag{5}$$

$e(T',C_2) = e(t\,(H(k_z)\alpha P + \beta P), rP)$

$= e(P, P)^{(rt(H(kz)\alpha+\beta))}$ (6)

Using challenge ciphertext,

$e(C',ST_2) = e(c(\gamma_\sigma PK_1+PK_2), tP)$

$= e(c(\gamma_\sigma \alpha P+\beta P), tP)$

$= e(P,P)^{(ct(\gamma\sigma)\alpha+\beta)}$ (7)

$e(T',C_2) = e(t\,(H(k_z)\alpha P + \beta P), aP)$

$= e(P, P)^{at\,(H(kz)\alpha+\beta)}$ (8)

From the above Eq. (5), (6), (7), (8)

$e(P,P)^{(brt(\gamma z)\alpha+\beta)}/e(P,P)^{(rt\,(H(kz)\alpha+\beta))} = e(P,P)^{(ct\,(\gamma\sigma)\alpha+\beta)}/e(P,P)^{at\,(H(kz)\alpha+\beta)}$

$e(P,P^{)(brt(\gamma z)\alpha+\beta)} \cdot e(P,P)^{at(H(kz)\alpha+\beta)} = e(P,P)^{(ct(\gamma\sigma)\alpha+\beta)} \cdot e(P,P)^{(rt(H(kz)\alpha+\beta))}$

$e(P,P)^{(abt(\gamma z)\alpha+\beta)} \cdot e(P,P)^{rt(H(kz)\alpha+\beta)} = e(P,P)^{(ct(\gamma\sigma)\alpha+\beta)} \cdot e(P,P)^{(rt\,(H(kz)\alpha+\beta))}$

$e(P,P)^{(abt(\gamma z)\alpha+\beta)} = e(P, P)^{(ct(\gamma\sigma)\alpha+\beta)}$

$ab = c$ (9)

Furthermore, the given challenge (aP,bP,cP) can't be proved as DDH tuple in case of b'=0 since the challenge ciphertext includes random element at position ***i*** and so the Eq. (9) can't confirm.

The simulations for **B'**sthe advantage are: (i)S1 where **B**responses with the token queries for n keyword sent by **A** and (ii) S2 where **B** is not aborting in a challenge phase.

The probability of S1 and S2(for large enough q) is defined as

$Pr[S1]=1/e^n$ and $Pr[S2]=1/(nq)$

Thus, advantage for **B**to solve DDH problem is $\epsilon' = \epsilon \cdot Pr[S1 \cap S2] = \epsilon/(nqe^n)$.

As per the propositions discussed in(Golle et al., 2004), if an adversary with non-negligible advantage in winning ICC(Indistinguishability of Ciphertext from Ciphertext) game is available, then there exists an another adversary with the non-negligible advantage in winning ICLR game. However, **B**'s advantage is $\epsilon/(nqe^n) \in [0,1/2(nqe^n)]$ which is negligible. Thus, the MKSE-MWMRS is at least $(1\text{-}1/2\,(nqe^n))$ secure in ICLR game providing DDH assumption holds. This proves **Theorem 2**.

PERFORMANCE ANALYSIS

The performance analysis of MKSE-MWMRS as compared to the existing multi-writer, multi-reader searchable schemes especially designed for secure sharing of medical records(Jingzhang et al., 2018; Sharma & Jinwala, n.d.; Wu et al., 2016) is given in this section.

Theoretical Comparison

The theoretical comparison of MKSE-MWMRS with the schemes(Jingzhang et al., 2018; Sharma & Jinwala, n.d.; Wu et al., 2016) is given in Table 3. Focusing on the characteristics offered by each of the listed schemes, it could be determine that though the scheme (Jingzhang et al., 2018)provides multi-keyword search similar to the proposed MKSE-MWMRS, it can search for the exact set of keywords only. On the other hand, though the scheme (Sharma & Jinwala, n.d.)performs conjunctive keyword search, it doesn't offer controlled environment for multiple writers and readers as that in MKSE-MWM. More precisely, in (Sharma & Jinwala, n.d.)any data owner knowing public key can write ciphertexts onto the storage server. With such an uncontrolled environment, the storage server could be overloaded by malicious writers. Furthermore, the scheme (Sharma & Jinwala, n.d.)doesn't discuss encryption/decryption mechanism to securely share medical records.

In addition, Table 3demonstrates the storage overhead (ciphertext size) on the server for the scheme (Wu et al., 2016)which is O(n+R) where n=total number of keywords in ciphertext and R=total number of readers in system. Such scheme is impractical in comparison with MKSE-MWMRS and (Sharma &

Table 3. Comparative Analysis

Analysis Parameters		**(Wu et al., 2016)**	**(Jingzhang et al., 2018)**	**(Sharma & Jinwala, n.d.)**	**MKSE-MWMRS**
Characteristics	**Type of Search**	Single-Keyword	Multi-Keyword (Exact)	Multi-Keyword (Conjunctive)	Multi-Keyword (Conjunctive)
	Controlled MWMR	No	Yes	No	Yes
	Record Encryption/ Decryption	Yes	Yes	No	Yes
Storage Overhead	**On Cloud Server**	$(1+n+R)G_1+G_2$	$(3+2A)G_1+G_2$	$(1+n)G_1+C$	$(2+n)G_1+C$
Computational Overhead	**On Writer (during Encryption())**	(2+n+R)E+P+ Enc	(5+3A)E+P+ Enc	(1+2n)M+ Enc	(3+2n)M+ Enc
	On Reader (during TokGen())	2E	(8+2A)E	$(1+\ell)$M	$(1+\ell)$M
	On Server (during Search())	2nP	(3+2A)P+AE	1M+2P	1M+2P
	On Reader (during Decryption()	1E+1P+Dec	(1+2A)P+Dec	-	1X+Dec

n: Total keywords in ciphertext, ℓ: Total keywords in a query, **R**: Total readers in system, **A**: Number of attributes in policyused in CPABE ciphertext, **C**: Size of ciphertextoutput by symmetric encryption, G_1, G_2: Element Size for bilinear groupsG_1 and G_2, **P**: Pairing, **E**: Exponentiatio, **X**: Ex-OR,**M**: Scalar Multiplication, (**Enc, Dec**): Computationaloverhead incurred by Encryption and Decryption algorithm of the used symmetric cipher

Jinwala, n.d.)where storage overhead is optimal (independent from R) i.e. O(n). On the other hand, the storage complexity for the scheme (Jingzhang et al., 2018) is O(|A|) where |A|=total number of attributes involved in policy associated with CPABE ciphertext. Though the scheme offers optimal overhead, the size ofA affects the storage overhead. More precisely, increasing the size of A offers strong access control whereas setting |A|=1, any reader possessing only a single attribute satisfying access policy would be able to get ciphertexts from the server.

Furthermore, the ciphertext computational overhead suffered by the writers in the proposed MKSE-MWMRS is optimal i.e. O(n) as compared to the schemes- (Wu et al., 2016) with complexity O(n+R) and(Jingzhang et al., 2018)with complexity O(A). Such optimal complexity makes the proposed scheme more acceptable in practical applications. In addition, the scheme(Wu et al., 2016)puts constant overhead O(1) on reader during token generation. However, with such an overhead, the reader could generate a token for a single keyword search. On the other side, though the scheme (Jingzhang et al., 2018)generates a token for multi-keyword search with overhead O(A), the server with such a token can search for the exact match of keyword sets. For the proposed scheme, a reader could construct a token with computational complexity O(ℓ)same as the scheme (Sharma & Jinwala, n.d.). However, the server receiving such a token would be able to perform conjunctive search. Furthermore, with the proposed scheme, the server could perform conjunctive search across available ciphertexts with constant computational complexity O(1). With such optimal search complexity, the proposed scheme performs much more better than the scheme (Wu et al., 2016) with search overhead O(n) and the scheme (Jingzhang et al., 2018) with overhead in O(A). Additionally, from Table 3 one woulddetermine that the result processing cost besides actual decryption in proposed scheme is optimal (involving only one EX-OR operation). With such a cost, MKSE-MWMRS performs record decryption faster than the schemes(Jingzhang et al., 2018; Wu et al., 2016) involving exponentiation and pairing operations besides actual decryption.

EXPERIMENTAL EVALUATION

For empirical analysis, the authors simulate the proposed MKSE-MWMRS as well as the existing schemes (Jingzhang et al., 2018; Sharma & Jinwala, n.d.; Wu et al., 2016) on Windows 7 machine having 32-bit, Pentium Core 2 Duo,2.10 GHz CPU. The implementation is done through Java language with JPBC (java pairing based cryptographic) library (De Caro & Iovino, 2011). To build the cryptographic environment, an elliptic curve of***Type A***having160-bit group order and 512-bit field order from JPBC is used.

Since there does not exist any public EHR dataset, the Enron email dataset (Cohen, 2009)is used as test dataset. The randomly chosen emails from this dataset are used as M to simulate the algorithms. Furthermore from javax.crypto package, the standard AES algorithm (128-bit key) is usedfor encryption/decryption of payload message M. Additionally, a keyword space K of size n=100 keyword fields relevant to the Email system is prepared. The experimental results are demonstrated with respect to three parameters: the number of keywords in ciphertext (n), number of Keywords in a query (ℓ), and the number of readers in the system (R). The simulation is performed at least 10 times for different values of each of these parameters (Table4) and the average value is considered as the final result. The simulation results for *Encryption(), TokGen(),Search()* algorithms are shown in Figure3, 4, 5respectively. Note that to simulate the scheme (Jingzhang et al., 2018), the authors consider the number of attributes A=10 for average case analysis.

Table 4. Simulation Parameters

Parameter	Simulation Values
n	{10,25, 50, 75, 100}
ℓ	{10,20,30,40,50}
R	{100, 200, 300, 400, 500}

Figure 3. Simulation Results: Encryption()

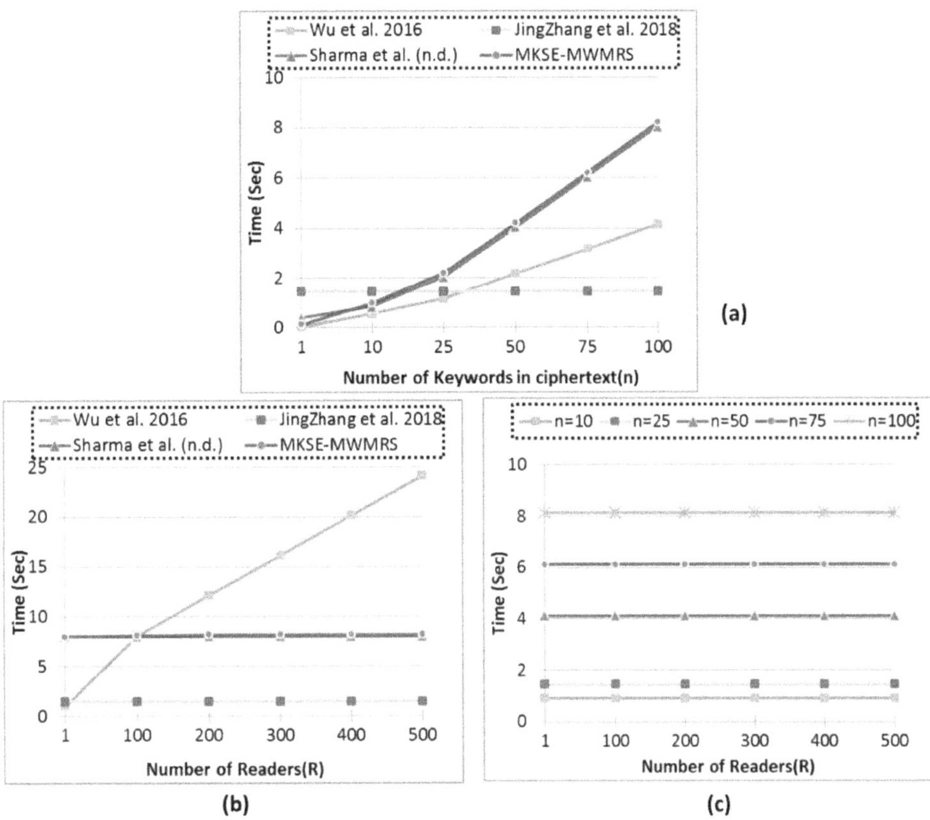

The results in Figure 3(a) show that the encryption costinMKSE-MWMRSis linearly proportional to **n** as that in the other MWMR schemes (Sharma & Jinwala, n.d.; Wu et al., 2016). However, the provision of the controlled multi-writer multi-reader support makes the proposed *Encryption()* algorithm slower than the encryption algorithms of(Sharma & Jinwala, n.d.; Wu et al., 2016). On the other hand, though the scheme (Jingzhang et al., 2018)reflects efficiency with constant encryption cost on writer, the computed ciphertext can't support conjunctive keyword search which is more desirable in real-life applications. Additionally, Figure 3 (b) shows that the *Encryption()*is not affected by the existence of multiple readers in system. The scheme (Wu et al., 2016) on the other hand offers multi-reader support with the computational overhead linear to **R** on the writer. Besides this, the Figure 3 (c) represents that though the encryption complexity is constant even in case of more than one reader, the increasing values of keywords(n) in ciphertextindeed affects the performance of *Encryption()* algorithm.

Figure 4. Simulation results: TokGen()

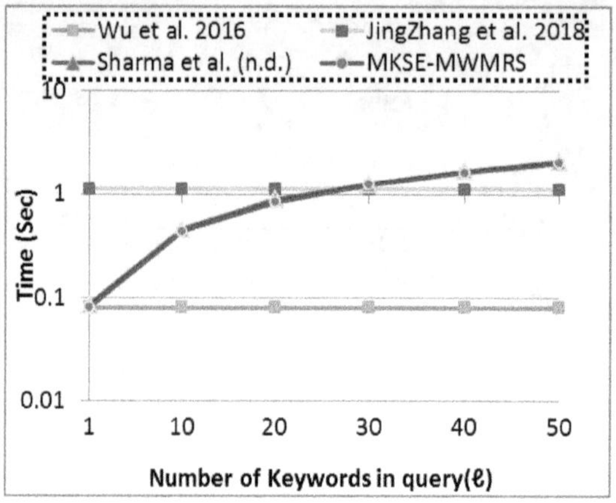

The results in Figure 4 show that the computational complexity of the proposed *TokGen()* algorithm is linear to the keywords in query, unlike the schemes (Jingzhang et al., 2018; Wu et al., 2016). However, with this overhead, the proposed algorithm generates a search token for the given conjunctive query that ultimately supports conjunctive search. The other remarkable point is that with the almost same token generation cost as that in (Sharma & Jinwala, n.d.), the proposed scheme offers the controlled environment for multiple readers and writers in system.

Figure 5. Simulation results: Search()

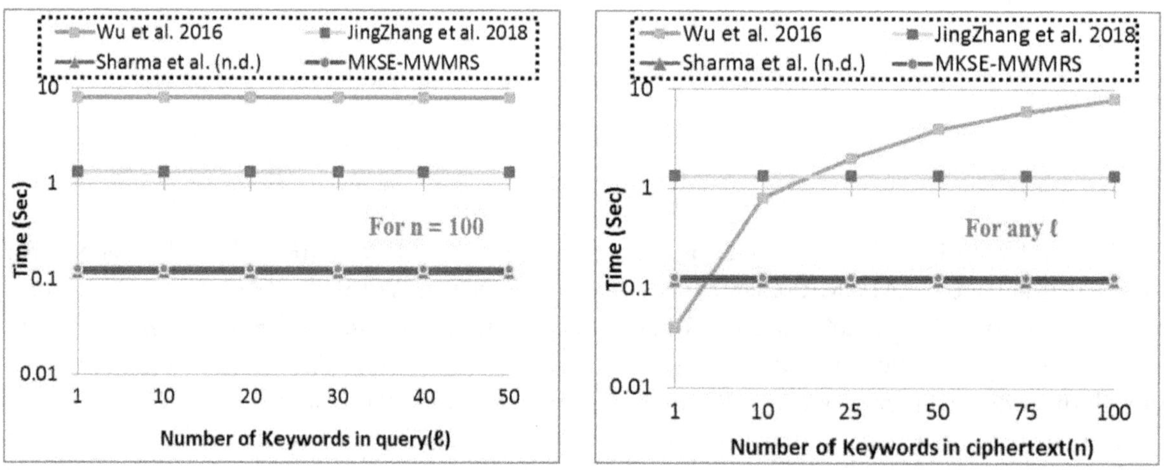

(a) (b)

The Figure 5 (a) demonstrates that the computation cost for the *Search()* is constant regardless of the number of keywords in query as that in (Jingzhang et al., 2018; Sharma & Jinwala, n.d.; Wu et al., 2016). More specifically, the proposed scheme provides conjunctive search with reduced computational overhead on server in comparison to the overhead incurred by the schemes(Jingzhang et al., 2018; Wu et al., 2016) and so it is more practical.

Furthermore, Figure 5 (b) shows that efficiency of *Search()* is never affected by the number of keywords in ciphertexts(n). However, it indeed degrades the search performance of (Wu et al., 2016).

Finally, it could be infer that that with the almost same computational overhead as in (Sharma & Jinwala, n.d.), MKSE-MWMRS offers controlled multi-reader, multi-writer settings where only the registered writer can share medical records and registered reader can search for the records.

Figure 6. Simulation results: Decrypt()

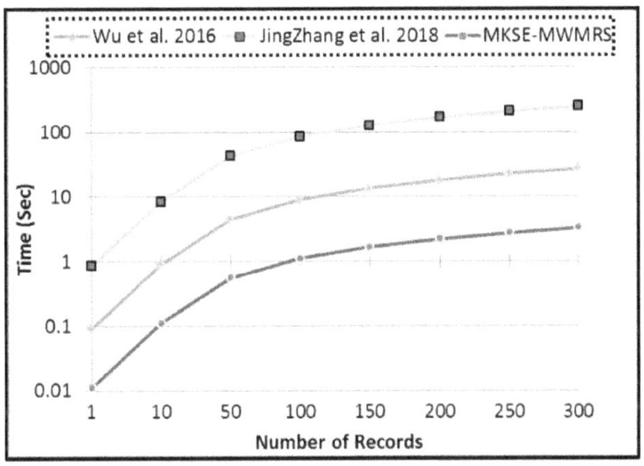

Furthermore, as shown in Figure6, the decryption cost incurred by the proposed *Decryption()*is much less than the schemes supporting record decryption(Jingzhang et al., 2018; Wu et al., 2016). Hence, the proposed solution is more suitable for the environment where readers have resource constrained devices.

CONCLUSION

The authors in this work define a multi-keyword public key searchable encryption (MKSE-MWMRS) especially for Medical Records. Precisely, the MKSE-MWMRS is a practical approach for E-Health system with multi-writer,multi-reader settings. In the proposed work, the registered data writer can securely share medical records with the registered data readers. For such a data sharing, the proposed scheme incurs optimal computational burden on writers and optimal storage overhead on the cloud server. Furthermore, each registered reader can easily search for records based on the conjunctive query of the chosen keywords. With the security analysis, it is proved that the proposed scheme ensures medical record security as well as searchable ciphertext security against chosen keyword attack. With the extensive performance analysis, including theoretical comparisons and experimental evaluation, the authors show the efficiency of MKSE-MWMRS in terms of storage and computationalcomplexity as compared to the existing searchable schemes designed for medical data. Since the proposed work offers cost-effective solution, it would indeed be acceptable by the real-word E-Health applications.

REFERENCES

Akinyele, J. A., Pagano, M. W., Green, M. D., Lehmann, C. U., Peterson, Z. N. J., & Rubin, A. D. (2011). Securing electronic medical records using attribute-based encryption on mobile devices. *Proceedings of the 1st ACM Workshop on Security and Privacy in Smartphones and Mobile Devices*, 75–86. 10.1145/2046614.2046628

Baek, J., Safavi-Naini, R., & Susilo, W. (2008a). Public key encryption with keyword search revisited. *Computational Science and Its …*, 1–15. Retrieved from https://link.springer.com/chapter/10.1007/978-3-540-69839-5_96

Baek, J., Safavi-Naini, R., & Susilo, W. (2008b). Public key encryption with keyword search revisited. In *Computational Science and Its Applications—ICCSA 2008* (pp. 1249–1259). Springer. doi:10.1007/978-3-540-69839-5_96

Ballard, L., Kamara, S., & Monrose, F. (2005). Achieving efficient conjunctive keyword searches over encrypted data. In *Information and Communications Security* (pp. 414–426). Springer. doi:10.1007/11602897_35

Bao, F., Deng, R. H., Ding, X., & Yang, Y. (2008). Private query on encrypted data in multi-user settings. *International Conference on Information Security Practice and Experience*, 71–85. 10.1007/978-3-540-79104-1_6

Boneh, D., Di Crescenzo, G., Ostrovsky, R., & Persiano, G. (2004). Public key encryption with keyword search. *Advances in Cryptology-Eurocrypt, 2004*, 506–522.

Boneh, D., & Waters, B. (2007). Conjunctive, subset, and range queries on encrypted data. In *Theory of cryptography* (pp. 535–554). Springer. doi:10.1007/978-3-540-70936-7_29

Byun, J. W., Lee, D. H., & Lim, J. (2006). Efficient conjunctive keyword search on encrypted data storage system. *European Public Key Infrastructure Workshop*, 184–196. 10.1007/11774716_15

Chen, T.-L., Chung, Y.-F., & Lin, F. Y. S. (2012). A study on agent-based secure scheme for electronic medical record system. *Journal of Medical Systems, 36*(3), 1345–1357. doi:10.100710916-010-9595-8 PMID:20857325

Chen, Z., Wu, C., Wang, D., & Li, S. (2012). Conjunctive keywords searchable encryption with efficient pairing, constant ciphertext and short trapdoor. *Pacific-Asia Workshop on Intelligence and Security Informatics*, 176–189.

Cohen, W. W. (2009). *Enron email dataset*. Academic Press.

De Caro, A., & Iovino, V. (2011). jPBC: Java pairing based cryptography. *Proceedings of the 16th IEEE Symposium on Computers and Communications, ISCC 2011*, 850–855. 10.1109/ISCC.2011.5983948

Ding, M., Gao, F., Jin, Z., & Zhang, H. (2012). An efficient public key encryption with conjunctive keyword search scheme based on pairings. *2012 3rd IEEE International Conference on Network Infrastructure and Digital Content*, 526–530.

Dutta, R., Barua, R., & Sarkar, P. (2004). Pairing-based cryptography: A survey. *Cryptology Research Group, Stat-Math and Applied Statistics Unit, 203.*

Eom, J., Lee, D. H., & Lee, K. (2016). Patient-controlled attribute-based encryption for secure electronic health records system. *Journal of Medical Systems*, *40*(12), 253. doi:10.100710916-016-0621-3 PMID:27714562

Golle, P., Staddon, J., & Waters, B. (2004). Secure conjunctive keyword search over encrypted data. *Applied Cryptography and Network Security*, 31–45.

Goyal, V., Pandey, O., Sahai, A., & Waters, B. (2006). Attribute-based encryption for fine-grained access control of encrypted data. *Proceedings of the 13th ACM Conference on Computer and Communications Security - CCS '06*, 89. 10.1145/1180405.1180418

Huang, H., Du, J., Wang, H., & Wang, R. (2016). A Multi-keyword Multi-user Searchable Encryption Scheme Based on Cloud Storage. *Trustcom/BigDataSE/I SPA, 2016 IEEE*, 1937–1943.

Hwang, M.-S., Hsu, S.-T., & Lee, C.-C. (2014). A new public key encryption with conjunctive field keyword search scheme. *Information Technology and Control*, *43*(3), 277–288. doi:10.5755/j01.itc.43.3.6429

Hwang, Y. H., & Lee, P. J. (2007). Public key encryption with conjunctive keyword search and its extension to a multi-user system. In *Pairing-Based Cryptography—Pairing 2007* (pp. 2–22). Springer. doi:10.1007/978-3-540-73489-5_2

Jingzhang, S., Chunjie, C., & Hui, L. (2018). Searchable Encryption Scheme Based on CPABE with Attribute Update in a Cloud Medical Environment. *International Conference on Cloud Computing and Security*, 265–276. 10.1007/978-3-030-00012-7_25

Kiayias, A., Oksuz, O., Russell, A., Tang, Q., & Wang, B. (2016). Efficient encrypted keyword search for multi-user data sharing. *European Symposium on Research in Computer Security*, 173–195. 10.1007/978-3-319-45744-4_9

Lai, J., Zhou, X., Deng, R. H., Li, Y., & Chen, K. (2013). Expressive search on encrypted data. *Proceedings of the 8th ACM SIGSAC Symposium on Information, Computer and Communications Security*, 243–252.

Lee, C.-C., Hsu, S.-T., Hwang, M.-S., & ... (2013). A Study of Conjunctive Keyword Searchable Schemes. *International Journal of Network Security*, *15*(5), 321–330.

Li, H., Yang, Y., Dai, Y., Bai, J., Yu, S., & Xiang, Y. (2017). *Achieving secure and efficient dynamic searchable symmetric encryption over medical cloud data. IEEE Transactions on Cloud Computing.*

Li, J., & Chen, X. (2013). Efficient multi-user keyword search over encrypted data in cloud computing. *Computer Information*, *32*(4), 723–738.

Liu, J., Huang, X., & Liu, J. K. (2015). Secure sharing of personal health records in cloud computing: Ciphertext-policy attribute-based signcryption. *Future Generation Computer Systems*, *52*, 67–76. doi:10.1016/j.future.2014.10.014

Löhr, H., Sadeghi, A.-R., & Winandy, M. (2010). Securing the e-health cloud. *Proceedings of the 1st Acm International Health Informatics Symposium*, 220–229. 10.1145/1882992.1883024

Lv, Z., Zhang, M., & Feng, D. (2014). Multi-user searchable encryption with efficient access control for cloud storage. *2014 IEEE 6th International Conference on Cloud Computing Technology and Science*, 366–373.

Macdonald, A. J. R. (1986). An introduction to medical manipulation. *Pain*, *24*(1), 124. doi:10.1016/0304-3959(86)90035-7

Pohlig, S., & Hellman, M. (1978). An improved algorithm for computing logarithms over GF (p) and its cryptographic significance (Corresp.). *IEEE Transactions on Information Theory*, *24*(1), 106–110. doi:10.1109/TIT.1978.1055817

Rau, H.-H., Hsu, C.-Y., Lee, Y.-L., Chen, W., & Jian, W.-S. (2010). Developing electronic health records in Taiwan. *IT Professional*, *12*(2), 17–25. doi:10.1109/MITP.2010.53

Schabetsberger, T., Ammenwerth, E., Andreatta, S., Gratl, G., Haux, R., Lechleitner, G., Schindelwig, K., Stark, C., Vogl, R., Wilhelmy, I., & Wozak, F. (2006). From a paper-based transmission of discharge summaries to electronic communication in health care regions. *International Journal of Medical Informatics*, *75*(3-4), 209–215. doi:10.1016/j.ijmedinf.2005.07.018 PMID:16112892

Sharma, D., & Jinwala, D. C. (2017). Multi-User Searchable Encryption with Token Freshness Verification (MUSE-TFV). *Security and Communication Networks*, *2017*, 16. doi:10.1155/2017/6435138

Sharma, D., & Jinwala, D. C. (n.d.). Multi-Writer Multi-Reader Conjunctive Keyword Searchable Encryption. *International Journal of Information and Computer Security*.

Sunyaev, A., Kaletsch, A., Mauro, C., & Krcmar, H. (2009). Security Analysis of the German Electronic Health Card's Peripheral Parts. *ICEIS*, (3), 19–26. doi:10.5220/0001854000190026

Wang, S., Zhang, X., & Zhang, Y. (2016). Efficiently Multi-User Searchable Encryption Scheme with Attribute Revocation and Grant for Cloud Storage. *PLoS One*, *11*(11), e0167157. doi:10.1371/journal.pone.0167157 PMID:27898703

Wu, Y., Lu, X., Su, J., & Chen, P. (2016). An efficient searchable encryption against keyword guessing attacks for sharable electronic medical records in cloud-based system. *Journal of Medical Systems*, *40*(12), 258. doi:10.100710916-016-0609-z PMID:27722976

Xu, L., Xu, C., Liu, J. K., Zuo, C., & Zhang, P. (2019). Building a dynamic searchable encrypted medical database for multi-client. *Information Sciences*.

Ye, J., Wang, J., Zhao, J., Shen, J., & Li, K.-C. (2016). Fine-grained searchable encryption in multi-user setting. *Soft Computing*, 1–12.

Yu, Y.-C., Huang, T.-Y., & Hou, T.-W. (2012). Forward secure digital signature for electronic medical records. *Journal of Medical Systems*, *36*(2), 399–406. doi:10.100710916-010-9484-1 PMID:20703711

Zhang, B., & Zhang, F. (2011). An efficient public key encryption with conjunctive-subset keywords search. *Journal of Network and Computer Applications*, *34*(1), 262–267. doi:10.1016/j.jnca.2010.07.007

Zhang, R., Xue, R., & Liu, L. (2017). Searchable encryption for healthcare clouds: A survey. *IEEE Transactions on Services Computing*, *11*(6), 978–996. doi:10.1109/TSC.2017.2762296

Zhang, Y., Liu, L., & Wang, S. (2016). Multi-User and Keyword-Based Searchable Encryption Scheme. *Computational Intelligence and Security (CIS), 2016 12th International Conference on*, 223–227.

Chapter 8
A Comparative Study on Symmetric and Asymmetric Key Encryption Techniques

Sneha Padhiar
Charotar University of Science and Technology, India

Kuldip Hiralal Mori
Charotar University of Science and Technology, India

ABSTRACT

With the rise in use of internet in various fields like education, military, government sector, banking, the security and privacy of the info has been the foremost concern. As in today's era, most of the handling of data and transactions are done online. When the data is transferred from the one end of sender to the other end of receiver online, it's eavesdropped by an intruder and thus could be a threat to the secrecy or confidentiality of the info. The hottest technique that protects the confidentiality of the data is cryptography which converts the plain text into scrambled form which is unreadable. Then the receiver applies a reverse mechanism to decrypt the unreadable data to readable form. This mechanism is known as encryption-decryption process or cryptography. Cryptography can be both symmetric and asymmetric. Here the authors discuss symmetric and asymmetric algorithms.

1. INTRODUCTION

Everyday an outsized amount of Security related Data/Information is transferred and accessed across the online (over the internet) in world application. In today's corporate world where access to Information in lesser time is required. With the goal of running the enterprise smoothly and efficiently it is vital to supply right information to right people at right time. There might be a scenario where one that's sending an important file to the other person, who is sitting at another site office then the message crosses through an insecure channel and will be possible that somebody within the center can retrieve the message and modify it then passes it to destination. which may cause many undesirable side effects to the

DOI: 10.4018/978-1-7998-6988-7.ch008

corporate and thus the company may suffer a huge loss in economic terms. So, to make data of company in secure manner we required some desired techniques to protect data. Cryptography plays a very big role keep the message safe because the info is transit. It ensures that the message being sent at one end remains confidential and can be received only by the intended receiver at the other end. Cryptography converts the primary message in to non-readable format and sends the message over an insecure channel. Cryptography are often stated because the art of writing which involves the transition of message to a concealed form or unintelligible data. the data in unintelligible within the sense that understanding of the data is difficult or impossible. The terms encryption and decryption are mainly used with the cryptographic process. Encryption is that the method of converting an original message to a random / non-readable message which is known as cipher text. Encryption has its reverse process mentioned as decryption. Where cipher text is converted back to the form of original message. Plain text Cipher Text Plain Text Encryption Decryption.

Figure 1. Encryption decryption process

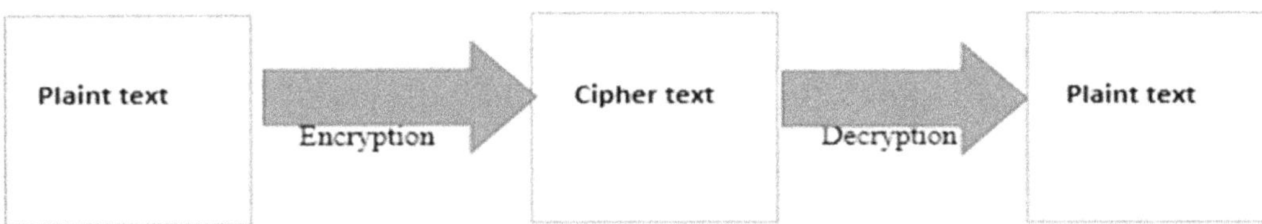

in fig 1. the primary message, plaintext is converts into the cipher text by applying the encryption procedure and cipher text is converted back to plain text through the decryption process. The cryptographic process makes use of an algorithm and a secret(key). The key's often same for encryption decryption process or are often a special one relying on the type of encryption algorithm used. Depending upon the type of key used, cryptography techniques are often divided into two different categories.

- **Symmetric Key encryption**
- **Asymmetric Key encryption**

Symmetric key encryption is that the type of encryption method where the sender and receiver make use of the same secret key for encryption also as decryption process. Because one keys used for both functions, secret key cryptography is additionally called symmetric key encryption. Samples of symmetric key encryption-algorithms.

There are two main applications of asymmetric cryptography: authentication and secrecy. Messages can be signed with a private key using asymmetric cryptography, and then someone with the public key can verify that someone who has the corresponding private key has generated the message. This can be combined with an identity proof scheme in order to understand what individual (person or group) actually owns.

Figure 2. Classification of encryption methods

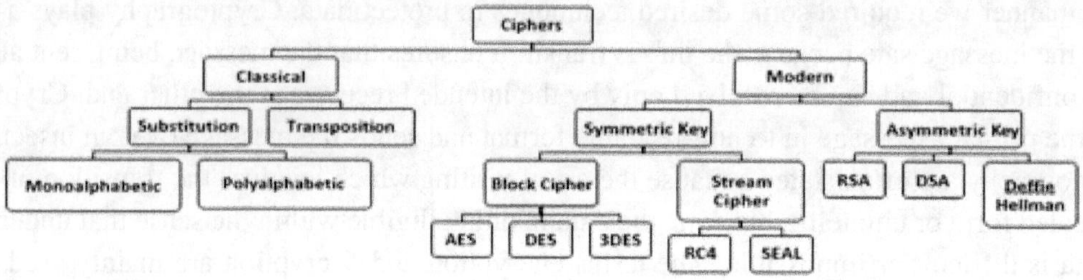

2. LITERATURE REVIEW

Seth, S. M., & Mishra, R. (2011) has shown many cryptography algorithms that can used to secure information, such as DES, 3DES, Blowfish, AES, RSA, ElGamal and Paillier. All of these algorithms are compared by Seth, S. M., & Mishra, R. (2011) unique on their way. However, the problem is that how to find the best security algorithm which provides the high security and also take less time for a key generation, encryption, and decryption of information. Security algorithms will depend on pros and cons of each algorithm, requirement and suitable for different application.

Nadeem, A., & Javed, M. Y. (2005), has evaluated performance of two algorithms DES and Blowfish on basis of certain parameters such as encryption speed, power consumption, and security analysis. Experiment result showed that performance of Blowfish is fastest than DES and AES algorithm by Nadeem, A., & Javed, M. Y. (2005). However, Tamimi, A. (2008) showed that AES performance is good than Blowfish.

some of the cryptography algorithms details are given by Mandal, A. K., Parakash, et.al. (2012), such as AES, DES, 3DES, RC6, Blowfish and RC2. Furthermore, the performance of these security algorithms is also evaluated and experiment is performed on text file and image. The result is showed that all algorithms slow in performance as compare to Blowfish as increased the packet size. However, selecting the image as the type of data instead of text file then Blowfish, RC6, and RC2 the algorithm has consumed more time than AES, DES and 3DES algorithms. Nadeem, A., & Javed, M. Y. (2005), has conclude that DES is still faster in performance than 3DES.

3. ALGORITHMS IMPLEMENTED

Encoding Standard as described by Nadeem, A., & Javed, M. Y. (2005), DES was the primary encryption algorithm to be published by NIST (National Institute of Standards and Technology). it's a widely used method of knowledge encryption which uses the private key. DES applies a 56-bit key to every 64-bit block of knowledge and maps 64-bit input block into a 64-bit output block. This process involves 16 rounds or operations and may run in several modes. the dimensions of the key involved in DES encryption is really 64 bits although the key size used is merely 56 bits because the least significant little bit of each byte is either used as a parity. DES was considered as an insecure block cipher thanks to its vulnerability to the brute-force attack and comparatively small key size.

Table 1. Symmetric Encryption algorithms

	DES	TDES	AES	BLO WFISH
Block Size	64bit	64bit	128bit	64bit
Key Size	56bit	168bit	128,192,256 bit	32-448 bit
Algorithm Structure	Feistel	Feistel	Substitution Permutation	Feistel
Rounds	16	48	9,11,13	16

Triple Encryption Standard

Seth, S. M., & Mishra, R. (2011) has discussed triple encoding Standard. Triple DES was developed as an alternate to affect the problems in DES without designing a replacement cryptosystem. 3DES preserves the prevailing investment in software and equipment by using multiple encryptions with DES and multiple keys. Triple DES simply extends the key size of DES by applying the algorithm 3 times in succession with three different keys. It uses as input 64-bit plaintext to provide 64-bit cipher text, almost like DES. But unlike DES the combined key size is thus 192 bits with actually key-size usage of 168 bits (3 times 56).3DES is thus slower than other block cipher method because it essentially applies the DES algorithm 3 times. during this paper, two keys are used for 3DES encryption (K1, K2, K1) describing encrypt(k1)-decrypt(k2)- encrypt(k1) mode and for decryption decrypt(k1)- encrypt(k2)- decrypt(k1) mode. There are two flavors of Triple DES available one is 2 Key Triple DES and another is 3 Key Triple DES algorithm. In 2 Key Triple DES only two 56 bits keys are used with EDE sequence by using K1 for encryption, K2 for decryption and again K1 for encryption. On other hand if you use 3 Key Triple DES three 56 bits keys are used with EDE sequence K1for encryption, K2 for decryption & K3 for encryption. Thus 3 Key Triple DES provides additional layer of security. Triple DES provide resistance towards most of the attacks against which DES is not safe as Triple DES uses three 56 bits keys. But Triple DES is computationally costly to implement on general purpose computers compared to DES. Triple DES is slower compared to DES.Triple DES is still used with electronic payment industry. According to Nadeem, A., & Javed, M. Y. (2005), Earlier versions of MS office also used Triple DES for password protected file. Moreover, Firefox and Mozilla used Triple DES with CBC mode to save login credentials and passwords.

Advanced Encryption Standard

Tamimi, A. (2008) has shown two major problems faced by the earlier algorithms were the small key size (in case of DES algorithm) and slow speed (Triple des algorithm). to beat these shortcomings; NIST (National Institute of Standards and Technology) published a replacement encryption algorithm mentioned as AES (Advanced Encryption Standard) in 2001. AES may be a crucial symmetric block cipher that has replaced DES for the big choice of applications. The AES encryption algorithm could also be a block cipher because it works on one block of data at a time. AES makes use of an encryption key and variety of other rounds of encryption. The AES takes as input the block length of 128 bits and thus the key length of 128,192 or 256 bits like 10, 12 or 14 rounds.

The three main parameters suggested by the cryptographic specialists were as follows:

- **Security**

The most important thing was to protect the data from brute-force attacks that failed in the Data Encryption Standard (DES). AES is able to secure the data better than DES. Lots of experiments and functional implementations have shown this [8].

- **Costs**

AES has been nominated by NIST because it has high computational efficiency and can be used in high-speed broadband connections.

- **Features of Algorithms and Implementation**

Flexibility, simplicity and suitability of the algorithm for the variety of hardware and software implementation are the requirements for the algorithm to be simple to use.

Diffie-Hellman Algorithm

One of the important developments in public-key cryptography is the Diffie-Hellman key exchange. The primary aim of using the Diffie-Hellman protocol is to exchange a protected shared key with each party to derive keys that can be used for message encryption and decryption. Protection protocols such as TLS, IPsec, SSH, PGP and many others follow this protocol. Diffie-Hellman key exchange is complex and it would be difficult to get in your head about its working because it involves large numbers and a lot of math.

Blowfish

Tamimi, A. (2008) has discussed Blowfish. Which uses the secret key that is standard during both the processes performed in this algorithm to perform encryption as well as its decryption. It is also a type of a block cipher, that is, for its working method, First, it divides the message given to encrypt into a number of blocks that are all fixed in length, while both encrypting and decrypting processes are performed.

In this algorithm, S-box and P-array tables are involved. Feistel cipher composition is the basis of this algorithm. The 16 round function F is used here, which makes the concepts of DES (Data Encryption Cipher) more understandable. The protection is therefore the same as DES with the speed of success boosted.

The Blowfish algorithm mainly consists of two key components:

- Entry Generation Sub-key and S-box
- Data encryption and decryption

For Blowfish, around 5KB of memory is needed. It can perform encryption/decryption of a 64-bit message in more or less 12 clock cycles with careful implementation on a 32-bit processor.

Messages with more lengths make the computation time more linear; for example, a 128-bit message requires more or less (2*12) clocks.

RSA Algorithm

RSA is an asymmetric cryptographic algorithm used by modern computers for an encryption and decryption of messages.RSA involves a public key and private key. The public key can be known to everyone it is used to encrypt messages. Messages encrypted using the public key can only be decrypted with the private key.

The keys for the RSA algorithm are generated the following way:

- Choose two different prime numbers p and q.
- Calculate **n = pq**. n is the modulus for public key and private key.
- Calculate the **Φ(n)= (p-1) (q-1)**
- Choose an integer e such that **1< e < Φ(n),** and e is co-prime to **Φ(n).**

i.e.: **e and Φ(n)** share no factors other than 1, in short **gcd(e, Φ(n))=1**.
- Calculate **d = (1+x Φ(n))/e. d** is kept as private key component.

Digital Signature Algorithm

It becomes essential to identify his/her authenticity for security and safety purposes when a person sends data via a document. For this identification, digital signatures are used. Document authentication means being aware of who has created it and that it has not interfered with its transmission. Using certain algorithms, these signatures are generated. The Algorithm for Digital Signature (DSA).The DSA algorithm is standard for digital signature which is based on the algebraic properties of discrete logarithm problem and modular exponentiations and is based on the on public-key cryptosystems principal.

Figure 3. Digital signature algorithm

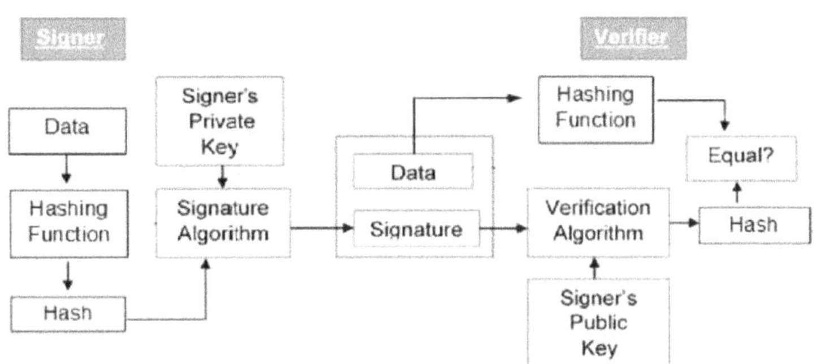

Nadeem, A., & Javed, M. Y. (2005), has state that DSA algorithm is a regular digital signature based on the algebraic characteristics of the problem of separate logarithms and modular exponentiations and is based on the main public-key cryptosystems. Digital signatures operate on the premise that two cryptographic keys are mutually authenticated. Signatures are based on key pairs that are public/private. You can construct a mathematically based algorithm with a public key algorithm like RSA.

The private key and the public key are connected. With his private key, one can sign a digital message. Through the use of a private key, signature related data can be encrypted by an individual. The private key should always be with a person who wants a digital signature to be produced. Both the public key and the private key can still be extracted from each other since they are mathematically related. The only way to decrypt this data is by using Signer's public key. One can give the public key to someone who needs the signature of the signer to be checked. Holding private key secrets is important as you can create your signature on a document with the aid of this. Digital signature authentication is achieved in this way. In a digital signature, only public and private keys are validly guaranteed. The digital signature algorithm, on the other hand, does not use a private key to encrypt data. A digital signature algorithm uses a public key to decrypt this information, too. DSA operates on the concept of a special mathematical function to render a digital signature with two 160-bit numbers. By using the private key and the message digest, these two numbers are generated. Verma, O., Agarwal, et.al. (2011) shown that authentication process is complicated as the public key is not used to authenticate the signature. In a special digital signature algorithm for further security assurance, both keys are used to protect data. Now, for generating a message digest, a hash function is used. Together with the DSA algorithm, the created message digest is what gives the digital signature.

ElGamal Algorithm

ElGamal key exchange is public-Key cryptography. It uses in asymmetric key encryption for communicating between two parties. This algorithm is based on discrete logarithm in a cyclic group.

- ElGamal is non-determinism-encrypting algorithm. Same plain text multiple times will result in different ciphertexts, since a random k is chosen each time.
- ElGamal encryption is used in the free GNU privacy Guard software, recent version of PGP and other cryptosystem.

Elliptical Curve Cryptography

ECC is a public-key technology that offers performance advantages at higher security levels. It includes an Elliptic Curve version of Diffie-Hellman key exchange protocol [DH1976] and an Elliptic Curve version of the ElGamal Signature Algorithm [E1985]. The elliptic curve versions of these algorithms are referred to as ECDH and ECES, respectively. The adoption of ECC has been slower than had been anticipated, perhaps due to the lack of freely available normative documents and uncertainty over intellectual property rights. The primary advantage of ECC is that, for key sizes in use today, it is clearly stronger than RSA. The standard 256-bit ECC key size equals a 3072-bit RSA key and is 10,000 times stronger than a 2048-bit RSA key! RSA keys have to be longer in order to remain ahead of the processing power of an intruder. After 2013, the CA/Browser Forum and leading browser vendors formally terminated support for 1024-bit RSA keys. The ECC's other protection advantage is clearly that it offers an alternative to RSA and DSA. If a major RSA weakness is found, the best alternative is likely to be ECC, especially if the RSA weakness suddenly needs a sharp increase in the main size to compensate. For a variety of factors, the ECC is often quicker.

4. EXPERIMENTAL SETUP

In order to compare the performance of various algorithms, the simulation is done in C language, compiled with Borland C++ Compiler.

We evaluate the performance of symmetric and asymmetric algorithms by using parameters such as encryption time, decryption time, throughput and memory utilization.

The various performance metrics shown by Verma, O., Agarwal, R et.al. (2011) that are evaluated here: -

Throughput: - The throughput of an encryption or decryption scheme defines the speed of encryption. The throughput of the encryption can be calculated as in equation:
Encryption time: -Encryption is the time required by any encryption function to convert plaintext into ciphertext.
Decryption time: -Decryption is the time required to convert again cipher text into plain text.

All these functions generate different times according to the size of text files and key length in any algorithm.

Throughput = Tp (Kilobytes)/Et (Second)

where Tp: Total plain text (Kilobytes) Et: Encryption time (second)

Figure 4. Comparison of Throughput for various Symmetric algorithm

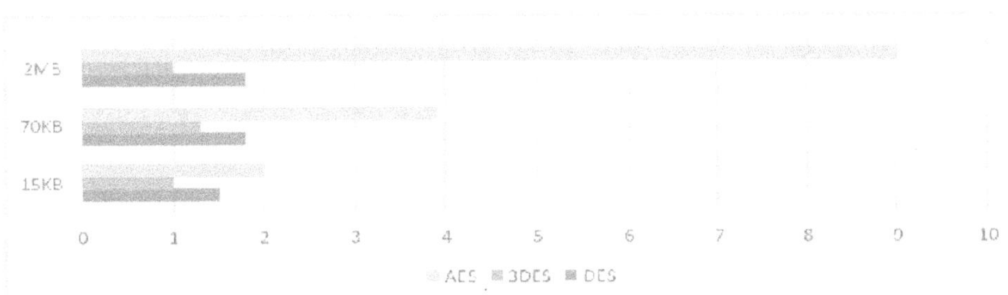

The following table describes the performance of **Diffie-Hallman algorithm** on files of different sizes.

Table 2. Diffie-Hellman Algorithm Result

Size of Files	5KB	20KB	50KB
Encryption time(second)	0.00200	0.00900	0.01900
Decryption time(second)	0.00100	0.00800	0.02200
Throughput(kb/sec) =>Throughput=size/encryption time	2500	2222.22	2631.5789
Memory utilization(kb)	5.123	20.361	50.483

The following table describes the performance of **RSA algorithm** on files of different sizes.

Table 3. RSA Algorithm Result

Size of Files	5KB	20KB	50KB
Encryption time(second)	0.01519	0.04643	0.09339
Decryption time(second)	0.01724	0.04629	0.10927
Throughput(kb/sec) =>Throughput=size/encryption time	367	477	592
Memory utilization(kb)	5.581	22.178	55.369

The following table describes the performance of **Elgamal algorithm** on files of different sizes.

Table 4. Elgamal Algorithm Result

Size of Files	5KB	20KB	50KB
Encryption time(second)	0.002020	0.009400	0.056837
Decryption time(second)	0.003622	0.012963	0.031949
Throughput(kb/sec) =>Throughput=size/encryption time	2475.25	2127.65	879.70
Memory utilization(kb)	5581	22178	55369

The following table describes the performance of Elliptical Curve Algorithm on files of different sizes.

Table 5. Elliptical Curve Cryptography Result

Size of Files	5KB	20KB	50KB
Encryption time(second)	0.002020	0.009400	0.056837
Decryption time(second)	0.003622	0.012963	0.031949
Throughput(kb/sec) =>Throughput=size/encryption time	2475.25	2127.65	879.70
Memory utilization(kb)	5581	22178	55369

Comparison of Encryption Time

A depiction of the comparison of encryption time for various asymmetric algorithms is shown in Figure 5.

Figure 5. Comparison of Encryption time for various Asymmetric algorithm

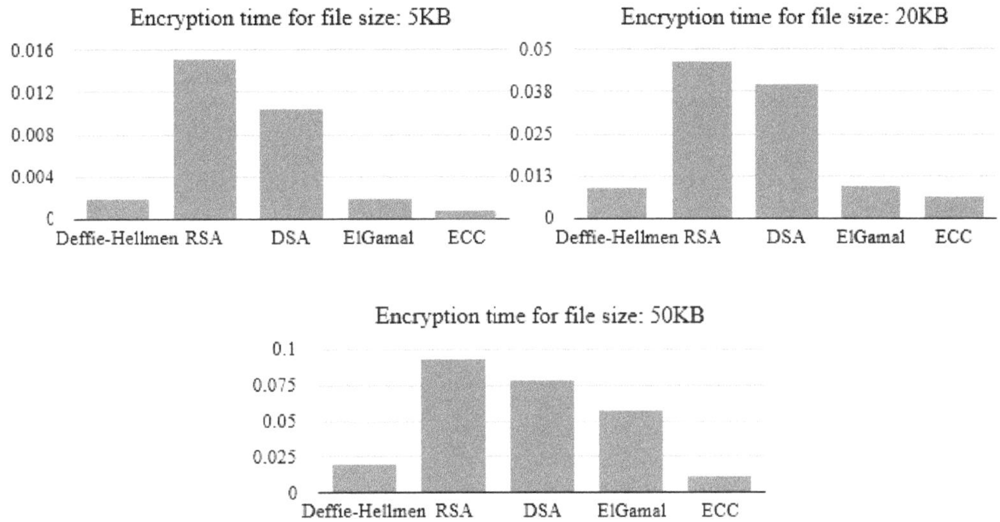

Comparison of Decryption Time and Throughput

A depiction of the comparison of decryption time and throughput for various asymmetric algorithms is shown in Figure 6.

Comparison of Memory Utilization (in KB)

A depiction of the comparison of memory usage for various asymmetric algorithms is shown in Figure 7.

5. FUTURE SCOPE

Cryptography is the perfect solution for effective data transmission. Many algorithms have been developed. So far, we've built a system that is based on both Symmetric and Cryptography of asymmetric keys. The algorithms have been developed. efficient in maintaining data privacy, honesty, and confidentiality Non-repudiation and authenticity. There are, however, several places that are still accessible. Quantum cryptography is thought to be a good substitute for the Diffie-Hellman algorithm since data transmitted via it is extremely stable. However, it will not shield you from traditional bucket brigade assaults. Methods for resolving this issue may be devised. Data is highly secure when messages are scrambled using two primary factors. Methods could be built to make it easier to produce large prime numbers.

CONCLUSION

The paper presents the performance evaluation of various symmetric key encryption algorithms DES,3DES and AES for text. The result show that encryption and decryption time of AES algorithm is a smaller amount than other algorithms because the number of rounds is relatively less just in case of AES while

Figure 6. Comparison of Decryption time & Throughput for various Asymmetric algorithm

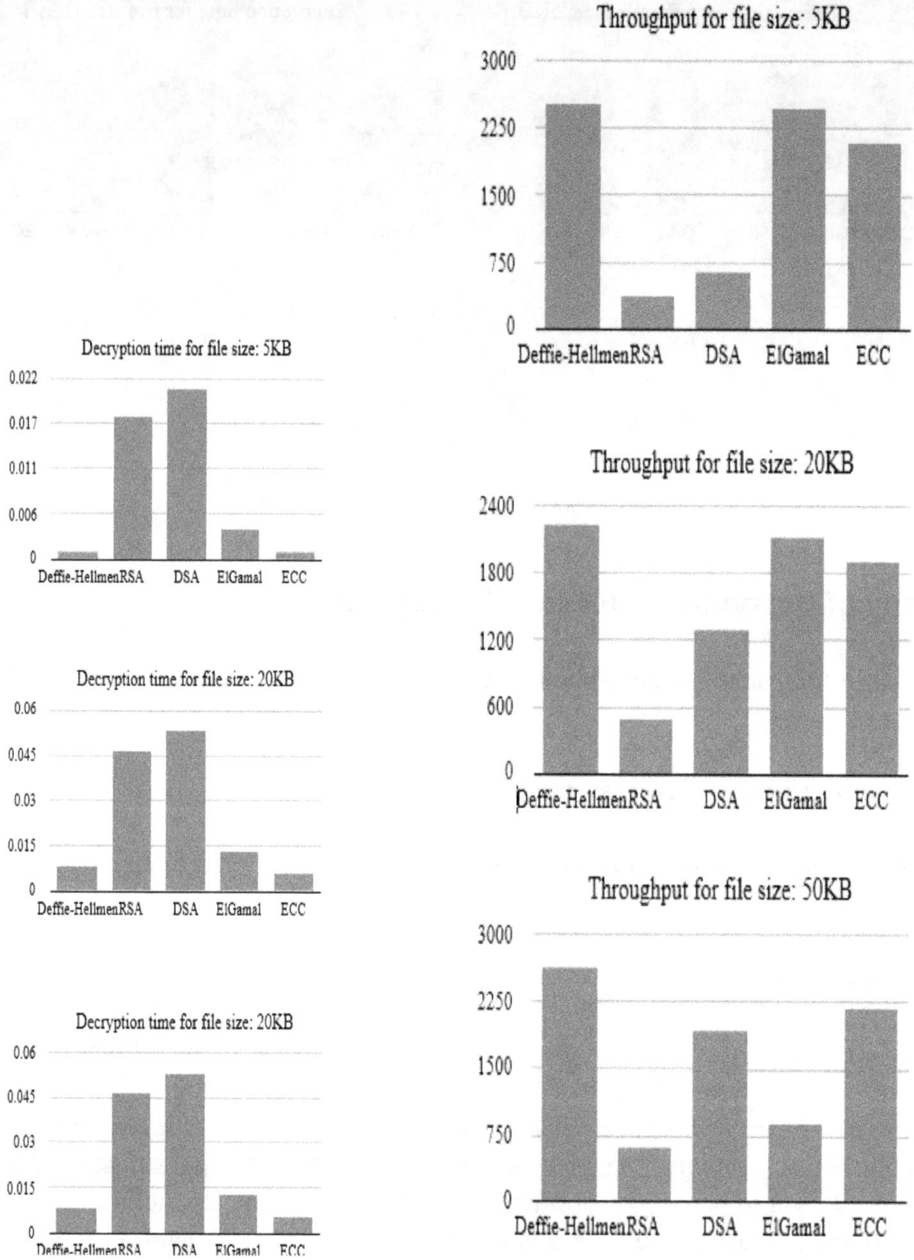

3DES has more encryption-decryption time because it applies the algorithm 3 times. Throughput varies inversely to the encryption or decryption time, Therefore AES has more and 3DES has less throughput than the opposite algorithms. The detailed comparative analysis for Asymmetric Algorithm is presented in graphical format for better understanding of each algorithms in terms of encryption time, decryption time, calculated throughput and memory utilization when running is presented.

Figure 7. Comparison of Memory Usage for various Asymmetric algorithm

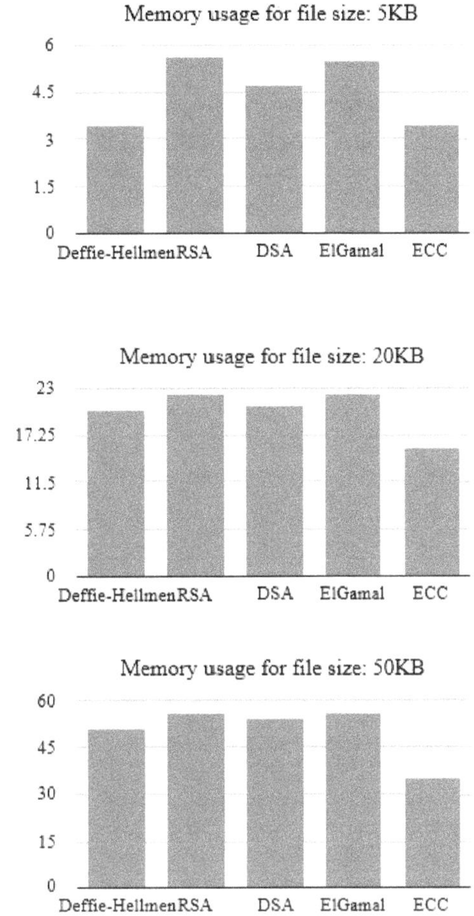

REFERENCES

Arora, P., Singh, A., & Tiyagi, H. (2012). Evaluation and comparison of security issues on cloud computing environment. *World of Computing and Knowledge Technology Journal*, *2*(5), 179–183.

Elminaam, D. S. A., Abdual-Kader, H. M., & Hadhoud, M. M. (2010). Evaluating the Performance of Symmetric Encryption Algorithms. *International Journal of Network Security*, *10*, 216–222.

Forouzan. (n.d.). *Data Communications and Networking*. McGraw-Hill.

Kumar, A., Jakhar, S., & Makkar, S. (2012, July). Comparative analysis between des and RSA algorithms. *International Journal of Advanced Research in Computing and Software Engineer in G*, *2*(7), 386–391.

Mandal, A. K., Parakash, C., & Tiwari, A. (2012). Performance evaluation of cryptographic algorithms: DES and AES. In *IEEE Students' Conference on Electrical, Electronics and computing (SCEECS), 2012* (pp. 1–5). IEEE.

Nadeem, A., & Javed, M. Y. (2005). A performance comparison of knowledge encryption algorithms. In *Information and communication technologies. ICICT. (2005). First international conference on, 2005* (pp. 84–89). Academic Press.

Seth, S. M., & Ishra, R. (2011, June). Comparative Analysis of encryption algorithms for data communication. *International Journal of Computers and Technology*, 2(2), 292–294.

Seth, S. M., & Mishra, R. (2011). *Comparative Analysis of Encryption Algorithms for digital communication*. Academic Press.

Tamimi, A. (2008). *Performance analysis of knowledge encryption algorithms*. Academic Press.

Verma, O., Agarwal, R., Dafouti, D., & Tyagi, S. (2011). Performance analysis of knowledge encryption algorithms. In *Electronics Technology (ICECT) 3rd International Conference on, 2011* (pp. 399–403). Academic Press.

William, S., & Stallings, W. (2006). *Cryptography and network security*. Pearson Education.

Chapter 9
Demystifying Multi-Tier Cost Model for Scheduling in Fog Communication Networks

Jagadesh T.
KPR Institute of Engineering and Technology, India

Jaishankar B.
KPR Institute of Engineering and Technology, India

ABSTRACT

In this chapter, the authors explore a cost model and the come about cost-minimization client booking issue in multi-level mist figuring organizations. For an average multi-level haze figuring network comprising of one haze control hub (FCN), different fog access nodes (FANs), and user equipment (UE), how to model the cost paid to FANs for propelling assets sharing and how to adequately plan UEs to limit the cost for FCN are still issues to be settled. To unravel these issues, multi-level cost model, including the administration delay and a straight backwards request dynamic installment conspire, is proposed, and a cost-minimization client planning issue is defined. Further, the client planning issue is reformulated as an expected game and demonstrated to have a Nash equilibrium (NE) arrangement.

1. INTRODUCTION

With the blast of savvy gadgets and the prevalence of low-inertness applications, for example, web based recordings, current remote organizations have been experiencing information traffic burst and tough requests on help delay. To adapt to this test, fog processing has arisen as a promising engineering for Internet of Things (IoT) and future remote organizations (X.Chen et al, 2016). Fog processing shifts part of the correspondence, calculation, and storing assets from the far-off cloud to the organization edge, along the cloud-to things continuum. It enables end client types of gear (UEs) with multi-level figuring or administration (Yang et al,2018; Liuet al,2018). In such an engineering information can be prepared,

DOI: 10.4018/978-1-7998-6988-7.ch009

or administrations can be given, deftly at various levels, which are nearer to UEs. In this way, both the traffic load and the administration deferral can be adequately diminished.

Giving QoS certifications to multi-level administrations isn't a direct assignment due to two primary reasons: One is that the remaining task at hand designs are eccentric and persistently change over the long haul, and the other is that the complex communication between levels expands the trouble in recognizing the bottlenecks and settling them naturally (Bi et al,2018;Liu et al,2018). Along these lines, the Cloud Providers (CP) needs to receive a unique asset provisioning and improvement way to deal with satisfies the commitment to the administration proprietors concerning for to the Service Level Agreements (SLA) necessities (Kitanov et al,2016). Given that the said administrations run on a common framework, the CP necessities to advance the asset provisioning between the diverse running administrations when the total asset requests surpass the CP asset pool ability to build the CP administration provisioning benefits.

Without loss of consensus, let us consider a multi-level fog figuring network comprising of one fog control network (FCN), numerous fog access networks (FANs), and UEs, as appeared in Fig. 1. With the assistance of FANs, UEs can be presented with decreased administration delay and upgraded nature of administration(QoS). For model, delay-open- minded administrations can be given by distant FCN, while delay-delicate applications can be prepared at neighboring FANs (Zheng et al,2019;Penget al,2014). Through compelling client planning, the traffic burden and administration postponement can be significantly decreased (Xiong et al,2017).

Albeit various parts of client planning in multi-level fog processing networks have been talked about in written works, a viable client planning plan actually faces difficulties, particularly when the cost model is thought of (Romana et al,2016). For the most part, the FCN is worked by a telecom administrator, who signs an assistance contract with UEs, while the FANs are have a place with various people. To all the more likely persuade the FANs to share assets and foresee in storing, the cost model, particularly for FANs, should be contemplated.

In this chapter, a brought together multi-level cost model, including the administration delay and a straight converse interest dynamic installment plot, and the came about cost-minimization client booking issue, are researched, in a multi-level fog registering network comprising of one FCN, various FANs and UEs(Tikhvinskiy et al,2018).

This chapter means to eliminate the impediments of current asset provisioning approaches for cloud multi-level administrations. This examination proposes a unique asset enhancement and provisioning system (ROP) and working model framework based on a cloud stage(Ai et al,2017). The model utilizes multi-level internet business applications to act as an illustration of cloud multi-level help. The model intermittently screens the exhibition as far as a start to finish delay, gathers central processor uses of every level, identifies the bottlenecks, and employs the proposed system to change and upgrade the asset provisioning strategy.

(Romana et al,2016) present a completely actualized two-way validation security conspire for IoT dependent on existing Internet norms, particularly the DTLS convention. They assess the proposed approach in regards to execution and handshake. They showed that the proposed approach gives message trustworthiness, classification, and credibility with moderate energy, start to finish slowness, and memory overhead.

(Corminardi et al, 2017) proposed a DTLS header pressure conspires that expects to diminish energy utilization by utilizing the 6LoWPAN norm. Creators assessed DTLS in regards to execution, overhead, and handshake.

(Wang et al, 2015) give an assessment of DTLS in distinctive obligation cycled networks. They examined overhead and handshake when utilizing three obligation cycling join layer conventions: preface testing, the IEEE 802.15.4 reference point empowered mode, furthermore, the IEEE 802.15.4e Time Slotted Channel Hopping mode.

(Rubertis et al,2013) built up a client booking calculation to boost every UE's nature of involvement, based on the likely game hypothesis. It also inferred a raised streamlining-based joint client booking and asset portion calculation to limit the absolute framework cost for a cloud-mist registering network with non-symmetrical different access. (Bi et al,2018) used the Lyapunov improvement procedures to plan an online joint client booking and asset allotment calculation to boost the normal organization throughput for a mist empowered substance conveyance organization

The remainder of this chapter is coordinated as follows. The framework model of multi-level fog processing networks is given in Section II, along with the numerical plan of the expense model and came about the cost-minimization client planning issue. Given because of the potential game, the client planning game is created and examined in Section III. The NE of this game is demonstrated to exist and the comparing conveyed client planning calculation, i.e., COUS calculation, is proposed. At that point, Section IV assesses the presentation of created calculation and the highlights of the proposed cost model through re-enactment. At last, Section V closes this chapter.

II SYSTEM MODELING

A multi-level fog figuring network comprising of one FCN, M FAN's, and N UEs is thought of. The FCN is worked by a telecom administrator, which gives administrations to N UEs, i.e., administration supporters, while the FANs are having a place with various people(Vucinic et al,2015). To diminish administration delay furthermore, improve QoS, the FCN is eager to pay cash to FANs if they offer types of assistance to UEs. For the simplicity of articulation, we take reserving as an illustration in the accompanying setting 1. In the fog-empowered storing network, the FCN can assign documents to FANs during an off-top time, i.e., record position, and in this way the UEs can be related with appropriate FANs or FCN to download records during the top time, i.e., client planning(Rubertis et al,2013).

Asset provisioning in multi-level cloud administrations is an essentially mind-boggling issue in light of their unpredictable conduct what's more, powerful changing of traffic designs. Numerous asset provisioning approaches have been acquainted with recognize the on-request assets that can meet help SLA execution necessities. Even though these methodologies chiefly point to give QoS to running administrations, they expand on various asset provisioning methods. Our past study separated the asset provisioning approaches into rule-based also, model-based methodologies.

Like numerous past works (Han et al,2014;Mi et al,2010) a quasistatic situation, wherein the UEs stay unaltered during a client booking stretch, is expected. Plus, the client planning issue is the focal point of this work, and accordingly, the document portion issue is overlooked in this chapter.

Interestingly, the standard-based methodologies utilize a predefined rule to recognize when and how many measure of assets are expected to meet SLAs of running help. In this unique circumstance, a static predefined CPU edge is utilized to scale an assistance up if there should arise occurrence of SLAs infringement. Also, fluffy rationale regulator is utilized to catch the administration execution dispersion and to distinguish on-request assets dependent on SLAs necessity. Without model neural fluffy regulator (NFC) is utilized to offer QoS ensures dependent on percentile start to finish delay(Liu et al,2018).

If UE n is associated with FAN m, the downloading delay of a file can be expressed as

$$t_{m,n} = \frac{L}{R_{m,n}} \tag{1}$$

Where $R_{m,n}$ is the transmission rate from FAN m to UE n.

Much the same as the instalment conspire for online commercial, i.e., cost-per-click (Ghetas et al,2015), the FANs are expected to charge by utilization sums or downloads. To rouse more UEs to download records from it, and along these lines acquire more incomes, the FANs set their cost as an opposite interest work. Expect a direct converse interest work, and the cost for single download or the instalment for downloading a document from FANs is given by,

$$\alpha_m - \beta_m \sum_{n=1}^{N} \alpha_{n,m}, m \in M \tag{2}$$

$$min \sum_{n=1}^{N} O_n \left(a_n, A_{-n}\right) \tag{3}$$

At the point when various multi-level administrations share a typical foundation, accomplishing the administration level QoS objectives becomes an altogether intricate issue on account of the multi-level design of the administrations notwithstanding the intricate connections between singular levels. The total on-request assets from all administrations sharing a typical framework normally surpass the CP asset pool limit. The ROP system handles this circumstance by giving execution confinement and separation among co-facilitated administrations to increment the CP administration provisioning benefits and fulfill the administration level QoS targets if conceivable.

III PERFORMANCE EVALUATION

Information downloading task will be infused after every open cloud allotted undertaking when its kid is dispensed in the private cloud. As in information transferring errands, the same downloading errands will be likewise infused previously each assignment dispensed in the private cloud when its parent is allotted in the public one while downloading information burns-through the two veils of mist assets. The downloaded information won't be traded locally any longer .Accordingly, in instance of downloading information, neighbourhood information move costs in the bunch will be zero. The performance comparison of various UE's is shown in Figure.1 which shows that fog communication network yields better results. Information transferring task will be infused before every open cloud distributed errand when its parent is allotted in the private cloud. The equivalent transferring assignments will be likewise infused after each undertaking assigned in the private cloud when its youngster is apportioned in the public one, while transferring information burns-through the two mists assets. The transferred information won't be traded locally any longer. If there should arise an occurrence of transferring information, neighborhood information move costs in the bunch will be zero.

Figure 1. Overall cost comparison

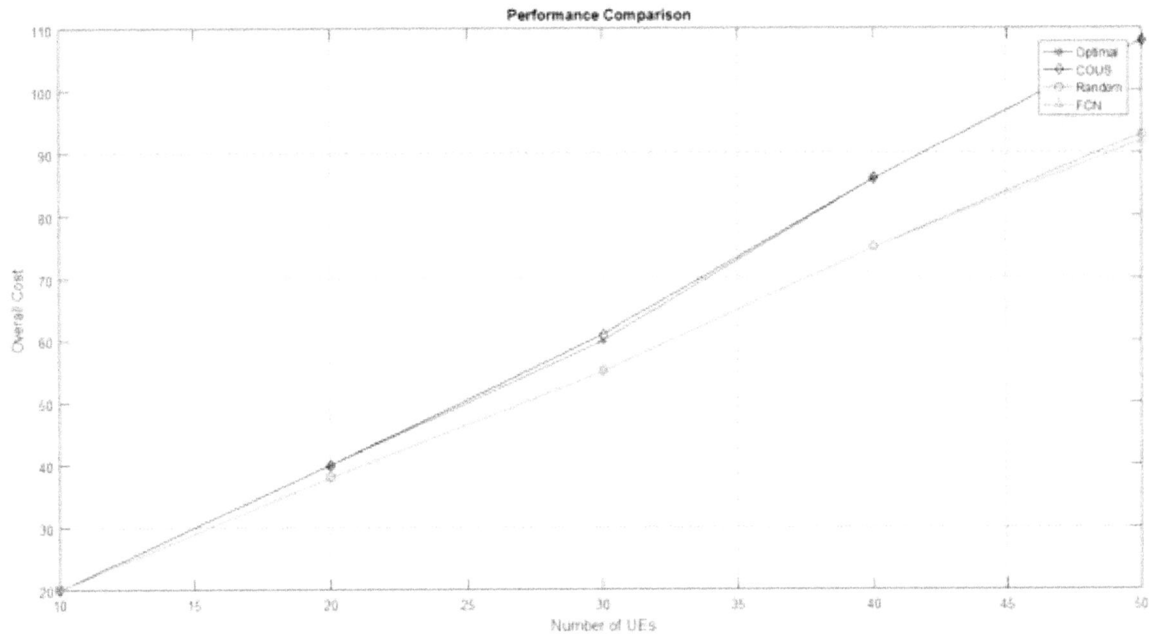

A Markovian investigative model is utilized to distinguish the start to finish defer dependent on assistance deals and the a number of apportioned VMs(Wang et al,2018). Specifically, these methodologies receive normal start to finish postponement to assess the deliberate assistance execution. Be that as it may, the normal start to finish delay can't catch the state of the reaction time conveyance also, prompts SLA infringement. These middle-of-the-road schedulers are considered as a dark box right now, a considerable lot of the cutting edge executed schedulers can be applied. In the wake of reproducing both assignment designs, the execution reports of the two recreations, which contain the supreme execution season of each errand on its facilitating processor, will be shipped off the covering stage. In this stage, the genuine execution season of assignments will be determined for example in the event that M1, M2, and M3 took 50 minutes and transferring information to the public cloud took 5 minutes, at that point the R1 and R2 will be begun at minute 55, etc.

Initially, the work processes will be created and consolidated into one by infusing sections and leave assignments. The significant level scheduler chooses to plan R1 and R2, for instance, in the public cloud and the rest in the private one. The private and public portion plans contain infused transferring furthermore, downloading errands. Every information transferring activity costs the two veils of mist, the private one while sending, and the public one while getting.

Albeit extraordinary exertion has been given to contemplate work process planning for mists for quite a while, the vast majority of the examinations try not to consider booking work processes that comprise of different map-reduce occupations. Likewise, albeit logical work processes are characterized by space specialists, numerous improvements and tasks can be performed through them to use organization also, figuring assets. Also, the vast majority of studies accept that processors are inactive following assignments execution without thinking about that trading information devours processors' time. Additionally, a considerable lot of mixture cloud task booking approaches don't contemplate the web association among private and public mists as a basic restricted asset which should be shared and used admirably Also, these methodologies try not to consider the expenses for transferring and downloading information volumes.

IV CONCLUSION

In this chapter, we explored a bound- together multi-level cost model also, the came about cost-minimization client booking issue in multi-level fog registering networks comprising of one FCN, various FANs, and UEs. A bound-together multi-level cost model, including the administration delay and a direct reverse interest dynamic installment conspire, was proposed, and an expense minimization client booking issue was detailed. Further, the client booking issue was reformulated as an expected game and in this way demonstrated to have a NE arrangement. Likewise, an appropriated calculation called COUS was created to accomplish a NE of the game. Investigative and reenactment results indicated that the COUS calculation could offer close ideal execution regarding generally speaking expense. Plus, the proposed dynamic installment plot could accomplish a mutually beneficial result for the two FANs and FCN, however, an unreasonable remaining burden conveyance among FANs, looked at with the fixed installment plot.

REFERENCES

Ai, Y., Peng, M., & Zhang, K. (2017). *Edge cloud computing technologies for Internet of Things: A primer*. Digital Communications and Networks.

Bi, S., & Zhang, Y. J. (2018). Computation rate maximization for wireless powered mobile-edge computing with binary computation offloading. *IEEE Transactions on Wireless Communications*, *17*(6), 4177–4190.

Chen, X., Jiao, L., Li, W., & Fu, X. (2016). Efficient multi-user computation offloading for mobile-edge cloud computing. *IEEE/ACM Transactions on Networking*, *24*(5), 2795–2808. doi:10.1109/TNET.2015.2487344

Chiang, M., & Zhang, T. (2016). Fog and IoT: An overview of research opportunities. *IEEE Internet of Things Journal*, *3*(6), 854–864. doi:10.1109/JIOT.2016.2584538

Corminardi, L. (2018). Opportunities and Challenges of Joint Edge and Fog Orchestration. *IEEE Wireless Communication and Networking Conf.*

European Telecommunications Standards Institute. (2016). *Network Functions Virtualisation (NFV); Acceleration Technologies; VNF Interfaces Specification*. ETSI GS NFV-IFA 002.

Ghetas, M., Yong, C. H., & Sumari, P. (2015). Harmony-based monarch butterfly optimization algorithm. *International Conference on Control System, Computing and Engineering (ICCSCE), 2015 IEEE International Conference on. IEEE*, 156–161.

Han, R., Ghanem, M. M., Guo, L., Guo, Y., & Osmond, M. (2014). Enabling cost-aware and adaptive elasticity of multi-tier cloud applications. *Future Generation Computer Systems*, *32*, 82–98. doi:10.1016/j.future.2012.05.018

Kitanov, S., Monteiro, E., & Janevski, T. (2016). 5G and the Fog – Survey of Related Technologies and Research Directions. *Proceedings of the 18th Mediterranean IEEE Electrotechnical Conference MELECON*, 1-6.

Lama, P., & Zhou, X. (2013). Autonomic provisioning with self-adaptive neural fuzzy control for percentile-based delay guarantee. *ACM Transaction. Autonomous. Adapt. Syst. (TAAS), 8*(2). doi:10.1109/ICNISC.2015.91

Liu, T., Li, J., Kim, B., Lin, C.-W., Shiraishi, S., Xie, J., & Han, Z. (2018). Distributed file allocation using matching game in mobile fog-caching service network. *IEEE INFOCOM 2018-IEEE Conference on Computer Communications Workshops (INFOCOM WKSHPS)*, 499–504. 10.1109/INFCOMW.2018.8406854

Liu, Y., Yu, F. R., Li, X., Ji, H., & Leung, V. C. (2018). Hybrid computation offloading in fog and cloud networks with non-orthogonal multiple access. *IEEE INFOCOM 2018-IEEE Conference on Computer Communications Workshops (INFOCOM WKSHPS)*, 154–159. 10.1109/INFCOMW.2018.8406940

Liu, Yang, Yang, Wang, & Mao. (2018). DATS: Dispersive stable task scheduling in heterogeneous fog networks. *IEEE Internet of Things Journal.*

Mi, H., Wang, H., Yin, G., Zhou, Y., Shi, D., & Yuan, L. (2010). Online self reconfiguration with performance guarantee for energy-efficient large-scale cloud computing data centers. In Services *Computing (SCC), 2010 IEEE International Conference on 2010*, (pp. 514–521). IEEE. 10.1109/SCC.2010.69

Peng, M., Li, Y., Jiang, J., Li, J., & Wang, C. (2014). Heterogeneous cloud radio access networks: A new perspective for enhancing spectral and energy efficiencies. *IEEE Wireless Communications*, *21*(6), 126–135.

Romana, R., Lopeza, J., & Mambob, M. (2016). Mobile Edge Computing, Fog et al: A Survey and Analysis of Security threats and challenges. In Future Generation Computer Systems. Elsevier.

Rubertis, A. D., Mainetti, L., Mighali, V., Patrono, L., Sergi, I., Stefanizzi, M. L., & Pascali, S. (2013). Performance evaluation of end-to-end security protocols in an internet of things. *International Conference Software, Telecommunications and Computer Networks*, 1–6.

Taleb, T., Samdanis, K., Mada, B., Flinck, H., Dutta, S., & Sabella, D. (2017). On Multi-Access Edge Computing: A Survey of the Emerging 5G Network Edge Cloud Architecture and Orchestration. *IEEE Communications Surveys and Tutorials*, *19*(3), 1657–1681. doi:10.1109/COMST.2017.2705720

Tikhvinskiy, V., & Bochechka, G. (2015). Prospects and QoS Requirements in 5G Networks. *Journal of Telecommunications and Information Technologies*, *1*(1), 23–26.

Vucinic, M., Tourancheau, B., Watteyne, T., Rousseau, F., Duda, A., Guizzetti, R., & Damon, L. (2015). DTLS performance in duty-cycled networks. *International Symposium on Personal, Indoor, and Mobile Radio Communications,* 1333–1338.

Wang, F., Xu, J., Wang, X., & Cui, S. (2018). Joint offloading and computing optimization in wireless powered mobile-edge computing systems. *IEEE Transactions on Wireless Communications*, *17*(3), 1784–1797.

Xiong, K., Chen, C., Qu, G., Fan, P., & Letaief, K. B. (2017). Group cooperation with optimal resource allocation in wireless powered communication networks. *IEEE Transactions on Wireless Communications*, *16*(6), 3840–3853.

Yang, X., Liu, Z., & Yang, Y. (2018). Minimization of weighted bandwidth and computation resources of fog servers under per-task delay constraint. *2018 IEEE International Conference on Communications (ICC)*, 1–6. 10.1109/ICC.2018.8422318

Ye, X. (2015). A Survey on Scheduling Workflows in Cloud Environment. In *Network and Information Systems for Computers (ICNISC), 2015 International Conference on*. IEEE.

Zheng, Xiong, Fan, Zhong, & Letaief. (2019). Fog-assisted multi-user SWIPT networks: local computing or offloading. *IEEE Internet of Things Journal*.

Chapter 10
Analysis of Bluetooth Versions (4.0, 4.2, 5, 5.1, and 5.2) for IoT Applications

S. D. Padiya
https://orcid.org/0000-0002-7462-0187
Shri Sant Gajanan Maharaj College of Engineering, Shegaon, India

V. S. Gulhane
Sipna's COET, India

ABSTRACT

IoT includes many sensors that have to collect the data and send it to the superior nodes; for such interaction between the IoT devices, various wireless technologies are available, like infrared, Li-Fi, WI-Fi, Zigbee, Bluetooth, etc. Among all the available, Bluetooth proved the most promising short-range wireless communication technology due to various factors. To fulfil the increasing demand for wireless connectivity, the Bluetooth SIG must continuously perform up-gradation. Here, analysis of Bluetooth versions are discussed based on the characteristics such as speed, bandwidth, range, power, message capacity, beacon provision, compatibility, reliability, errors detection, correction capability, advertisement packets, duty cycle, slot availability masks, and many more. This analysis concluded that all the versions have their own set of merits and limitations. For the basic IoT applications (limited functionalities), Bluetooth 4.0/4.2 is a good choice, while for the complex IoT applications (advance functionalities), Bluetooth 5/ 5.1/ 5.2 is better.

INTRODUCTION

The Internet of Things (IoT) involves various wireless communication technologies to makes devices capable of interacting with each other. Nowadays, IoT with various dedicated sensors, devices and wireless communication technologies making a human lifestyle easier and smarter. Therefore, in our personal lives, IoT devices are becoming more prevalent and pervasive. Due to the IoT era, sensors are everywhere

DOI: 10.4018/978-1-7998-6988-7.ch010

in the smart society, and the trend will continue in the future. Today, every industry is changing towards automation under the vision of Robotic Process Automation (RPA). In such an industry, nowadays, machines/equipment are powered by some advanced technologies, like wireless sensor networks (WSN), artificial intelligence (AI), machine learning (ML), cloud computing (CC), etc.

As IoT use increases, new challenges, requirements and research needs have arisen to fulfil the demands. Among all the IoT application requirements, wireless communication technology(low power requirement and high data security)is the most viable demand. IoT includes many sensors that collect the data from an environment and send the collected data to their superior node. For the data transmission from the sensors to its superior node, many wireless communication technologies are available such as radio-frequency identification (RFID), Infrared, Li-Fi, Wi-Fi, Zigbee, Bluetooth, LoRa, etc.

A general IoT architecture includes hardware, software, communication medium and application layer. The communication layer is a critical bridge between all the other layers and consists of a multi-layer stack, including the data link, network or transport, and session protocols. Bluetooth is at the data link layer to establish sensor to sensor connection or sensor to the gateway connection. Aguilar, Vidal and Gomez (2017), analytically and experimentally proved that among all the available short-range wireless technologies, Bluetooth has become the most adopted technology. Bluetooth is the most promising because of appearing everywhere or of being a very common technology. After reviewing all the other available wireless communication technologies, it has been considered that Bluetooth becomes the ultimate technology for many IoT applications. Following are the main reasons:

- Bluetooth has been embedded massively in the latest smartphones.
- It is less costly to deploy and maintenance.
- It is supported by all the operating systems including, iOS, Android, Linux, OSX, and Windows.
- It has low interference and a standardized protocol compatible with IPv6.

The day-to-day nature of IoT applications becoming more complex as it adds many new sensor nodes, devices and communication technologies. In the WSN, devices and sensor nodes regularly have to interact with each other. This interaction must be in minimum time and energy. Therefore, for the IoT, fast but lightweight short-range communication technology is needed. To fulfil the increasing demand for high speed, lightweight and secure wireless connectivity, Bluetooth Special Interest Group (Bluetooth SIG) continuously upgrading Bluetooth technology. In this chapter, Bluetooth versions 4.0, 4.2, 5, 5.1 and 5.2 are analyzed based on their characteristics such as data speed, network bandwidth, communication range, power requirement, message capacity, beacon provision, support for IoT, compatibility, reliability, errors detection, and correction capability, advertisement packets, coexistence with Wi-Fi devices, duty cycle, slot availability masks, periodic advertising synchronous transfer capability and many more.

For the development of any application, the developer has a responsibility to select the appropriate resources. The selection of resources must be according to the need. The resources must be simple, cheap, low energy consumable, high output and durable. This analysis for the various Bluetooth versions will help the IoT developer for the selection of the correct Bluetooth version as per the need.

Background

In the IoT environment, all the devices have connections with each other. Nowadays, IoT system involves wireless communication technologies known as Wireless Sensor Network. For such an interaction between IoT devices, many wireless communication technologies are available. The WSN must provide a high speed and secure data transmission in minimum energy.

After reviewing all the available wireless communication technologies, it has considered that Bluetooth is the ultimate technology for many IoT applications. Bluetooth is a lightweight and secured short-range communication technology. The foundation of protocol and hardware developments of Bluetooth to implement IoT applications and services has a strong nature. Bluetooth has many advantages like small-size, lightweight-data-transmission, low-cost, low-maintenance, and less-energy-consumption. To improve the efficiency for high speed, lightweight and secure wireless connectivity, 'Bluetooth SIG' continuously upgrading Bluetooth technology.

Cognizant observed that as seek-to-reap the benefits of IoT, enterprises facing many challenges such as selection, implementation, customization and support to the new technologies across the IoT continuum. For the development of any application, the developer has a responsibility to select the appropriate resources. The selection of resources must be according to the need. The resources must be simple, cheap, low energy consumable, high output and durable. The analysis for the various Bluetooth versions will help the IoT developer for the selection of the correct Bluetooth version as per the need.

INTERNET OF THINGS (IoT)

The IoT is a network of physical objects or 'things' embedded with electronics, software, sensors and connectivity, enabling objects to exchange data over the internet. In 1982, Carnegie Mellon University has discussed the concept of an Internet-connected "Coca-Cola Vending Machine" with the ability to report its inventory and temperature status of drink to the system. In 1999 at MIT's Auto-ID Center the term "Internet of Things" was coined by Mr. Kevin Ashton of P&G with the role of RFID to allow computers to manage all individual things.

The IoT ecosystem involves physical devices like sensors and actuators with the Internet. A sensor is a device that detects both physical and environmental conditions. (Estrin et al., 1999) the output of the sensors is an electrical signal that must transmit to a controller for further processing. As per the requirement, various types of sensors are available to sense the environment. Sensors make IoT capable of smarter decisions based on the collected data. Many sensor nodes are available to sense the environmental condition for the different parameters, such as i) seismic, magnetic, thermal, visual, infrared, acoustic & radar, ii) measurements including temperature, noise level, speed, size of an object, pressure & humidity and iii) conditions of vibration, radiation, vehicular movement, soil erosion, the presence or absence of objects, mechanical stress levels on objects and direction. Sensors are also capable of continuous sensing, event detection, location sensing and local control.

The IoT has various wireless communication technologies to make devices capable of interacting with each other. Today's most IoT systems are with smart-sensor networks mainly connected wirelessly through WSN. WSN involves many sensor nodes to sense the environmental conditions. Therefore, for the transmission of collected huge data, it requires an energy-efficient, flexible and low-cost wireless communication medium. Hence sensor network routing protocols have been designed by considering the limitations of power, processor's capacity, and memory of sensor devices.

HIS Technology forecasts (2016) that in the last decade, the IoT importance has notably improved up to 15.4 billion IoT devices installed in 2015, 30.7 billion in 2020 and targeted to 75.4 billion in 2025. In the "The Five Essential IoT Requirements and How to Achieve Them-2019", Cognizant Digital Business

(2019) observed that enterprises that adopted IoT have decreased supply chain costs by more than 20%, increased productivity by 10% to 20% and reduced design-to-market times by 20% to 50%.

Importance of Bluetooth in IoT

As the demand for IoT increasing, new challenges, requirements, and research needs have also arisen. Devices capable of interaction with low power and high data security are the most viable demand for diverse IoT applications. At the heart of the explosion of the IoT and the Industrial IoT (IIoT) is wireless technology. This technology enables devices to communicate with each other without being physically connected.

Technological enhancements that support incredible growth include the speed and bandwidth of the underlying networks, extended battery life of IoT devices, broader capabilities of wireless communication protocols, and more secure management of devices and networks. These advancements have allowed many industries to replace existing expensive and unreliable wired communication with wireless communication. However, to realize IoT's potential standardized communication protocol is needed. Recent developments in Bluetooth make the technology 'the communication protocol of choice' for IoT. According to Bluetooth SIG (2020), more than one-third of all installed IoT devices will be Bluetooth-enabled. Bluetooth enabled devices with the ability to connect diverse devices through short-range technology transformed the way of interaction with each other.

Today, Bluetooth technology has evolved from classic Bluetooth to smart Bluetooth stage, including the latest version Bluetooth 5. For instance, compared to its earlier versions, Bluetooth 5 has four times the range, double the speed and 800% more data broadcasting frequency. These enhanced features will increase the number of applications for which Bluetooth will be a good choice. Many applications with large and critical infrastructure can have 100% uptime and cost-effective solutions using Bluetooth 5 enabled IoT devices. In addition, the latest Bluetooth versions have been designed for low-powered devices so that IoT devices conserve less energy by maintaining devices in sleep mode until they are connected. The new specifications make Bluetooth ideal for IoT applications because it does not require devices paring for the interaction. Any device can broadcast the messages into a network, and any other device willing can receive it. This feature helps in enhancing operational efficiency and also improving device availability.

BLUETOOTH TECHNOLOGY

Bluetooth is a short-range wireless communication technology for exchanging data between fixed and mobile devices using ultra-high frequency (UHF) radio signals in the industrial, scientific and medical (ISM) radio bands. The technology provides a short period connection without any fixed infrastructure hence also called an Ad Hoc network. In 1997 Jim Kardach of Intel has proposed the name "Bluetooth", who developed a system that would allow mobile phones to communicate with computers. Bluetooth managed by the Bluetooth SIG, established by Ericsson, IBM, Intel, Nokia and Toshiba, and later joined by more than 35,000 member companies. The IEEE standardized Bluetooth as IEEE 802.15.1, but no longer maintains the standard. 48 billion IoT devices are expected by 2021 and Bluetooth is predicted to be in nearly one-third of those devices. Figure 1presents the official logo for Bluetooth.

Figure 1. Official Bluetooth Logo (© Bluetooth SIG, Inc.)

Bluetooth operates at frequencies range 2.402GHz to 2.480GHz, or 2.400GHz to 2.4835GHz with guard bands 2MHz wide at the bottom end and 3.5MHz wide at the top. It is not a globally licensed ISM 2.4GHz short-range radio frequency band. Bluetooth uses a radio technology called frequency-hopping spread spectrum. Bluetooth divides data to transmit into packets and each packet on one of 79 designated Bluetooth channels. Each channel has a bandwidth of 1MHz. It usually performs 1600 hops per second, with Adaptive frequency-hopping (AFH) enabled.

Bluetooth specification supports multiple radio options that enable developers to build products meeting the unique connectivity requirements of their market. Bluetooth is packets-based protocol with master/slave architecture. One master can communicate with up to seven slaves. All devices use the clock provided by the master as the base for packet exchange.

Market Demand

In the latest forecast, ABI research (2016) predicted the annual Bluetooth devices shipment to reach 5 billion up to 2021. At this time, only smartphones have a share of 43% of total shipments. Bluetooth Low Energy (BLE) shows the highest notable growth with a predicted 34% CAGR (Compound Annual Growth Rate) between 2016 and 2021, driven by new opportunities in beacons. As a result, Bluetooth smart devices will contribute 27% of total Bluetooth shipments by 2021. The research on Bluetooth is a continuing process, which results in progressive improvement. Martin Woolley, Bluetooth SIG (2016), when released Bluetooth 5 specifications, stated that Bluetooth 5 improved the network from 50 meters for Bluetooth 4 to 200 meters with a doubled speed.

Bluetooth enabled IoT applications are being increasingly used in industries for Asset Management, Safety and Worker Health (Healthcare, Automotive, Home and Entertainment), and Security. Bluetooth-powered IoT may be a big differentiator for companies looking to reduce manual tasks and enhance operational efficiency. Organizations with a well-planned IoT strategy and a deep understanding of use cases and technology protocols can establish new forms of machine-to-machine and machine-to-human communication, thereby driving superior experience and customer delight.

Bluetooth Versions

Bluetooth have been evolved which include v1.0, v1.1, v1.2, v2.0+EDR, v2.1+EDR, v3.0+HS, v4.0 LE. This up-gradation is based on the improvement in speed, range and data capacity. The important thing to be noted is that the additional features supported by the higher versions or standards of Bluetooth are optional and also, they do not affect the encoding and transmission of audio. In other words, all the Bluetooth versions are compatible with their previous versions which, give the advantage that device with one version can interoperate with the other version.

Bluetooth 1.0 and 1.0B

These were initial versions, so faced many problems with interoperability. Manufacturers were faced problems and difficulty in making their products, being the drawback of these versions. They also included mandatory hardware device address (BD_ADDR) transmission in the connecting process which, was the main setback for certain services planned for use in Bluetooth environments.

Bluetooth 1.1

It overcomes the interoperability problem faced by p*revious* specifications by adding non-encrypted channels and signal strength indicators to it.

Bluetooth 1.2

It enhances the connection, discovery, AFH spread spectrum, and resistance to RF interference by avoiding the crowded frequencies in the hopping sequence. It also enhances the transmission speeds up to 721 Kbps, synchronous connections (eSCO), audio latency to provide better concurrent data transfer, and Host Controller Interface (HCI) operation with three-wire UART (universal asynchronous receiver and transmitter).

Bluetooth 2.0 + EDR

This version was released (2005) with the Enhanced Data Rate (EDR) for faster data transfer. The bit rate of EDR is 3 Mbps, although the maximum data transfer rate is 2.1 Mbps. EDR uses a combination of **Gaussian frequency-shift keying** (GFSK) and Phase-shift keying (PSK). EDR can provide lower power consumption through a reduced duty cycle. This version specification implies that EDR is an optional feature. EDR includes some other minor improvements to the previous.

Bluetooth 2.1 + EDR

This version by the Bluetooth SIG (2007) included a Secure Simple Pairing (SSP) to improve the pairing experience for Bluetooth devices while increasing the use and strength of security.

Bluetooth 3.0 + HS

The Bluetooth *v*3.0+HS by the Bluetooth SIG (2009) claimed theoretical data transfer speed up to 24 Mbps, though not over the Bluetooth link itself. Instead, the Bluetooth link is used for negotiation and establishment, and the high data rate traffic is carried over the co-located 802.11 links in the addition of 802.11 as a high-speed transport. The high-speed part is not mandatory.

BLUETOOTH LOW ENERGY

As the name introduced, Bluetooth Low Energy also, known as 'Bluetooth LE' or 'Bluetooth Smart', is a wireless communication technology designed to achieve secured and energy-efficient data transmission than the classic Bluetooth. Due to the low power requirement, the BLE allows small devices to operate on tiny batteries. Younis & Fahmy (2004) discussed that Nokia had realized a need for low power consuming wireless technology and published its solution after three years, named Bluetooth Low-End Extension. The design and development of BLE began in the first half of 2001. It had appealed to develop a small size device to perform secure data transmission.

Figure 2. Official Bluetooth Smart/ BLELogo(© Bluetooth SIG, Inc.)

After the failure of Wibree, Bluetooth SIG included it in the future Bluetooth, which is today known as BLE. Figure 2 represents the official logo for BLE specification. The low power latency nature and lower complexity structure make BLE perfect to perform at low-cost microcontrollers. BLE is a radio standard design with the lowest possible power consumption, optimized for low cost, low bandwidth, low power, and low complexity.

What Makes BLE Different?

While BLE is a better technology on its merit, what makes BLE genuinely exciting and what has pushed its phenomenal adoption rate so far so quickly is that it's the right technology, with the right compromises, at the right time. Relatively, BLE is a young standard (introduced in 2010) that has seen a rapid adoption rate. Many products already included BLE puts them self well ahead of other wireless technologies at the same point in time in their release cycles.

BLE is available in mostly all smartphones, tablets, and other mobile computing devices. Hence due to the incredible growth of Bluetooth equipped devices, the BLE standard becoming common and popular. Active adoption of BLE technology by the mobile industry (Apple & Samsung) opens the doors for a wider implementation of the BLE. All-in-one radio-plus-microcontroller BLE solutions are available today at fewer prices as compared to other similar wireless technologies.

Features and Market Demand

BLE protocol is a radical departure from Classic-Bluetooth protocol. The Bluetooth Basic Rate/Enhanced Data Rate (BR/EDR) is still the best for voice and music streaming but, BLE is better for wireless sensor and control applications. BLE version 4 performs error detection while fails to perform error correction.

After reviewed all other available wireless technologies for communication, we have decided to focus on BLE, which becomes the ultimate technology for many IoT applications due to the following reasons:

- The lowest power consumption, i.e., BLE energy utility is 2.5 times better than ZigBee.
- BLE expected an average latency and average energy consumption for it.
- Cost-efficient and compatible (Dual-mode with both BLE and classic Bluetooth)
- BLE has been embedded massively in all smartphones.
- It is supported by all the operating systems including, iOS, Android, Linux, OSX, and Windows.
- It has low interference and a standardized protocol compatible with IPv6.
- Achieved robustness by using the same AFH technology.
- It is fast to achieve speed between 1 Mbps to 2 Mbps.
- It performs with 2Mbps to 5Mbps bandwidth.
- It potentially supports the maximum Link Layer (LL) data throughput 70% higher than the maximum throughput possible for IEEE 802.15.4.
- The Bluetooth 5 range is enhanced; support 200m in outdoor and 40m in indoor.
- Bluetooth 5 with a capacity of 255 bytes capable to gives more bytes for actual data payload.
- Bluetooth 5 included eight new Protocol Data Units (PDUs) for advertising, scanning and connecting purpose.
- It is less costly to deploy and maintenance.
- Bluetooth 5 facilitates errors detection and correction to avoid data retransmission.

BLE Specifications

According to Goosen (2014), the Bluetooth specification covers both classic Bluetooth and BLE (the new, highly optimized wireless standard introduced in 4.0). These two communication standards are not directly compatible, and Bluetooth devices qualified on any specification version before 4.0 cannot communicate in any way with a BLE device. The on-air protocol, the upper protocol layers, and the applications are different and incompatible between the two technologies. Figure 3 shows the configuration possibilities between Bluetooth versions and device types, along with the protocol stacks that allow these devices to communicate with each other.

The Bluetooth specification (4.0 and above) defines two wireless technologies:

Figure 3. Configurations between Bluetooth versions and device types(© Bluetooth SIG, Inc.)

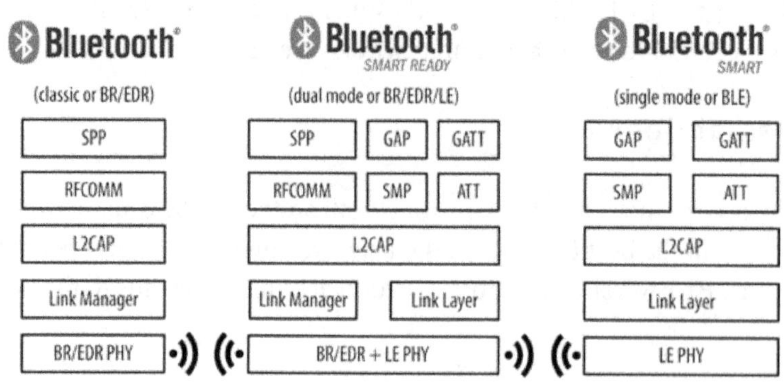

BR/EDR (Classic Bluetooth)is the wireless standard that has evolved with the Bluetooth specification since 1.0 and

BLE (Bluetooth Low Energy)is the low-power wireless standard introduced with version 4.0. BLE is not backwards compatible with BR/EDR protocols.

There are the two types that can use with the above-defined configurations:

Single-mode (BLE) devices implement BLE to communicate with single-mode but not with devices supporting BR/EDR only. A device with only BLE (single-mode) cannot communicate with a device with only Classic Bluetooth Protocol.

Dual-mode (BR/EDR/LE, Bluetooth Smart Ready) devices implement both the BR/EDR and BLE, which can communicate with any Bluetooth device.

Network Topologies

In the external world, BLE devices can communicate by following two approaches: Broadcasting or Connections. Both mechanisms have advantages and limitations.

Broadcasting and Observing

Figure 4 represents the broadcasting and observing topology. Broadcasting allows for data transmission to anyone or multiple receivers within a range, i.e., multiple receivers can receive the broadcasted data. Broadcasting defines two separate roles as broadcaster and observer.

Figure 4. Broadcast and Observer Topology(© Bluetooth SIG, Inc.)

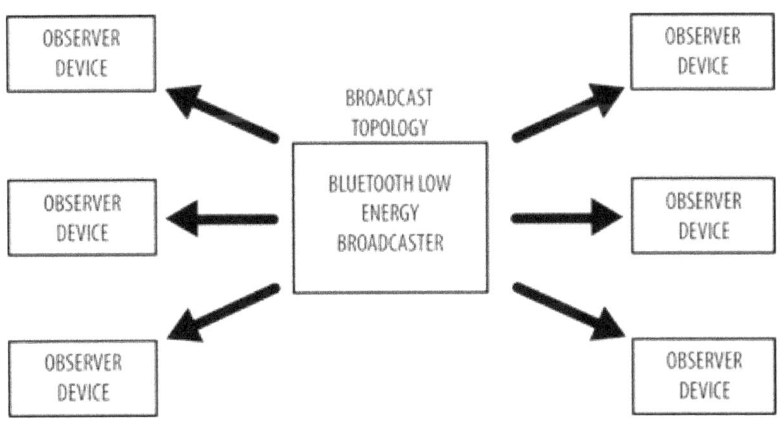

Broadcaster

Any device can send non-connectable *advertising* packets periodically to anyone willing to receive them. Broadcasting is only the way for devices to transmit data to multiple peers at a time. Users can broadcast data out by taking advantage of the advertising features of BLE.

The standard advertising packet contains a 31 bytes payload for the data that describes the broadcaster and its capabilities. A user may also include any custom information to broadcast. If a standard payload (31 bytes) is not enough for total data broadcasting, then optionally BLE also supports to second advertising payload (*Scan Response*). BLE allows devices that detect a broadcasting device to request a second advertising frame with another payload (31 bytes), i.e., payload for a total of 62 bytes. Since devices become in a sleep mode until an advertisement is initiated, this advertising feature enables BLE for minimum power consumption.

Broadcasting is fast and easy to use, and it's a good choice if the user has to push only a small data on a fixed schedule or to multiple devices. The limitation of broadcasting compared to regular connections is that it does not have any provision for data security and privacy (any observer can receive the data broadcasted in a network), so it might not be suited for sensitive data transmission.

Observer

It repeatedly performs the scanning in the environment to obtain the available non-connectable advertising packets currently being broadcasted. The slave device (broadcaster) performs "advertising" when it wants to broadcast a message in the environment. The client performs regular scanning for new devices (observer). When the observer finds a device that has to connect with it, initiate a connection. The advertisement may contain broadcasted data.

Connections

If any two devices have to interact with each other, i.e., if they have to share the data with each other then, the devices must have to use a 'connection topology'. A 'connection topology' is a permanent and

Figure 5. Connected Topology(© Bluetooth SIG, Inc.)

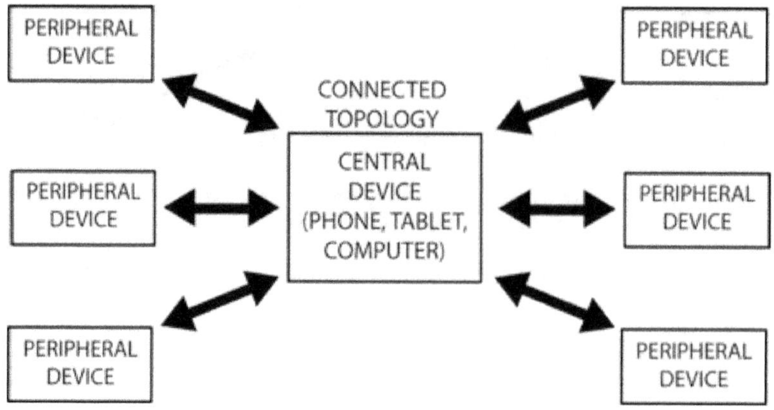

periodical data exchange between the two devices hence, it maintains privacy for data. The communication is only between the two peers involved in a connection so; no other device can access or enter in the connection. Figure 5 presents the connection topology for Bluetooth. Connection topology involves two separate roles:

Central (master): It repeatedly performs the scanning for available frequencies for the connectable advertising packets from an environment. After obtaining the desired frequency signal, it initiates a connection. Once a connection is established, the central manage the timing and initiate a periodical data exchange.

Peripheral (slave): It periodically sends the connectable advertising packets and accepts incoming connections. In an active connection, the peripherals follows central's timing and performs a data exchange regularly with it.

When connected, the Client/Central controls the communication by sending data and "polling" the Server/Peripheral for data at regular intervals. The selected interval is application dependent and can be specifically set. Once the connection is established peripheral stops advertising so devices can begin exchanging data in both directions.

Mixed Topology

In the previous version, a limitation has obtained regarding the connection of peripherals to a single central. The user can have multiple services and characteristics organized in a meaningful structure. The service may include many characteristics, like access rights, descriptive metadata, a higher throughput, a secure encrypted link establishment, and negotiation of connection to fit the data model. The previous two topologies can be combined in a single BLE network which is known as Mixed Topology. Figure 6 shows the Mixed topology architecture.

Figure 6. Mixed Topology(© Bluetooth SIG, Inc.)

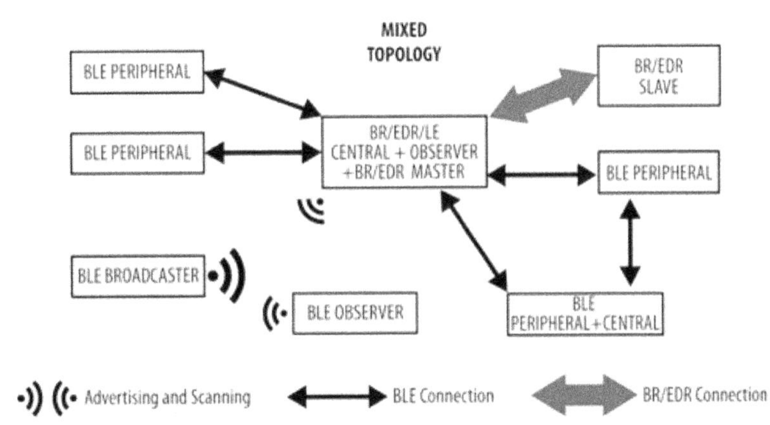

PROTOCOLS AND PROFILES

From its inception, the Bluetooth specification introduced a clear separation between the distinct concepts of *protocols* and *profiles*:

Protocols

The protocol defined the different building blocks used by all devices related to the Bluetooth specification. The protocols are the layers that may implement the various packet formats, routing, multiplexing, encoding, and decoding to allow data to be sent effectively between the peers.

Profiles

The profile defined the functionality to covers either basic modes of operation required by all devices (Generic Access Profile and Generic Attribute Profile) or specific use cases (Proximity Profile and Glucose Profile). Profiles essentially define, how protocol should be used to achieve a particular goal?

Generic Profiles

The generic profiles are defined in the Bluetooth specification. These profiles explain the fundamentals of ensuring interoperability between BLE devices from different vendors.

Generic Access Profile (GAP)

The GAP covers the used model of the lower-level radio protocols to define roles, procedures, and modes that allow devices to broadcast data, discover devices, establish connections, manage connections, and negotiate security levels. GAP is the topmost control layer of BLE. This profile is mandatory for all BLE devices, and all must comply with it.

Generic Attribute Profile (GATT)

The GATT deals with data exchange in BLE, it defines a basic data model and procedures for devices to discover, read, write, and push data between them. It is the topmost data layer of BLE.

BLE Working

BLE performs essentially one-way communication. Let us consider an example, BLE beacons for communication with a smartphone nearby - a BLE beacon periodically broadcasts data packets. These data packets are detected by app/pre-installed services on smartphones nearby. This BLE communication triggers actions such as pushing a message or promoting a mobile application. A BLE device operates on 40 channels spread in the 2.4GHz range. Three of these channels are dedicated to the advertisement and the rest are for data exchange.

There is only one general type of packet, specialized in advertisement and data types, which pretty much don't need to care about. Depending on the transmission power/reception sensitivity, the usable range will vary between 30 meters (a transmitter at 0.01mW and receiver with -70dBm sensitivity) and 100 meters (tx at 10mW and rx -90dBm). To save energy and provide higher data transfer speed, the entire BLE communication framework consists of 40 frequency channels, separated by 2MHz. Three of these channels are the primary advertisement channels while the remaining 37 channels are secondary channels, also known as data channels. The Bluetooth communication starts with the 3 primary advertisement channels and then offloads to the secondary channels.

BLE advertisement packets are broadcast periodically (fixed interval of 20 ms to 10.24 seconds, or random interval from 0 ms to 10 ms) on the three advertising channels and included the capabilities and services provided by the BLE node. Note that BLE operates in either an advertising or connected mode, which means that if a BLE node is in the connected mode it cannot send advertisement packets. Figure 7 presents the BLE Bands structure (37, 38 and 39 reserved for advertisement).

BLE differs from standard Bluetooth and employs the GAP for advertising a node's capabilities. When in a connected mode, BLE use a GATT profile to exchange data, it considers a simple structured list format. BLE beacons operate in only an advertising mode using GAP to send information in the advertisement packets, this is a simple one-to-many broadcast that enables BLE beacons to consume very little power. BLE is capable of much longer-range transmission (even at the initial version) than is popularly believed to be the case.

Figure 7. BLE Bands (37, 38 and 39 reserved for advertisement)

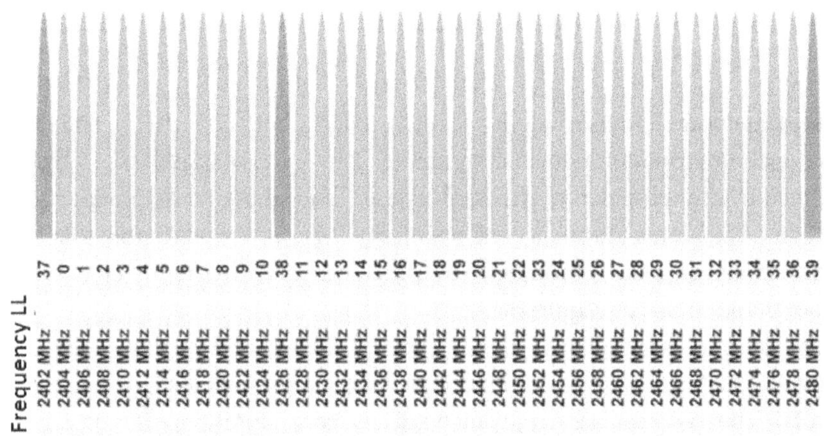

Demonstration using a standard smartphone and 'Mobile Control Unit' (MCU) shows a successful message transmission up to 350m in a sub-optimal environment. The commercial Bluetooth achieved up to 500m. The data transmission range for any wireless device depends on the operating environment, antenna, enclosure, noise, device orientation, power stability, etc. Transmit power (dBm) is usually configurable over a certain range, but the increase in the transmit power for more range cause more energy consumption, i.e., reducing the usable battery lifetime.

BLE Advertising Packets

A BLE beacon is a simple device that uses BLE node advertisements to broadcast its data to clients that can read the information offered. In BLE beacon applications, the BLE node operates in only an advertising mode using GAP to send messages in the advertisement packets. BLE beacons do not accept connection requests from clients. It is a simple one-to-many broadcast system so, BLE beacons consume very little power. Figure 8 presents BLE advertisement data packets.

Figure 8. Advertising Data Packets

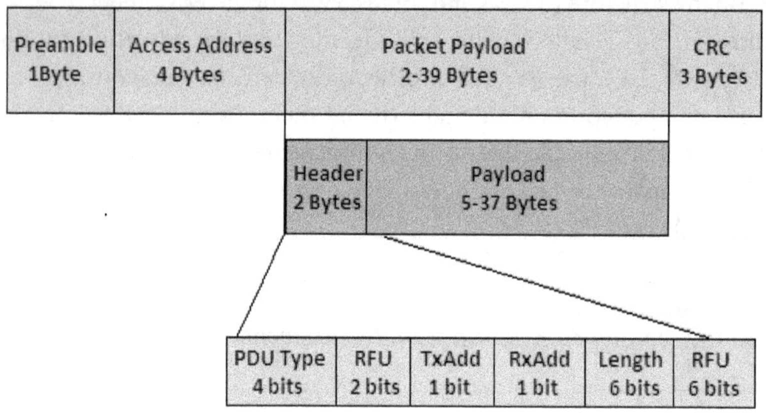

For ease of use, Bluetooth defines a single packet format for both advertising and data transmissions. The packet format includes four components: preamble (1 byte), access address (4 bytes), protocol data unit (2-39 bytes) and cyclic redundancy check (3 bytes). In all the advertisement packets, contents of the data payload can be customized for specific operators. It is important to note that the Bluetooth SIG has not specified a standard for BLE beacons. The Bluetooth specification is open for all the operators and consortiums to define it as per requirement.

Many organisations implemented the protocols with similar packet formats such as Apple iBeacon, Google Eddystone and the open-source AltBeacon. All the packet formats have different advertisement packet formats. A BLE beacon requires a relevant dedicated receiver/client (maybe a mobile application) that can recognize the beacon advertisements, extract the information and then take actions based on the content. In many cases, BLE beacons do not even send a direct message in advertisements; they only identify the beacon and its location. The mobile application itself looks up the beacon in a database (on a server platform or cloud platform) before suggesting actions to the user.

The data unit of an advertisement packet is called a 'Protocol Data Unit' with a 2 bytes header. The PDU specifies type and data payload length, which can be up to 37 bytes and includes 6 bytes for the advertisement address up to 31 bytes for data. The advertising header included the six segments, but the focus is only on the '*PDU Type*' and 'Length' fields. The Length field is 6 bits to define payload size to define the amount of data that can be advertised. The length may be between 6-37 bytes defined by PDU Type. BLE beacons operate within a limited zone so, when a mobile user enters the zone, their mobile application receives the beacon message and checks the UUID to determine if it is a service that the application recognizes. If the application recognizes the beacon service, then other messages can be extracted and actions can be triggered to engage the user.

The data unit of an advertisement packet is called a 'Protocol Data Unit' which has a 2 bytes header. The PDU specifies type and data payload length, which can be up to 37 bytes and includes 6 bytes for the advertisement address up to 31 bytes for data. The advertising header includes six segments, but the focus is only on the '*PDU Type*' and 'Length' fields. The Length field is 6 bits to define payload size to define the amount of data that can be advertised. The length may be between 6-37 bytes defined by PDU Type. BLE beacons operate within a limited zone so, when a mobile user enters the zone, their mobile application receives the beacon message and checks the UUID to determine if it is a service that the application recognizes. If the application recognizes the beacon service, then the other message data can be extracted and actions triggered to engage the user.

Table 1. Difference between Bluetooth Classic and Bluetooth Low energy

Specification	Bluetooth Classic	Bluetooth Low Energy (BLE)
Frequency Band	2.4GHz ISM Band (2.402 – 2.480 GHz Utilized)	2.4GHz ISM Band (2.402 – 2.480 GHz Utilized)
Channels	79 channels with 1 MHz spacing	40 channels with 2 MHz spacing (3 advertising + 37 data channels)
Channel Usage	Frequency-Hopping Spread Spectrum (FHSS)	Frequency-Hopping Spread Spectrum (FHSS)
Modulation	GFSK, π/4 DQPSK, 8DPSK	GFSK
Data Rate	EDR PHY (8DPSK): 3 Mb/s EDR PHY (π/4 DQPSK): 2 Mb/s BR PHY (GFSK): 1 Mb/s	LE 2M PHY: 2 Mb/s LE 1M PHY: 1 Mb/s LE Coded PHY (S=2): 500 Kb/s LE Coded PHY (S=8): 125 Kb/s
Network Topologies	Point-to-Point (including piconet)	Point-to-Point (including piconet) Broadcast or Mesh
Range	100 m (330 ft)	<100 m (<330 ft)
Data Rate	1–3 Mbps	125 Kbps – 1Mbps – 2Mbps
Application throughput	0.7–2.1 Mbps	0.27-1.37 Mbps
Active Slaves	7	Implementation dependent
Security	56/128-bit and application layer user-defined	128-bit AES in CCM mode and application layer user-defined
Robustness	Adaptive fast frequency hopping, FEC, fast ACK	AFH, Lazy Ack., 24-bit CRC, 32-bit Message Integrity Check
Latency	Typically, 100 ms	6 ms
Time to send data	0.625 ms	3 ms
Voice Capable	Yes	No
Power Consumption	1 (reference value)	~0.01x to 0.5x of reference
Peak Current	<30 mA	<15 Ma
Service Discovery	Yes	Yes
Profile Concept	Yes	Yes
Primary use cases	Mobile phones, gaming, headsets, audio streaming, smart homes, wearables, automotive, PCs, security, proximity, healthcare, etc.	Mobile phones, gaming, smart homes, wearables, automotive, PCs, security, proximity, healthcare, sports, industrial, etc.

Bluetooth Classic v/s BLE

Bluetooth core specification includes two technologies; these are Bluetooth Classic and Bluetooth Smart (BLE). Table 1 presents the differences between Bluetooth Classic and Bluetooth Low Energy with various parameters. The difference between these two beacon technologies is in power consumption.

BLE Versions

Among all the available wireless technologies, the BLE has proved as a most promising short-range wireless communication technology because of appearing everywhere or very common technology. To meet the increasing demand for wireless connectivity in IoT, Bluetooth SIG continuously publishing upgraded versions of BLE. Here, an analysis of Bluetooth 4.0, 4.2, 5, 5.1, and 5.2, based on the data speed, network bandwidth, communication range, power requirement, message capacity, Bluetooth beacon, support for IoT devices, compatibility, reliability, errors detection, and correction capability, advertisement packets, coexistence with Wi-Fi devices, duty cycle, slot availability masks, periodic advertising synchronous transfer capability and many more.

BLUETOOTH 4.0

Bluetooth SIG (2016) defined to achieve high efficiency in wireless communication; Bluetooth 4.0 has been introduced with the name Bluetooth Low Energy. It is suitable for many IoT applications that only need the periodic transmission of small size data because it returns to a lower data throughput of 1 Mbps using the GFSK modulation scheme. For example, a smart wristwatch that has to send small amounts of Blood-Pressure value to the user smartphone only when requested. In applications that require a continuous data stream like wireless headphones, lower data throughput (1 Mbps) may not be suitable.

BLE is a wireless personal area network based on Bluetooth 4.0 framework, designed and marketed by the Bluetooth-SIG for effective communication between the devices in the agriculture, aerospace, home appliances, healthcare, robotics, fitness, automobiles, beacons, security, entertainment industries, etc. It has been designed to improve security and energy efficiency in Classic-Bluetooth. Due to the low power requirement, it allows small devices to operate on tiny batteries. BLE protocol is a radical departure from Classic-Bluetooth protocol. The Bluetooth BR/EDR is still the best for voice and music streaming but, BLE is better for wireless sensor and control applications. BLE version 4 performs error detection while fails to perform error correction.

BLE works on the 2.4 GHz ISM band and a frequency hopping technique. In a departure from Bluetooth BR/EDR instead of 79 1MHz-wide channels, it uses 40 2MHz-wide channels. The sets of channels are numbered from 0 to 39; therefore, these versions are fundamentally incompatible. A device that can support both Bluetooth BR/EDR and BLE is called "Bluetooth Smart Ready". Most handsets, tablets, and other devices belong to this.

Bluetooth 4 has advertising packets of 37 octets with 6 octet header and at most 31 octets payload. Advertising packets are transmitted up to 2MHz wide with three dedicated channels. All the channels carry equal payload i.e., at a time one packet. For Point-to-Point (1:1), BLE optimized for short burst data transmission, the required setup time is <6ms and allows for unlimited connections. It has a 125 Kbps to 2 Mbps data rate, 251-bytes max payload size, and security provision of 128-bit AES at the user-defined application layer. For Broadcast (1:m) BLE has max payload size: 31 bytes primary, 255 bytes secondary channel, and concept of chaining of packets for larger messages.

BLUETOOTH 5

Bluetooth SIG (2016) published a document of Bluetooth 5 with updates to significantly increase the operating range, data transmission speed, and message broadcasting capability of applications like smart home, smart parking, e-governance, enterprise, industrial markets, etc. Bluetooth 5 shows a step-change in Bluetooth technology like area coverage is provided for with the new long-range LE Coded PHY, the higher symbol rate of LE 2M helps to improve spectral efficiency and supports in many applications and advertising extensions makes a provision for next-generation beacons. Bluetooth 5 makes a real impact in IoT due to low power consumption.

A Choice of Three PHYs (2x speed and 4x Range)

Bluetooth is a full protocol stack. The bottom layer of the stack is a Physical Layer (PHY). Bluetooth 5 is with two new PHY variants, these three PHYs are LE 1M, LE 2M, and LE Coded. Designed of every variant is specific to achieve particular characteristics.

Bluetooth 5 consist of LE 1M for use with its support. The new LE 2M PHY makes the physical layer capable to operates at 2 mega symbols per second (Ms/s) i.e., at higher data rates than LE 1M and Bluetooth 4. The LE 2M PHY is characterized by using double the symbol rate to double that of the Bluetooth 4 PHY. The LE Coded PHY achieved a range four times than the Bluetooth 4 without affecting the transmission power. The range improvement without affecting transmitter power was a problem concerned with achieving the same maximum permitted Bit-Error-Rate (BER) at a greater distance from the transmitter i.e., at a lower Signal-to-Noise Ratio (SNR).

Dealing with Errors

It facilitates detecting errors at the receiver and also correcting them to avoid the data being re-transmitted. Error correction can be performed by using a new method to achieve correct data decode at a lower SNR i.e., at a greater distance from the transmitter. On this basis, Bluetooth 5's range has been increased. The LE Coded PHY uses Forward Error Correction (FEC) to correct errors. It additionally adds redundant bits with the transmitted packets to support the application of the FEC algorithm and to determine the correct value that erroneous bits should have.

Advertising Extensions

In Bluetooth 5 protocol stack for advertising, scanning and connecting purpose eight new PDUs added to the GAP. These new PDUs make it possible for broadcasting large data and advertising is possible to perform in different ways by multiple sets of data broadcasting. Due to this, improvement has been achieved for the contention and duty cycle. Beacons are major applications for advertising. Bluetooth 5 protocol stack is most suitable and provides a basis for the development of a new beacon application.

Instead of only ID or URL broadcasting, a newly developed beacon will be capable to allow much richer and multi-faceted sets of contextual data.

Bluetooth 5 can broadcast packets up to 255 bytes long.

This can be accomplished in section by offloading the payload to one of the other channels in the 0-36 channels, previously only used for connection events. Header data with a new field called AuxPtr has been transmitted on channel numbers 37, 38 and 39 known as the Primary Channels. The AuxPtr field refers to the advertisement packets to be transmitted on the secondary channels. These all the channels have unique channel numbers so that receivers can find them easily. The larger amount of data can be broadcast by making a chain of the packets and for each packet to contain a different subset of the whole data set. AuxPtr header field referencing the next in a chain hence each chained packet can be transmitted on a different channel.

Contention and Duty Cycle

In Bluetooth 5 radio channels are now used with primary advertising channels (11, 12, and 13) carries less data while secondary channels (0-36) carrier large data. With advertising data using all available channels, and only small headers using the primary channels, there will be less contention on those channels. Bluetooth 4 can transmit the same payload up to three times on three different channels but now Bluetooth 5 can transmit data only once with a small header to referencing it from the primary channels.

The total amount of data transmitted is thus less and so the duty cycle has been reduced. For non-connectable advertising, the minimum advertising interval is reduced from 100ms to 20ms. Due to this lower interval, rapid recognition of and response to advertising packets from devices like beacons is possible.

Slot Availability Masks

Bluetooth 5 has major changes that help to improve coexistence with other wireless devices such as smartphones. Bluetooth uses the 2.4GHz ISM band which is immediately adjacent to the Mobile Wireless Standard (MWS) bands. It has scope for interference between the two systems with transmissions from one and the receiving on the other. It is with a "Slot Availability Masks" to show the time slots availability status and to synchronize in the best way by the use of an adjacent MWS band.

Improved Frequency Hopping

Bluetooth in a connection uses AFH. It is an algorithm used to determine the radio channel for transmission and receiving also, to change the frequency of selected channels such that, data is transmitted over a wide selection of channels. It helps Bluetooth to perform well in busy radio environments. Bluetooth 5 included a new channel selection algorithm called "Channel Selection Algorithm #2". It allows for the use of a shared event counter, it guaranteed that the entire peer in the connection selects the same channel from the next available in a pseudo-random sequence.

BLUETOOTH 5.1

Bluetooth Core Specification v5.1 contains a series of updates to the Bluetooth core specification. Bluetooth SIG (2019) published new direction-finding features to find the Bluetooth signal transmission direction by Bluetooth devices. The direction-finding includes two methods for finding the angle that a Bluetooth signal transmitted from with a high degree of accuracy. These methods are i) Angle of Arrival (AoA) and ii) Angle of Departure (AoD). Every method requires any one of the two communicating devices for an array of multiple antennae, with the antenna array included in the receiving device when the AoA method is used and in the transmitting device when using AoD.

Bluetooth 5.1 makes the BLE controller in the receiving device capable to generate data that can be used to calculate the directional angle to the transmitting device. At the start of Bluetooth, the roadmap is to ultimately enable key enhancements to Bluetooth location services. At the release of associated profiles, developers will be able to develop the new direction-finding controller capability to create high accuracy and interoperable positioning systems such as real-time locating systems and indoor positioning systems. These new direction-finding systems also enhance the proximity by finding device direction, particularly in directional item finding and point of information interest.

GATT Caching Enhancements

All BLE devices used the GATT. GATT devices maintain a database known as 'attribute table', which have GATT service, characteristic, descriptor structural details, values, and are central to how GATT-based BLE devices work. Entries in the attribute table are identified by the attribute handles. The GATT clients must perform a "Service Discovery" procedure to obtain the detail of the attribute table on the GATT server to which the clients have to connect. This detail can be used with the identifying attribute to handles the interaction with a server in subsequent Attribute Protocol (ATT). Service discovery takes time and consumes more energy, therefore Bluetooth 5.1 has a provision to skip service discovery when nothing has changed.

Improved Caching Strategy

This new version makes changes to the attribute caching and cache synchronization approached by GATT clients and servers. It allows client without a trusted relationship with a server so that it retains their attribute cache across connections and resolves the race condition issue. A good user experience and energy efficiency improvements noted due to 'Database Hash' and 'Client Support' features.

Better State Management

State machine defines the status of the synchronization of the client view of the attribute table and the server view of its attribute table. From this status, it will clear that service discovery by a client is required or not. The new 'attribute caching' specification defines a strong caching mechanism that formalizes state machine.

Randomized Advertising Channel Indexing

In the advertising state, it is not required to select advertising channels (strict and fixed sequence) because it starts at the lowest used channel index and ends at the highest. In this, it is allowed for random selection of channel indices. The random selection of advertising channel indices further reduces the chances of advertising packet collisions. Applications with making this change to channel index selection achieved improved scalability and reliability in busy radio environments.

Periodic Advertising Sync Transfer

The new Periodic Advertising Sync Transfer (PAST) feature makes it possible for another less constrained device to synchronize and then passes the acquired synchronization details over a point-to-point BLE connection to the other constrained device. For example, a smartphone can scan packets from a TV and then pass them to an associated smart-watch. The watch can use periodic advertising and scanning to get data from TV.

BLUETOOTH 5.2

Bluetooth Core Specification v5.2 also contains a series of updates to the Bluetooth core specification. Bluetooth SIG at CES 2020 introduced Bluetooth 5.2 with the announcement of next-generation Bluetooth audio, named: LE Audio. The major changes are as follows:

Enhanced Attribute Protocol (EATT)

The Enhanced Attribute protocol (EATT) has been introduced along with some associated improvements to the GATT. It supports two transactions simultaneously like, a process for interleaving of L2CAP packets relating to ATT packets from different applications and a process to change the ATT Maximum Transmission Unit (MTU) during a connection. These changes improve user experience on devices where multiple applications using the BLE stack at the same time by reducing instances where one application's use of the stack temporarily blocks that of another. This protocol helps to reduce the end-to-end latency of applications and to improve the user's experience of responsiveness.

In support of EATT, a new L2CAP Enhanced Credit-Based Flow Control Mode has been defined to provide flow control to allow applications to regard the protocol as reliable. EATT has security advantages as it may only be used over an encrypted connection.

LE Power Control

It helps to dynamically optimize the transmission power used in communication between paired devices. To maintain optimal signal strength based on signal quality and low-power-use perspective BLE receiver has to monitor signal strength and request transmission power- level changes in connected devices. Bluetooth controller regularly monitors and using the concept of zones reports to the host about path loss changes.

Following are the benefits of LE Power Control:

- Due to dynamic power management overall power consumption is reduced.
- Due to active maintenance of receiver signal strength improvement in reliability is noted.
- Improvement in the frequency as compare to same frequency devices is noted.

LE Isochronous Channels

The new topologies are made possible by LE Isochronous Channels. A transmitter can share data with the synchronized device through small private groups of devices or multiple devices. The multiple devices may be of unlimited sizes in public spaces such as cinemas, stadiums, etc. audio sharing is a popular application, therefore LE Audio has been built on top of new LE Isochronous Channels. It offers a new standard for hearing aids and support assisted hearing systems in diverse locations like theatres, conference halls, lecture halls or airports. It is expected that multi-language audio systems will become possible. For the LE Isochronous Channels support, various changes have been performed by the developer like,

- Addition of isochronous adaptation layer to the controller,
- Modification in the data transport architecture to support connection-oriented and connectionless isochronous communication,
- Provided new LE security mode 3 based around the use of a shared broadcast code which allows encryption to be used in broadcast isochronous groups,
- Procedures for setting up isochronous communication in connections or via broadcasting,
- New commands and events at the HCI layer so that isochronous communication to be configured and used, and

New PDUs (connected isochronous PDU and broadcast isochronous PDU) were introduced at the link layer.

SUMMARY AND PERFORMANCE ANALYSIS OF BLE VERSIONS

Bluetooth SIG (2020) released a Bluetooth 5 achieved double the speed (throughput up to 1.4Mbps), lower power consumption (15-50%), four times improvement in range, new channel selection algorithm enables +20dBm TX, eight times advertisement capacity, advertisement payload growth from 31B to 255B, 37 new advertisement channels to help offload 3 primary channels, new advertisement schemes for advanced beacons and periodic advertisement.

Bluetooth 5.1 achieved direction finding (AoA and AoD), faster and lower power connections (i.e., GATT Caching), reduced interference for busy RF environments, periodic advertising synchronous transfer (transfer between devices) and other minor enhancements. Bluetooth 5.2 achieved LE isochronous channels for audio enablement over BLE, high data throughput, audio broadcasting to the multiple devices, time-bound data distribution multiple devices, enhanced attribute protocol to make concurrent ATT transactions for reducing overall latency, more reliable connections, lower power and better coexistence.

The BLE up-gradation is a continuing process Bluetooth SIG continuously working to enhance the capability of BLE. Here, the various versions of Bluetooth with the following parameters are compared.

Speed: Bluetooth 4 is slow; supports up to 1 Mbps while Bluetooth 5/5.1/5.2 is fast supports up to 2Mbps (twice the speed of Bluetooth 4).

Bandwidth: Bluetooth 4 bandwidth is with 2.1 Mbps while Bluetooth 5/5.1/5.2 bandwidth is with 5 Mbps.

Range: Bluetooth 4 covers a limited range (50m in outdoor and 10m in indoor) while Bluetooth 5 covers enhanced range (200m in outdoor and 40m in indoor).

Power requirement: Bluetooth 4 requires high power, the Bluetooth 5 is formulated to use less power, Bluetooth 5.1 facilitates to skip service discovery when nothing has changed so, time and energy consumption for service discovery has reduced while Bluetooth 5.2 facilitates dynamic power management conducted between connected devices so, reduction of overall power consumption by transmitters obtained.

Message capacity: Bluetooth 4 is for a small message of 31 bytes (17 to 20 bytes for actual data payload) while Bluetooth 5/ 5.1/ 5.2 are for a large message of 255 bytes to gives more bytes for actual data payload.

Bluetooth beacon: Bluetooth 4 beacons are with less speed and range and also low message capacity of 31 bytes, therefore less popular, while Bluetooth 5 beacons are with high speed and large range and also large message capacity of 255 bytes, therefore more popular.

Support for IoT devices: Bluetooth 4 does not support IoT devices due to its low speed and short working range, while Bluetooth 5 support improved range and speed.

Compatibility: Bluetooth 4 works best with only the version 4 series, but fails to work with Bluetooth 5/5.1/5.2 while Bluetooth 5 is backwards compatible with v4, v4.1, and v4.2.

Reliability: Bluetooth 4 is not reliable for communication but Bluetooth 5.2 achieved improvements in reliability by involving the Bluetooth controller for monitoring and reporting path loss changes to the host using the concept of zones.

Dealing with Errors: Bluetooth 4 only facilitates detecting errors at the receiver, while Bluetooth 5/5.1/5.2 facilitates detecting errors also correcting them so that the receiver does not need to perform data retransmitted. The LE Coded PHY uses FEC to correct errors.

Advertising Packets: Bluetooth 4 has advertising packets of 37 octets long with a 6-octet header and a payload of, at most, 31 octets while Bluetooth 5/5.1/5.2 has eight new PDUs for advertising, scanning, and connecting. It is capable to broadcast larger amounts of data, advertising to be performed in a deterministic fashion, and multiple distinct sets.

Coexistence with Wi-Fi devices (Robustness): Bluetooth 5 has improved wireless coexistence and interoperability, information that can easily be misunderstood as enhanced coexistence with a Wi-Fi device. The real improvement is with other nearby BLE devices the random frequency-hopping decreases the chance that a neighbouring device transmitting on the same RF channel at the same time, thus increasing the overall robustness of each connection.

Duty Cycle: Bluetooth 4 can transmit payload up to three times on 3 different channels while Bluetooth 5 can transmit such data only once, with small headers referencing it from the primary channels.

Slot Availability Masks: Bluetooth 5 help to improve coexistence with other wireless devices such as smartphones. It is with a "Slot Availability Masks" to show the time slots availability status and to synchronize in the best way by the use of an adjacent MWS band.

Periodic Advertising Synchronous Transfer: Bluetooth 5.1 includes a new feature that makes it possible for another less constrained device to synchronize and then pass the acquired synchronization details over a point-to-point BLE connection to the other constrained device.

IoT Applications with Bluetooth

Bluetooth Smart or BLE is the intelligent, low energy version of Bluetooth wireless technology. BLE technology is turning heads and finding more application areas in recent times.

BLE in Medical/ Health Care

BLE make health care monitoring in the medical field by sharing data sensed by the medical sensors in a network for various physiological parameters. The shared data can be utilized for the proper treatment. Several applications are possible for monitoring, blood glucose, blood pressure, heart rate, pulse, body temperature, respiration rate, fitness level, exercise capacity, etc.

BLE in Retail

BLE makes it possible for consumers and retailers to communicate with each other quickly and easily. Beacons track the route of people inside a shop, gather customer data, and help to offer a personalized experience. Media such as ads, coupons, and additional product information can be pushed through beacons.

BLE for Location Tracking

Application may be for passengers at the airport to find personalized directions within the airport, show them the way to restaurants, services, and baggage carousels. It may be at the museum to help the visitors for leading themselves around the exhibits and get additional information as they walk around the exhibits on their phones.

BLE for Communicators

Applications to takes call by doing the flip action and serve as wireless speakers when linked to music streaming devices.

BLE in Agriculture

Beacons can be used to collect and analyze data such as environmental and sub-tidal water temperature, chlorophyll values, and more. Combined with satellite imaging data and properly analyzed, this could bring about efficient and sustainable food production.

Smart Tags

Smart tags are a type of Bluetooth LE gadget that shows the technology's potential. The most interesting example of smart tags is Tile. Users can attach this small tag to anything and it'll communicate with the phone over Bluetooth LE, allowing the phone to keep track of its location.

Home Automation Systems

Applications for smart-controlling of home appliances like smart-stoves, smart ovens, smart locks, smart-water tanks, smart-lighting, etc. to control and operate Bluetooth smart ready gadgets that can receive and transmit Bluetooth signals.

Military Applications

Military applications are like armed command, control, communications, computing, intelligence, surveillance, reconnaissance & targeting systems, tracking & environment monitoring surveillance systems, enemy tracking, security detections, and so on.

FUTURE RESEARCH DIRECTIONS

Here, a comparison of Bluetooth 4.0, 4.2, 5, 5.1, and 5.2 on various parameters has been discussed. This performance-based comparison discussed all the requirements, working style, capacity, merits and limitations for the Bluetooth versions. It will help the system developer to select the correct Bluetooth version to fulfill the need. In future, this comparison may be continuing based on the practical results.

CONCLUSION

In today's world, everyone follows the new trends; in previous literature, it has proved that BLE is the most promising short-range wireless communication technology. It is suitable for IoT applications which only need to transmit bits of data periodically. It has a max data throughput of 1Mbps, therefore, may not be useful for applications that require continuous data streaming like the audio transmission.

The next version Bluetooth 5 has an improvement upon the previous BLE standards. It is geared towards low powered applications with BLE's data rate and range improvement. Bluetooth 5 provides 2x speeds, 15-50% lower power consumption, 4x ranges, a new channel selection algorithm, 8x advertisement capacity with a new scheme for advance beacon and periodic advertisement. It performs continuous signals scanning and search for a new device to connect when Bluetooth is enabled hence slowly drains the battery. The Bluetooth 5.1 provides an improved direction-finding method with AoA and AoD which provides <1m accuracy, faster and lower power connections, reduced interference for busy RF environments, periodic advertising synchronous transfer, and other minor enhancements, and the latest Bluetooth 5.2 provides LE isochronous channels for audio enablement over BLE, more reliable connections, lower power, and better coexistence. From the above comparison, it has concluded that Bluetooth 5/5.1/5.2 are much better than the Bluetooth 4.0/4.2 with high data speed, lower power consumption, large communication range, enhanced advertisement capacity, and many more advanced methods to improve a BLE standard and make them capable to use for communication between various smart home and IoT.

It is concluded, that all the versions have their own set of advantages and limitations. For the basic IoT applications (only with limited functionality) to broadcast small messages (UID, URL, EID, and TLM) Bluetooth 4.0/4.2 are a good choice due to its simplicity and small advertisement capacity, while in complex and advanced IoT applications latest Bluetooth 5/5.1/5.2 are a better choice due to high data speed, lower power consumption, large communication range, enhanced advertisement capacity, and many more advanced methodologies.

REFERENCES

Aguilar, S., Vidal, R., & Gomez, C. (2017). Opportunistic Sensor Data Collection with Bluetooth Low Energy. *MDPI Sensors*, *17*(12), 159. doi:10.339017010159 PMID:28124987

Cognizant Digital Business. (2019). *The Five Essential IoT Requirements and How to Achieve Them.* Whitepaper.

Davidson, R., Townsend, K., Cufí, C., & Akiba. (2014). *Getting started with Bluetooth Low Energy - Tool and techniques for low-power networking*. Academic Press.

Estrin, D., Govindan R., Heidemann, J., & Kumar, S. (1999). Next Century Challenges. *MobiCom '99 Seattle Washington USA,* 263-270.

Goosen. (2014). *Design and Implementation of a Bluetooth 4.0 LE Infrastructure for Mobile Devices* (Thesis). ULM University, California.

Nakamura, M., Nakamura, J., Lopez, G., Shuzo, M., & Yamada, I. (2011). Collaborative processing of wearable and ambient sensor system for blood pressure monitoring. *Sensors (Basel)*, *11*(7), 6760–6770. doi:10.3390110706760 PMID:22163984

Padiya, S., & Gulhane, V., (2020). Analysis of Data Aggregation Methods to avoid Data Redundancy in Wireless Sensor Network. *12th IEEE CICN 2020,* 25-26.

Research, A. B. I. (2016). *Bluetooth Smart Evolution Helps the Technology Break into Key IoT Market Verticals. PRNewswire*.

Sam Lucero IHS Technology. (2016). *IoT Platforms: Enabling the Internet of Things.* Whitepaper.

She, J., Soonsawad, P., & Ng, P. (2018). BLE Beacons for the Internet of Things Applications: Survey, Challenges, and Opportunities. *IEEE IoT Journal.*

Woolley. (2019). *Bluetooth SIG Bluetooth Core Specification v5.1.* Bluetooth SIG.

Woolley. (2020). *Bluetooth SIG Bluetooth Core Specification v5.2.* Bluetooth SIG.

Woolley, M. (2016). *Bluetooth 5 /Go Faster. Go Further*. Bluetooth SIG.

Yin, Z., & Gunnarsson, G. (2017). Received-Signal-Strength threshold optimization using Gaussian processes. *IEEE Transactions on Signal Processing*, *65*(8), 2164–2177. doi:10.1109/TSP.2017.2655480

Younis, O., & Fahmy, S. (2004). HEED: A Hybrid, Energy-Efficient, Distributed Clustering Approach for Ad Hoc Sensor Networks. *IEEE Transactions on Mobile Computing*, *3*(4), 366–379. doi:10.1109/TMC.2004.41

KEY TERMS AND DEFINITIONS

Actuators: It is a device such as controllers, lasers, robotic arms, etc. that translates an electrical signal from the IoT system into physical actions.

Beacon: It is an intentionally noticeable object to attract attention to a specific location.

BLE Beacon: It is similar to the lighthouse; the beacon is a small (coin size) wireless device that transmits signals to other nearby smart devices by using low-energy Bluetooth technology.

Bluetooth Advertising: It is a method of data transmission through Bluetooth.

Radio Frequency Identification (RFID): It is a digital identification in which the RFID tag (a small chip with an antenna) broadcast a unique value, which can be read by the RFID reader.

Sensors: It is a device such as probes, gauges, meters, etc. those sense the physical parameters like temperature, humidity, object presence, light, etc. from an environment and convert it into an equivalent electrical signal.

Chapter 11
Evalutation of Turbo Decoder Performance Through Software Reference Model

Manjunatha K. N.
Jain University (Deemed), India

Raghu N.
https://orcid.org/0000-0002-2091-8922
Jain University (Deemed), India

Kiran B.
Jain University (Deemed), India

ABSTRACT

Turbo encoder and decoder are two important blocks of long-term evolution (LTE) systems, as they address the data encoding and decoding in a communication system. In recent years, the wireless communication has advanced to suit the user needs. The power optimization can be achieved by proposing early termination of decoding iteration where the number of iterations is made adjustable which stops the decoding as it finishes the process. Clock gating technique is used at the RTL level to avoid the unnecessary clock given to sequential circuits; here clock supplies are a major source of power dissipation. The performance of a system is affected due to the numbers of parameters, including channel noise, type of decoding and encoding techniques, type of interleaver, number of iterations, and frame length on the Matlab Simulink platform. A software reference model for turbo encoder and decoder are modeled using MATLAB Simulink. Performance of the proposed model is estimated and analyzed on various parameters like frame length, number of iterations, and channel noise.

DOI: 10.4018/978-1-7998-6988-7.ch011

INTRODUCTION

The turbo codes were first presented by Berrou, Glavieux, and Thitimajashima in 1993 (Manjunatha & Vaibhav, 2017), (Biatek et al., 2016). It has become most popular in nowadays because of the turbo code approaches Shannon limit performance. The turbo codes are efficient in error correcting capabilities because of this characteristic these codes are used in some standard mobile communication like LTE and 802.16e (WiMax). Nowadays there is a demand for high computing applications leads to increase in the research of multi standard turbo decoder with high throughput.

The communication system has transmitter and receiver along with a channel. The transmitter has Turbo encoder with modulator and the receiver consists of Turbo decoder with demodulator. The turbo encoder consists of two encoders in which one is connected across the interleaver and the other one is encoded output with interleaver. Similarly the turbo decoder consists of two SISO decoders, in which the first decoder will receive the noise added data rsys and other decoder takes the parity bits rpar. The existing work shows that the calculation of soft values using soft-output vertebri algorithm and maximum a posteriori (MAP) algorithm (Biatek et al., 2016). The literature work on the Turbo coder shows a better performance through the MAP algorithm. The goal of this design into support all sign of block and any degree parallelism. This existing work summary also reveals an algorithm of meaning remapping to eliminate the contention free memory when the parallelism is not the power of two or more.

TURBO CODES

There is a trade-off between energy and bandwidth efficiency (Saito, 2016) while designing a channel code. The codes with bigger redundancy or lower rate can usually rectify maximum number of errors. The communication system is expected to communicate long distance with minimum transmit power. It uses the smaller antennas to withstand more interference to communicate at higher data rates. This leads to correct more number of errors. The code energy becomes more efficient from these properties. The low rate codes consume maximum bandwidth with high overheads. The high computational requirements develop the decoding complexities as the code length increases. In channel coding always process of encoding is easy and decoding in difficult.

For every received noise power (N), bandwidth (W), channel type and signal power (S), the theoretical upper bound on the data rate is represented by R and only at this limit the transmitted data is error free.This limit is refers to Shannon capacity or channel capacity.

The mathematical model for the additive white Gaussian noise (AWGN) channels is

$$R < Wlog_2\left(1+\frac{S}{N}\right) \tag{1}$$

An ideal error-free channel is very difficult to find in real time. The transmission of error-free data is estimated through probability of bit error to an arbitrary small constant. The Bit Error Rate (BER) or bit error probability (Vargas et al., 2015) used as standard of measure and is often chosen to be 10-5 or 10-6. Based on the Convolutional encoding, first turbo code was presented in 1993 by Berrou et al. To cover the block codes and convolutional (Shrestha & Paily, 2014) codes later the term "turbo codes" has been coined. Turbo codes are simply parallel concatenation of two codes isolated by an interleaver.

The below fig 1. explains the structure of a turbo encoder.

Figure 1. General representation of turbo encoder

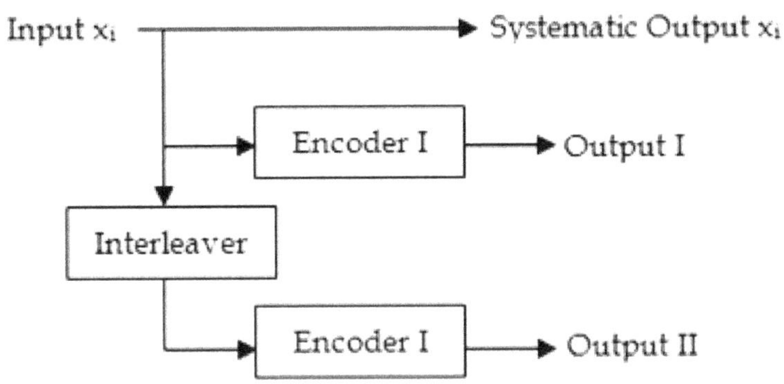

The turbo encoder design concept follows the choice of the encoders and interleaver (Kim & Kim, 2013), and most of the designs obey the ideas represented below:

- Normally two identical encoders are used.
- The input bits are also appeared across the output also called as systematic form.
- The pseudo-random order of the bit is read through interleaver.

The selection of interleaver in turbo encoder is a critical stage. The primary objective of the interleaver is to scramble the data bits in a pseudo-random order. The first encoder produces the output without scrambling of any data but the second encoder output is scrambled through the interleaver used. This causes the even if the output of the one encoder is low code weight also other does not show the same. For the benefit of the performance of the decoder higher weight code words are used.

To decode the turbo codes most of the algorithms uses the divide-and-conquer method, because these codes are parallel concatenated codes. In turbo decoder, the first decoder input is given by the output of the first encoder which is not scrambled where in the second decoder input is scrambled and this makes uncorrelated between two decoders. The two decoders will receive at maximum level by information exchange. The Convolution encoders with parallel concatenation are referred as pseudo-random technique.

A parallel concatenation of convolutional encoders via pseudo-random interleaver for turbo coding the informationbits, which need to be transmitted. It generates sequences of systematic bits as well as non-interleaved and interleaved parity bits.

On the other side, Fig. 1.2 shows a basic block diagram of turbo decoder which is an integration of constituent Soft-In Soft-Out (SISO) decoders with pseudo-random interleaver and de-interleaver. The soft-demodulated values of transmitted bits are referred as a-priori probability values and are fed to constituent SISO decoders, as shown in Fig.1.2 Such SISO (Calabuig et al., 2015) decoders are fundamentally based on BCJR(BahlCockeJelinekRaviv) algorithm that works on the principle of trellis graph and it processes a-priori probabilities of systematic and parity bits to produce a-posteriori probability values of the transmitted information bits. Thereafter, the extrinsic information is computed using a-posteriori probability values from the SISO (Calabuig et al., 2015) decoder, interleaved/non-interleaved a-priori probability values and interleaved/de-interleaved extrinsic information from another SISO decoder. Such extrinsic information values are taken between these SISO decoders and are iteratively processed along with a-priori probability values to produce error-free a-posteriori probabilities of the transmitted bits.

Figure 2. A typical Conventional turbo decoder

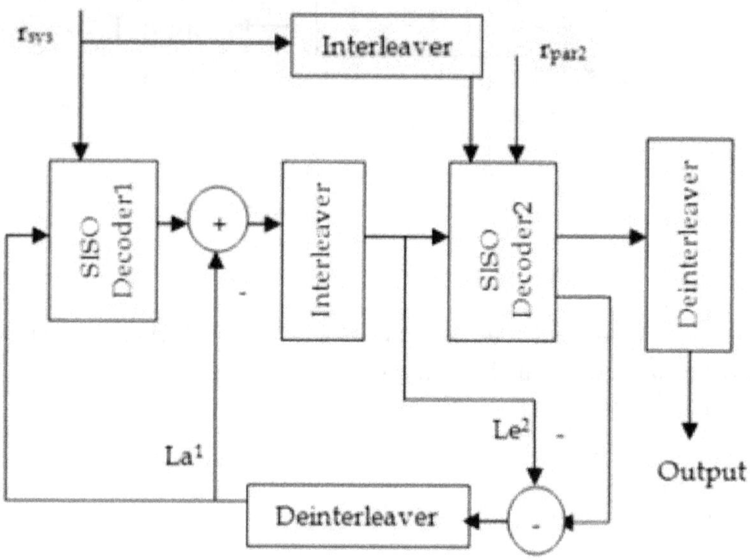

From fig. 2 The Soft In Soft Out decoders receive the two signals namely r_{sys} (noise added) and r_{par1} and r_{par2} (parity bits) from the AWGN channel. Based on the turbo decoding algorithm the two decoders repetitively exchange the extrinsic values (Le) with respect to interleaver or deinterleaver.

TURBO DECODING ALGORITHMS

Some of the important decoding algorithms are developed to decode (Dai et al., 2012)-(Reddy et al., 2010) the turbo codes are:

- Soft output Viterbi algorithm (SOVA),
- Maximum A Posteriori Probability (MAP),
- Log-MAP,
- Max-Log-MAP and

The SOVA is as Maximum Likelihood algorithm at medium and high SNR where ML is a MAP algorithm or maximum likelihood (ML) (Zhang & Li, 2011), (Li et al., 2013). The MAP algorithm and SOVA computations are comparable. The MAP algorithm computes the bit it has sent and SOVA computes the code sequence it has sent. Due to the way of computation in MAP algorithm minimizes the BER complexities, the same way SOVA minimizes the probability of word error. For MAP algorithm it is not required to return the data bits through the connected path through the Trellies but the SOVA returns through the connected track.

PROPOSED LOW-POWER TURBO DECODER

MAP algorithm is quite difficult to implement because of its design complexities. The major difference between the MAP and Max-Log-MAP algorithms is the computations of the values are in log domain and MAP algorithm will mainly perform the approximation. The approximated values are easy to implement. Approximations of MAP algorithm are reduced or avoided in Max-Log-MAP by applying

Figure 3. Log-MAP Decoder Computation Flow

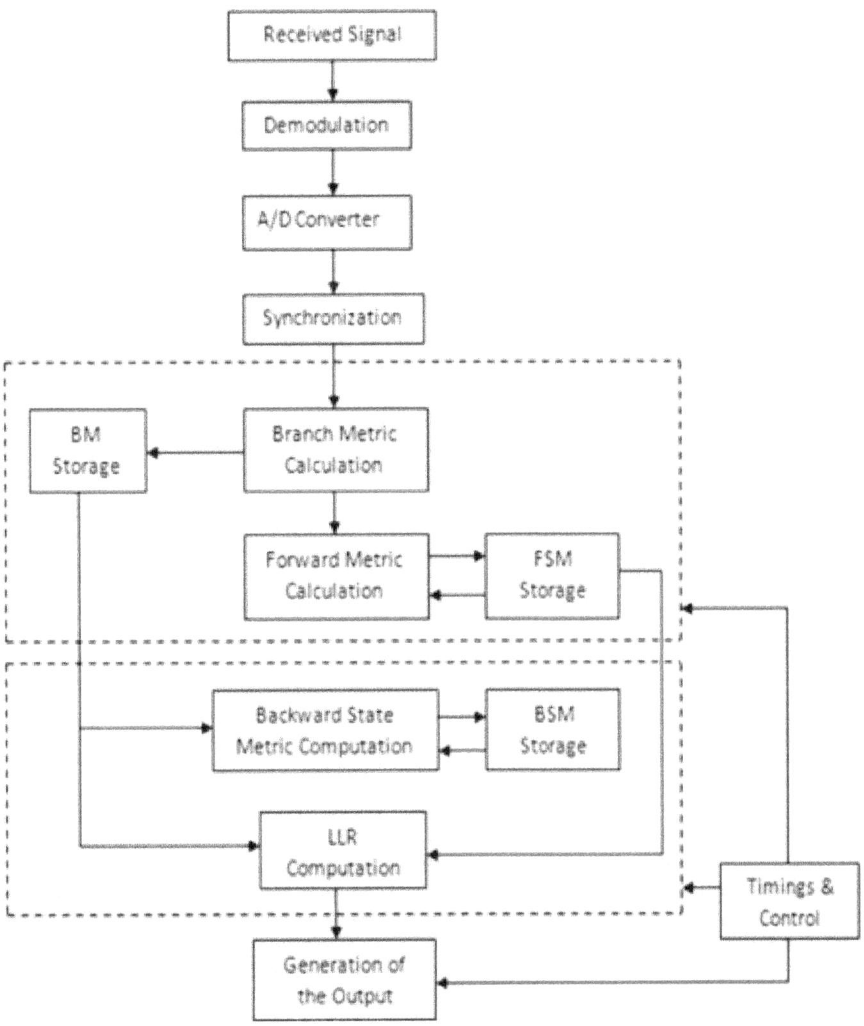

the small correction in each steps to maximize the operation. This small modification leads to effective computations near to the MAP algorithm.

Turbo codes are most popular and are generally used in parallel concatenated to iterative decoding techniques. The standard structure of turbo decoder has interleaver, deinterleaver and two decoders. Branch Metric Unit (BMU) to calculate, Forward State Metric Unit (FSMU) to verify, Backward State Metric Unit (BSMU) to correlate Log Likelihood Ratio (LLR) for evaluation: are the four major segments in Log-MAP decoder. Calculation steps of Log-MAP decoder are illustrated as in Fig 3.

The process of quantization is used to scale-up the received analog signal. The calculation is carried out in branch metric unit, this stage also has forward state metric unit along with the BMU. The MAP decoder expressions are derived from the standard decoding structure shown in Fig.4

Figure 4. Standard Decoding Structure of Turbo Codes

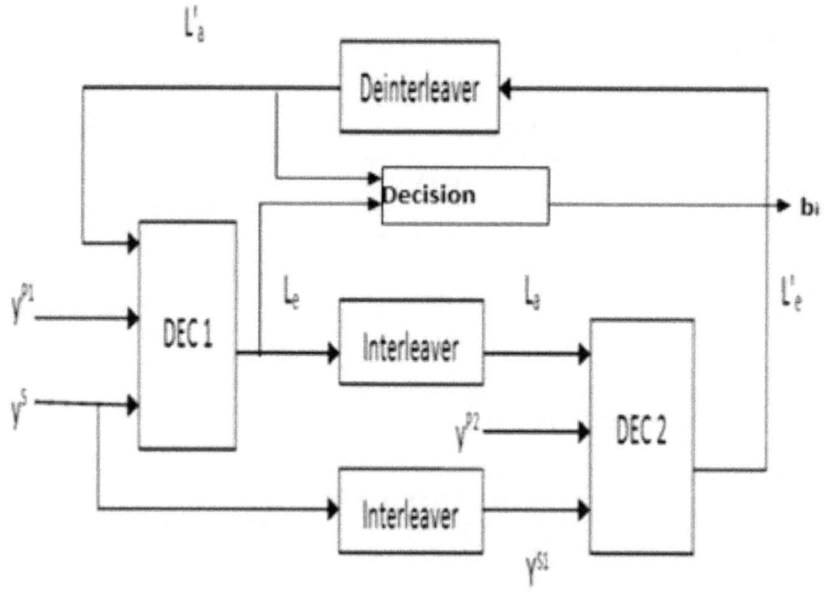

The stored branch metric values are computed throughThe Log likelihood Ratio (LLR) of log-MAP decoders with in-formation bit u_k can be written as

$$L(u_k) = \log \frac{P(u_k = 1 / y_k)}{P(u_k = 0 / y_k)} \tag{2}$$

Where $y_k = (y^s_k, y^p_k)$, assuming that the design uses the AWGN channel (Manjunatha & Vaibhav, 2017) for transmission of information and BPSK modulation is considered. The resultant expression for the operation becomes

$$L(u_k) = \log \frac{\sum_{(s',s)u_{k=1}} P(S_{k-1} = S', S_k = S, y_k) / P(y_k)}{\sum_{(S',S),u_{k=0}} P(S_{k-1} = S', S_k = S, y_k) / P(y_k)} \tag{3}$$

From BCJR (Bahl-Cocke-Jelinek-Raviv) algorithm (Reddy et al., 2010),

$$P(S', S, y_k) = \alpha_{k-1}(S') * \gamma_k(S', S) * \beta_k(S) \tag{4}$$

Forward recursive probable value represented as α_k(s), backward recursive functions are termed as β_k (s) and γ_k(s', s) is used to identify the notations probable values.The process of normalization is used to stop the overflow occurrence in computation and α_k(s) and β_k(s) values are computed through recursion.

$$\alpha' k(S) = \frac{\sum_{S'} \alpha'_{k-1}(S') \gamma_k(S', S)}{\sum_S \sum_{S'} \alpha'_{k-1}(S') \gamma_k(S', S)} \tag{5}$$

$$\beta'_k(S) = \frac{\sum_{S'} \beta'_{k-1}(S') \gamma_k(S', S)}{\sum_S \sum_{S'} \beta'_{k-1}(S') \gamma_k(S', S)} \tag{6}$$

From the above, it is derived that

$$\begin{aligned} \alpha(u_k) &= \ln \frac{\sum_{S+} \alpha'_{k-1}(S') * \gamma_k(S', S) * \beta'_k(S)}{\sum_{S-} \alpha'_{k-1}(S') * \gamma_k(S'.S) * \beta'_k(S)} \\ &= \ln \frac{\sum_{S+} \exp(\ln \alpha'_{k-1}(S') + \ln \gamma_k(S', S) + \ln \beta'_k(S)}{\sum_{S-} \exp(\ln \alpha'_{k-1}(S') + \ln \gamma_k(S', S) + \ln \beta'_k(S))} \\ &= \max{}^* \left\{ \sum_{S+} \exp(\ln \alpha'_{k-1}(S') + \ln \gamma_k(S', S) + \ln \beta'_k(S)) \right\} \\ &\quad - \max{}^* \left\{ \sum_{S+} \exp(\ln \alpha'_{k-1}(S') + \ln \gamma_k(S', S) + \ln \beta'_k(S)) \right\} \end{aligned} \tag{7}$$

Here max* function (Manjunatha & Kiran, 2012) can be represented as

$$\max{}^*(x, y) = \ln(e^x + e^y) \tag{8}$$

The Jacobean algorithm is applied to approximate and the same thing is simplified later, when x and y difference is too large.

$\ln(e^x + e^y) = \max(x, y) + \ln(\max(x, y))$ (9)

The expression (6) is simplified by applying the above equation and it is obtained with reference to Max-Log MAP algorithm. With the cost of performance degradation, algorithm simplifies the estimation from the series of expression above.

The modified version of the equation becomes

$$\left(\sum_{i=1}^{n} e^{x_i}\right)\left[n+(n-1)+\ldots+1\right] \le n*\left[ne^{x_1}+(n-1)e^{x_2}+\ldots+e^{x_n}\right] \tag{10}$$

The inequality is compared when x1³x2³.....³x_n,

$$\left(\sum_{i=1}^{n} e^{x_i}\right)\left[n(n+1)/2\right] \le ne^{x_1}*\left[n+(n-1)e^{x_2-x_i}+\ldots+e^{x_n-x_1}\right] \tag{11}$$

Apply the logarithm in the above equation of inequality on both sides, the equation becomes

$$\ln\left(\sum_{i=1}^{n} e^{x_i}\right) \le \ln\left[2/(n+1)\right]+x_1+\ln\left[n+(n-1)e^{x_2-x_1}++e^{x_n-x_1}\right] \tag{12}$$

In the right hand side of expression logarithm is considered for only two terms and after simplification leads to

$$\max^*(x_1,x_2,\cdots,x_n) \approx \max(x_1,x_2,\cdots,x_n)+\ln\left[2n/(n+1)\right]+\ln\left[1+(n-1)/n(n-1)e^{x_2-x_1}\right] \tag{13}$$

Second right term in the above equation (13) is constant and this does not affect sequence of computation. Final modified expression can be written as

$$\max^*(x_1,x_2,\cdots,x_n) \approx \max(x_1,x_2,\cdots,x_n)+\ln\left[1+(n-1)/n(n-1)e^{x_2-x_1}\right] \tag{14}$$

The subsequent extrinsic information is obtained after calculating the L (u_k)

$$L_e(u_k) = L(u_k) - L_c y_k^s - L_a(u_k) \tag{15}$$

Max-Log MAP algorithm performance increases gradually with scaling factor λ, when values are multiplied with Lc(u_k),

$$L_e^{'}(u_k) = \lambda\left(L(u_k) - L_c y_k^s - L_a(u_k)\right) \tag{16}$$

Clock Gating

To run multiple delayed cycles a control flag is developed to operate such components. Once the device stops execution, finally the combinational circuit linking registers are removed and the energy is saved due to the pipeline coordinates. The implementation of overall circuit requires additional circuit components to dissipate low power signals.

Group of techniques are used to reduce the power dissipation in digital circuits are called as Dynamic Power Management (DPM) strategies (Saito, 2016). A special type of circuits is used to disable the functionality when a circuit or portion of the circuit is not functioning at particular time duration to reduce the power consumption. A gate level circuit is used to achieve power reduction is called as ''gated clock'' technique, which stops the clock supply to the sequential elements when output is same as input, as it is shown in Fig. 5 and activating the FF only when the input signal is different from the actual output value.

Figure 5. Gated clock based on synchronization respect to input and output value of the FF

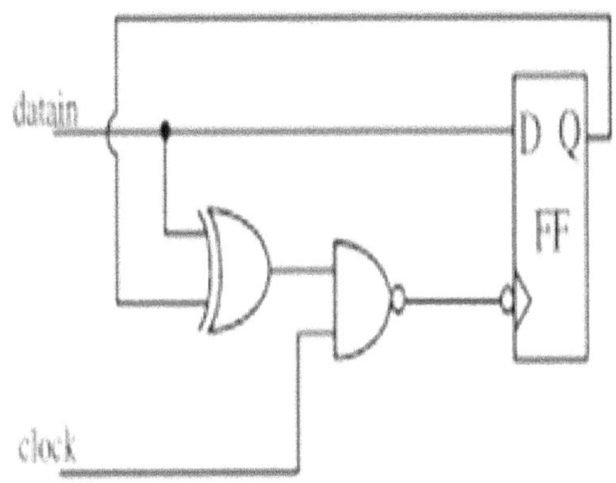

Adjustable Number of Iterations

The process of decoding will stop soon after the completion due to the adjustable number of iterations because iterations are not preset.This technique reduces the time taken to decode after the decoding process gets over results in reduced decoding latency and achieves less power dissipation.

Blocking of Floating Inputs

The RAMs are used to read the data very quickly but the reduced access time expects the higher performance to go along with compose and read function.Consequently it should be stopped to minimize the scattering of state power.When floating inputs are occurred, AND or OR logic is utilized to restrict the entry of inputs by applying the command sign as 0.

SOFTWARE REFERENCE MODELS

The software referencemodel consists of transmitter and receiver components with a Additive White Gaussian Noise (AWGN) channel. Bernoulli random binary generator is used to generate the data for reference model. Across the transmitter data is encoded by turbo encoder then it is converted from unipolar to bipolar before putting on to the channel. The channel data is decoded using turbo decoder ad finally the Bit Error Rate (BER) is calculated by error calculation unit taking two inputs one from input data is being sent and the received decoded data across the receiver. The complete software reference model for the conventional turbo decoder is as shown in the Fig.6

Figure 6. Conventional Turbo Decoder

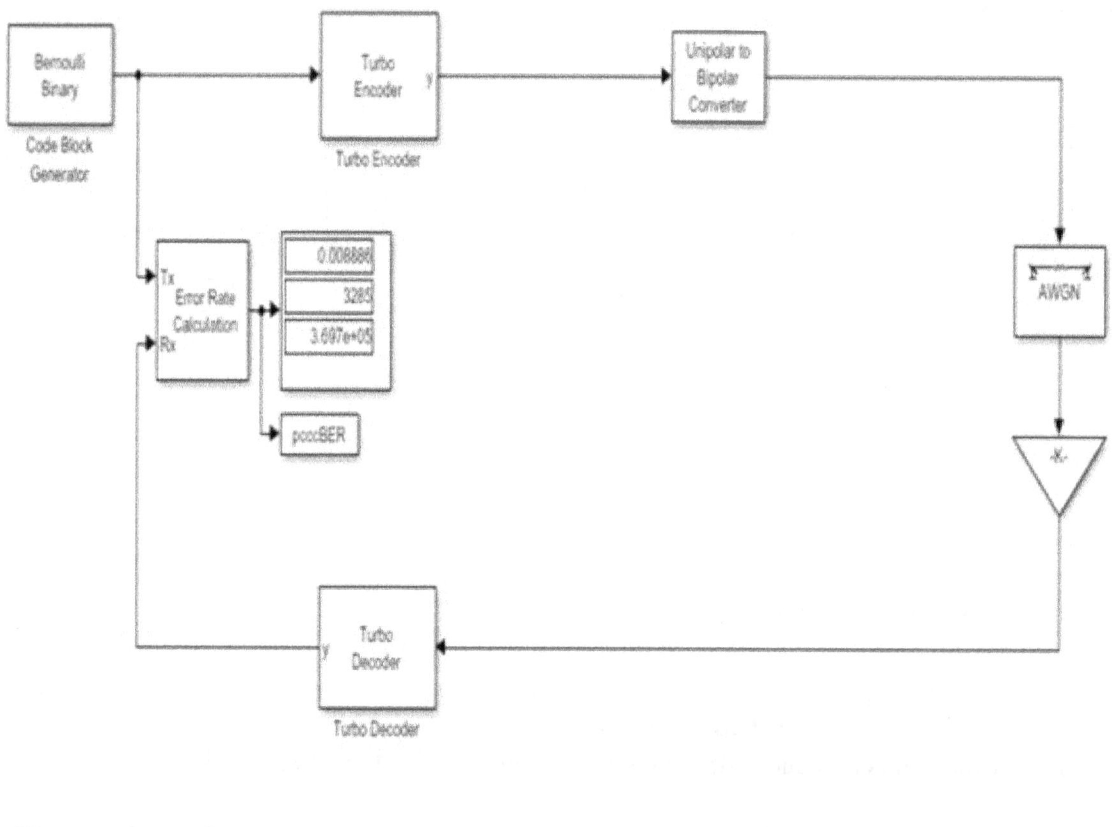

The rate of error from the conventional turbo decoder is high due to fixed number of iterative calculations and also increases the decoding latency. To overcome this, a modified software reference model is developed using Code Block Cyclic Redundancy Check (CBCRC) syndrome detector. As like previous wireless communication software reference model the modified version also has transmitter and receiver sections, but this model takes the input from three different sources as shown in Fig. 7. The input data generator consists of two code block generator and one control signal connected along with the switch to read the data in multiplexed way. Input is encoded using CBCRC generator through turbo encoder with maximum encoding size of 4096 in a single frame length. The encoded data is modulated using Binary Phase Shift Keying (BPSK) modulator before passing it on to the wireless channel. Compared to the conventional wireless communication model this modified model looks complex in structure but works very effectively in reducing the error rate. BPSK demodulator demodulates the channel data and allows data to decode using variable length iterative turbo decoder. Here the size of the frame length is not fixed and number of iterations as well. This helps in early termination of decoding process to reduce the overall latency and unwanted calculation in an iterative loop. Finally from the decoded data CRC syndrome is detected before calculating the BER. The decoder used in this reference model meets the specifications of LTE version so referred as the variable sized LTE decoder.

Figure 7. Variable-sized LTE Turbo Coding

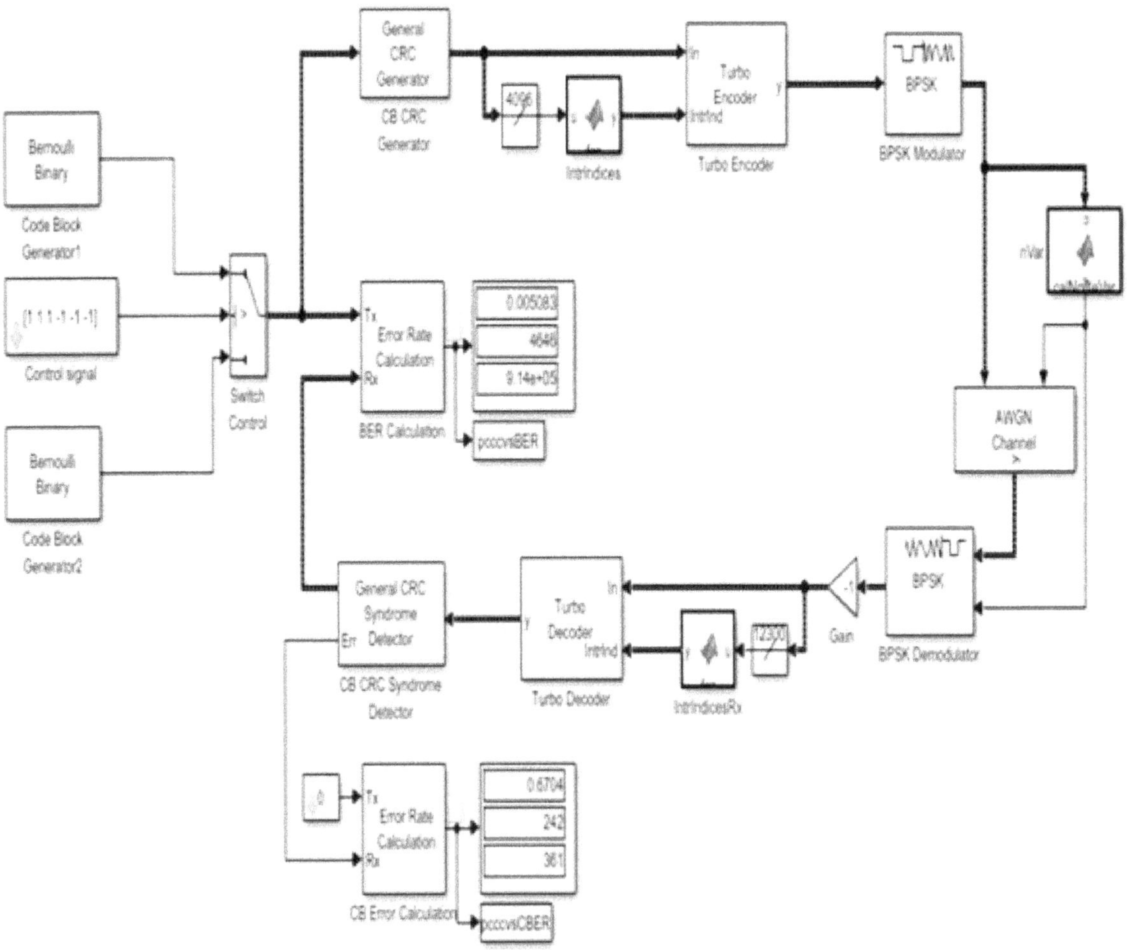

Basically there are 3 different channels in data communication namely AWGN, Rayleigh and Rician. The modulation scheme of phase shift keying (PSK) is kept same for all the execution and BER is observed with variations of different channels. The Fig. 8 shows that, the variation of Eb/N0 (dB) on X-scale and respective variation of Bit Error Rate (BER) for all 3 channels are plotted and shown in fig.1.8 It is observed that Rayleigh and Rician show moderately similar results when compared to AWGN. This proves that AWGN gives the ideal results on BER upto 10-8.

Figure 8. BER with different channels

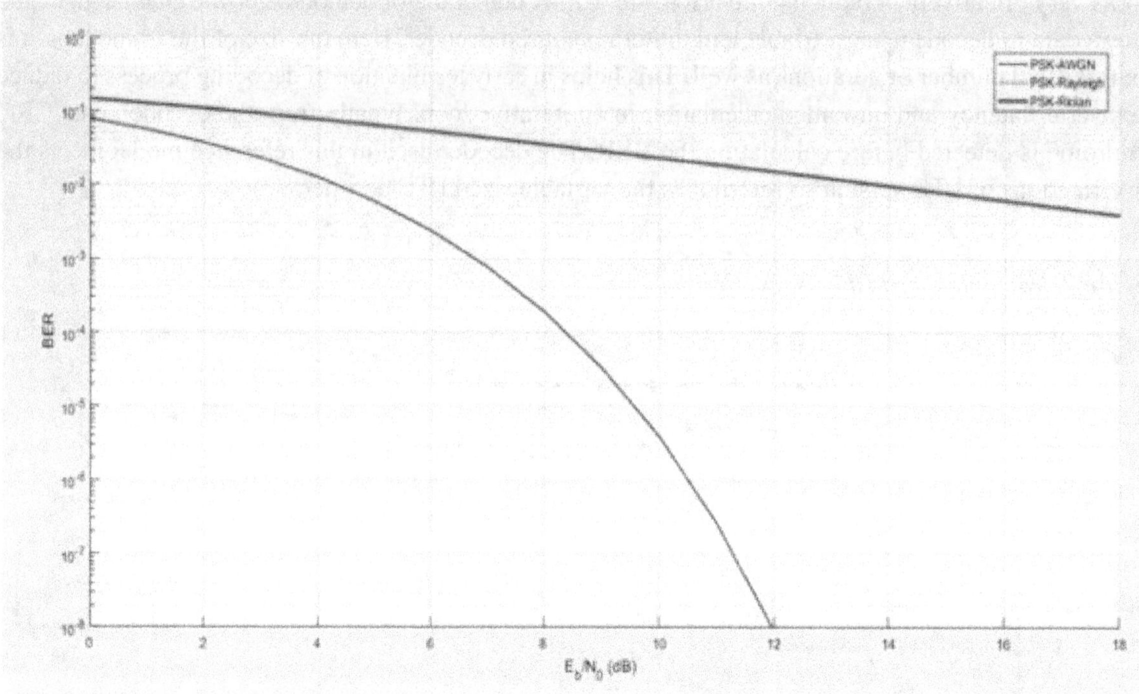

Figure 9. Number of errors bits for 10e8 bits

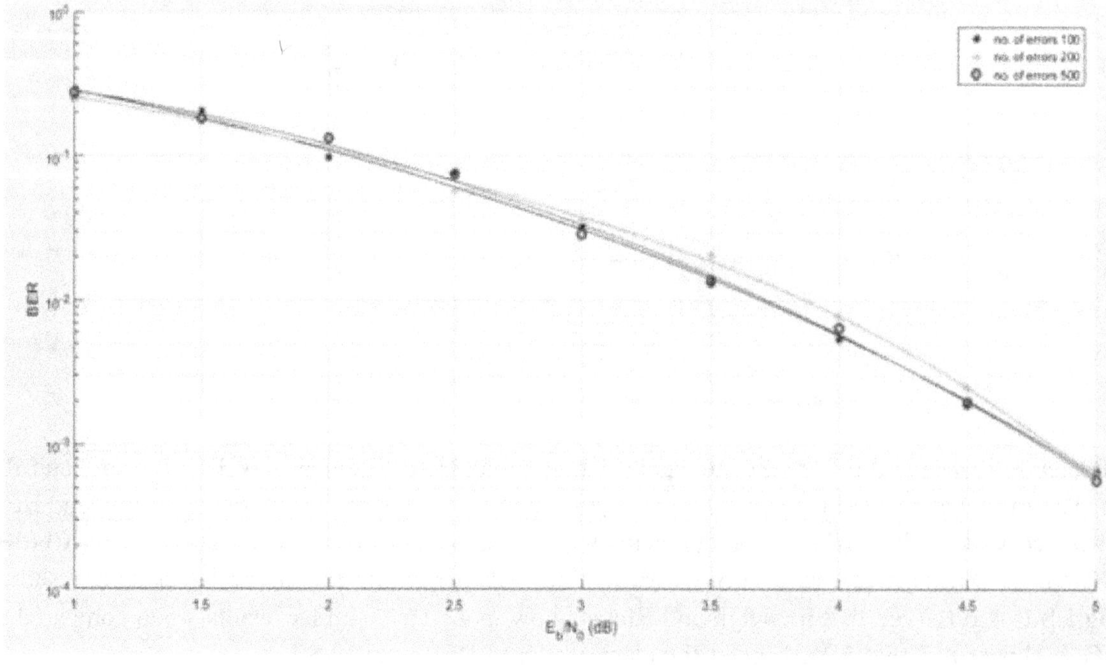

The plot represents the variation of BER for 100, 200 and 500 bits of error on 10e-8 bits which are transmitted from source to destination. This plot shows curves which almost resemble the executed errors. Here the Max algorithm for turbo decoder and PSK modulator and demodulator are kept same for all these execution. The minimal deviation can be observed when it is set for the maximum error bits and the same is plotted in fig. 9

The selection of modulation order also effects on the error rate and it is plotted in the Fig. 10 with different modulation orders for the PSK modulator and demodulator. Here also max algorithm is kept same for the Turbo decoder for all the executions. It is observed that as the orders of modulation increases, the received signal E_b/No (dB) also increases. The BER for the order four is relatively good when compared to the order 64.

Figure 10. PSK modulation with different orders

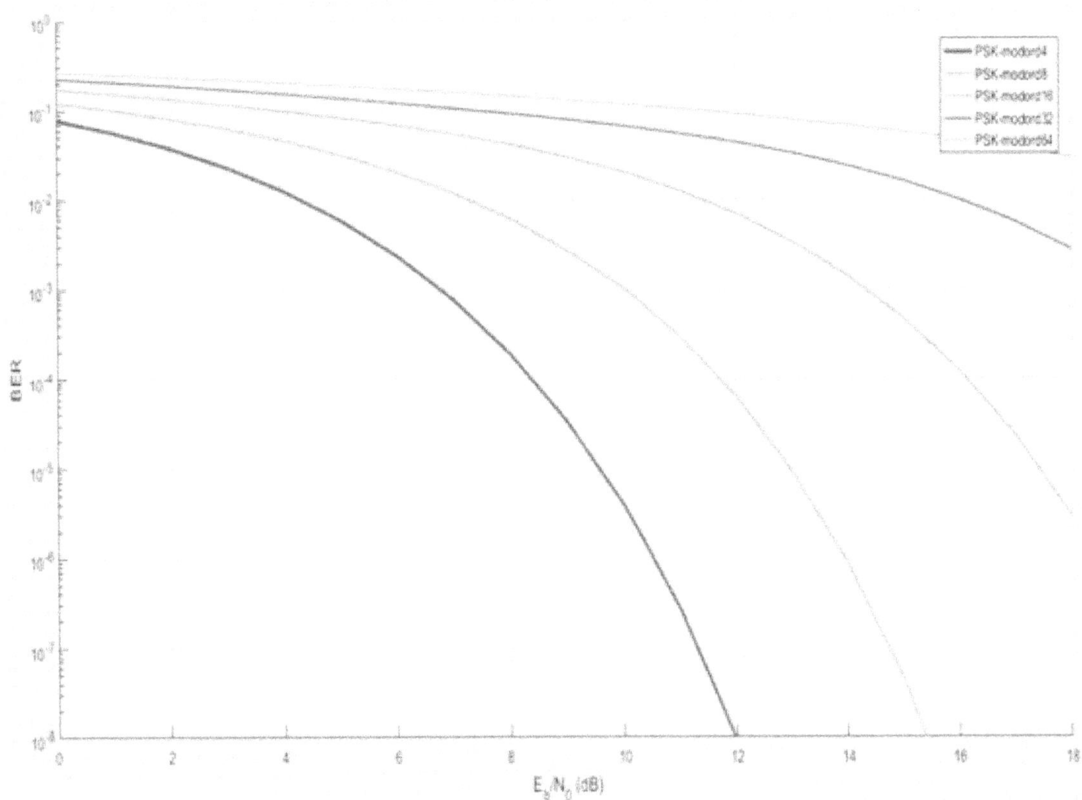

This plot shown in fig. 11 gives the variation of channel coding on two different techniques namely convolution and block coding techniques. Compared to the Block channel coding, convolution coding gives the better result. There are many coding techniques like MAP and SOVA, which proves the better accuracy on data coding.

Figure 11. BER with different channel coding techniques

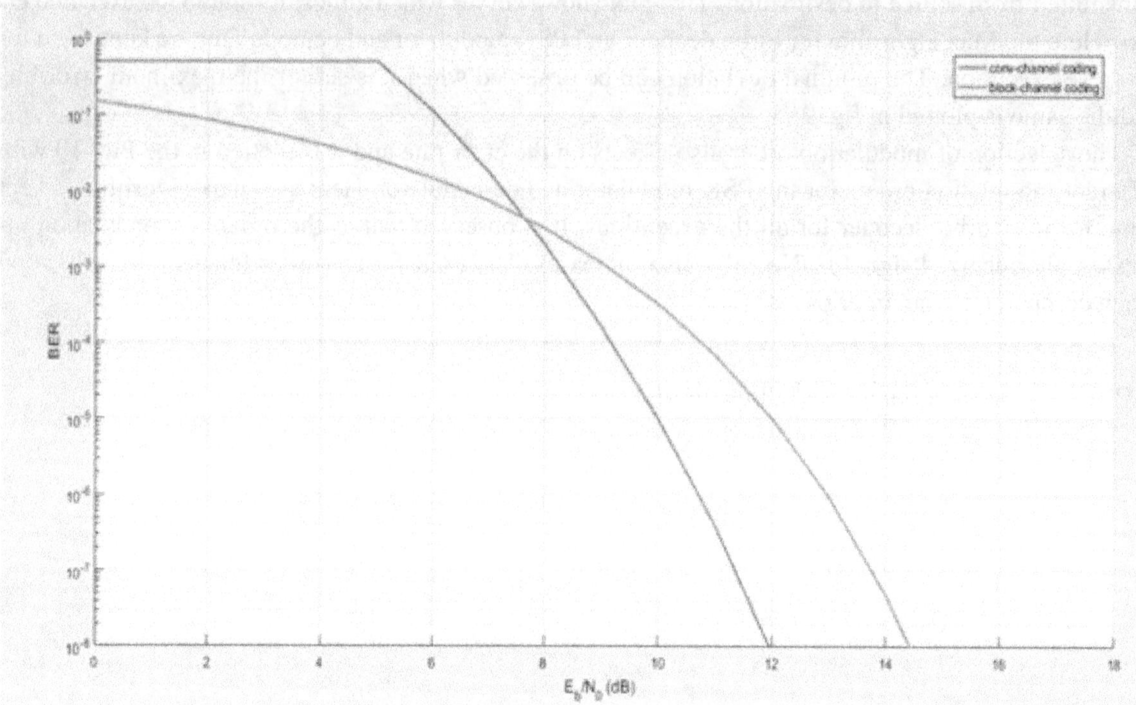

Figure 12. Iterative turbo decoder with variable frame length (N)

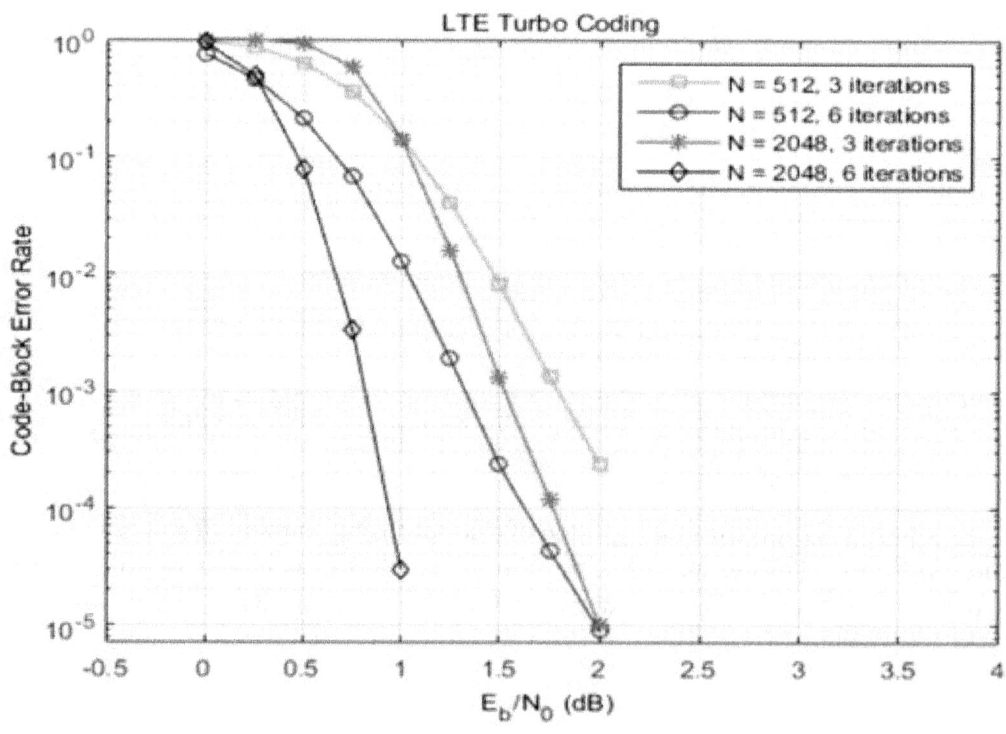

Table 1. BER measurements for frame length N=512

	IT=3	IT=6	IT=9
0	0.0986328125	0.0418526785	0.1513671875
0.5	0.0480468750	0.0161132812	0.0071806066
1	0.0054086538	0.0003363715	
1.5	0.0002387152	0	
2	0	0	

Table 2. BER measurements for frame length N=1024

	IT=3	IT=6	IT=9
0	0.09309895833	0.10498046875	0.0712890625000000
0.5	0.02761501736	0.00909090909	0.00205156880040323
1	0.00101725260	9.22309027777778e-05	1.08506944444444e-05
1.5	1.89887152777778e-05	0	

Table 3. BER measurements for frame length N=2048

	IT=3	IT=6	IT=9
0	0.104817708333333	0.0399693080357143	0.0674804687500000
0.5	0.0168619791666667	0.00187800480769231	0.00853650323275862
1	0.000425889756944444		

Figure 13. LTE Turbo Decoding latency in sec Vs Block length

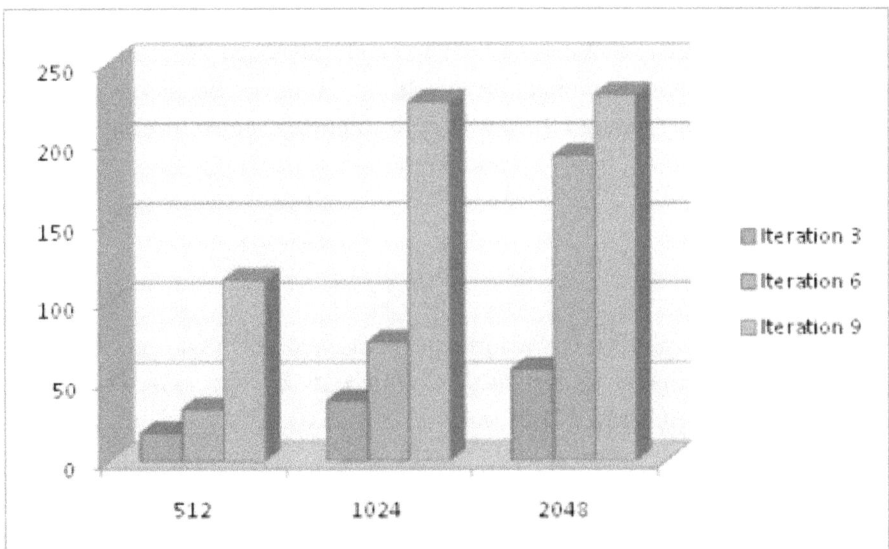

Figure 14. Encoding latency in sec Vs Block length

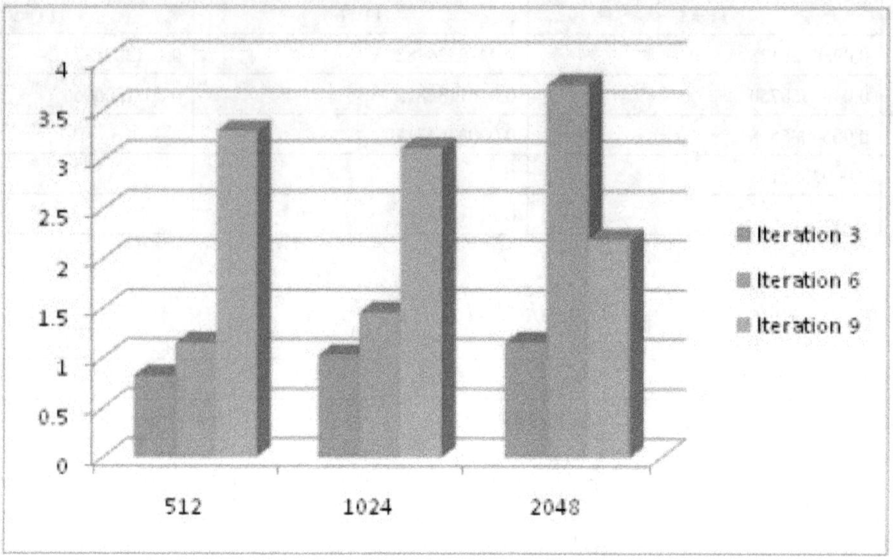

The turbo decoder is simulated on a MATLAB platform with frame length varied from 512 bits to 2048 bits at different iterations. The Fig. 12 Shows that Code- Bit Error Rate (CBER) along with the Eb/N0, this proves that, as the number of iterations increases there is a reduction in Eb/N0.

CONCLUSION

In this chapter, the main functional units of turbo decoder are analyzed. The significance of interleaver and types of interleaving blocks are demonstrated. Turbo decoder is simulated on MATLAB platform for various frame length and the iteration and the results are plotted.

A novel form of computation for Turbo decoder is proposed with a combination of clock gating and adjustable iteration in decoding to meet the LTE and LTE-advanced standards. The max-log-MAP algorithm is applied to the proposed architecture to achieve the power efficiency. The max-log-MAP based Turbo decoder has not only reduces the power consumption but also meet up the performance of LTE wireless standards.

REFERENCES

Abhishek, S. K., & Chakrabarti, S. (2011). Performance Evaluation of Asymmetric Turbo codes using Log-MAP decoding technique. *Proceedings of the IEEE International Conference on Devices and Communications*, 1-5. 10.1109/ICDECOM.2011.5738457

Ai-Mohandes, I., & Elmasry, M. (2004). A low-power 5mb/s turbo decoder for third generation wireless terminals. *Canadian Conference on Electrical and Computer Engineering*, 2387-2390.

Allan, G., & Simmons, S. (2001). A VLSI implementation of an adaptive-effort low-power Viterbi decoder for wireless communications. *Canadian Conference on Electrical and Computer Engineering (CCECE)*, 1183-1188. 10.1109/CCECE.2001.933609

Andrew, J. (2002). A 690-mW 1-Gb/s 1024-b, Rate-1/2 Low-Density Parity-Check Code Decoder. *IEEE Journal of Solid-State Circuits*, *37*(3), 404–412. doi:10.1109/4.987093

Andrew, J. V. (1998). An Intuitive Justification and a Simplified Implementation of the MAP Decoder for Convolutional Codes. *IEEE Journal on Selected Areas in Communications*, *16*(2), 260–264. doi:10.1109/49.661114

Ardakani, & Shabany. (2015a, June). A Novel Area-Efficient VLSI Architecture for Recursion Computation in LTE Turbo Decoders. *IEEE Transactions on Circuits and Wystems. II, Express Briefs*, *62*(6).

Ardakani & Shabany. (2015b). *An Efficient Max-Log MAP Algorithm for VLSI Implementation of Turbo Decoders*. IEEE.

Arun, C., & Rajamani, V. (2007). Minimized memory architecture for low latency Viterbi decoder using Zig-Zag algorithm. *International Journal on Wireless and Optical Communications*, *4*(3), 313–323. doi:10.1142/S0219799507000667

Aziz, Abdel-Kader, & Youssef. (2011). *Implementation of a Smart and Power Efficient Turbo Decoder Using SDR algorithm*. IEEE.

Belov, & Mosin. (2017). FPGA Implementation of LTE Turbo Decoder Using MAX-log MAP Algorithm. *6th Mediterranean Conference on Embedded Computing (MECO)*.

Biatek, Hamidouche, Travers, & Deforges. (2016). Optimal bit rate allocation in the scalable HEVC extension for the deployment of UHD services. *IEEE Trans. Broadcast.*, *62*(4), 826–841. . doi:10.1109/TBC.2016.2599266

Broich, M. (2014). *Optimal Data Path Widths for Energy- and Area efficient Max-Log-MAP Based LTE Turbo Decoders*. IEEE.

Calabuig, Monserrat, & Gómez-Barquero. (2015). 5th generation mobile networks: A new opportunity for the convergence of mobile broadband and broadcast services. *IEEE Commun. Mag.*, *53*(2), 198–205. . doi:10.1109/MCOM.2015.7045409

Dai, Wang, & Yang. (2012). Next-generation digital television terrestrial broadcasting systems: Key technologies and research trends. *IEEE Commun. Mag.*, *50*(6), 150–158. . doi:10.1109/MCOM.2012.6211500

Karim & Chakrabarti. (2010). An Improved Low-Power High-Throughput Log-MAP Turbo Decoder. *IEEE Transactions on Consumer Electronics*, *56*(2).

Karim & Chakrabarti. (2011). Design of Efficient High Throughput Pipelined Parallel Turbo Decoder Using QPP Interleaver. *International Conference on Multimeadia, Signal Processing and Communication Technologies, IEEE 2011*, 248-51. 10.1109/MSPCT.2011.6150486

Kim & Kim. (2013). Design of Early Stopping Unit in Parallel Turbo Decoder based on Galois Field Operation. *IEEE Proceedings ISOCC*.

Kim, H., Lee, Y., & Kim, J.-H. (2015, October 8). Low-complexity CRC-aided early stopping unit for parallel turbo decoder. *Electronics Letters, 51*(21), 1660–1662. doi:10.1049/el.2015.2262

Li, Maunder, Al-Hashimi, & Hanzo. (2013). A Low-Complexity Turbo Decoder Architecture for Energy-Efficient Wireless Sensor Networks. *IEEE Transactions on Very Large Scale Integration (VLSI) Systems, 21*(1).

Lin. (2015). An Area Efficient Radix-4 Reciprocal Dual Trellis Architecture for a High-Code-Rate Turbo Decoder. *IEEE Transactions on Circuits and Systems—II: Express Briefs, 62*(1). . doi:10.1109/TCSII.2014.2362733

Lin & Der-shieh. (2015). *Efficient highly parallel turbo decoder for 3GPP LTE-advanced*. IEEE.

Liu, J., Zhang, L.-M., & Zhong, Z.-G. (2014). Research on Low Latency Decoding Scheme of Turbo Codes. *2014 International Conference on Wireless Communication and Sensor Network*. 10.1109/WCSN.2014.10

Luo, H., & Zhang, Y. (2017). Low Latency Parallel Turbo Decoding Implementation for Future Terrestrial Broadcasting Systems. *IEEE Transactions on Broadcasting*, (Jan), 1–8.

Manjunatha, K. N., & Kiran, B. (2012). Design and ASIC Implementation of a 3GPP LTE Advance Turbo Encoder and Turbo Decoder. *International Journal of Engineering Research and Applications, 2*(4), 6-10.

Manjunatha, K. N., & Lohith Kumar, H. G. (2011). Design and Performance analysis of a 3GPP LTE/LTE-Advance turbo decoder using software reference models. *International Journal of Scientific & Engineering Research, 2*(7).

Manjunatha, K. N., & Vaibhav, A. (2017). Design and FPGA Implementation of Power Efficient Turbo Decoder for 4G LTE Standards. *International Journal of Applied Engineering Research, 12*(21), 10921-10925.

Reddy, Clermidy, Al Khayat, Bhagdadi, & Jezquel. (2010). *Power Consumption Analysis and Energy Efficient Optimization for Turbo Decoder Implementation*. IEEE.

Roth, C., Benkesery, C., & Huang, Q. (2014). Power-Efficient Turbo-Decoder Design based on Algorithm-Specific Power Domain Partitioning. *24th International Workshop on Power and Timing Modeling, Optimization and Simulation (PATMOS)*. 10.1109/PATMOS.2014.6951907

Saito. (2016). 8K terrestrial transmission field tests using dual polarized MIMO and higher-order modulation OFDM. *IEEE Trans. Broadcast., 62*(1), 306–315. . doi:10.1109/TBC.2015.2494853

Shrestha & Paily. (2014). High-Throughput Turbo Decoder with Parallel Architecture for LTE Wireless Communication Standards. *IEEE Transactions on Circuits and Systems—I: Regular Papers, 61*(9). . doi:10.1109/TCSI.2014.2332266

Sklar, B. (2001). *Digital Communications: Fundamentals and Applications with Fundamentals of Turbo Codes* (2nd ed.). Prentice-Hall.

Spasov, Gushev, & Ristov. (2015). *Max-Log-MAP Decoding with Reduced Memory Complexity*. IEEE.

Studer, Benskeser, Belfanti, & Huang. (2011). Design and Implementation of a Parallel Turbo-Decoder ASIC for 3GPP-LTE. *IEEE Journal of Solid-State Circuits, 46*(1). . doi:10.1109/JSSC.2010.2075390

Vargas, D., Kim, Y. J. D., Bajcsy, J., Gomez-Barquero, D., & Cardona, N. (2015, September). A MIMO-channel-precoding scheme for next generation terrestrial broadcast TV systems. *IEEE Transactions on Broadcasting*, *61*(3), 445–456. doi:10.1109/TBC.2015.2450431

Wang, G. (2014, May). Parallel Interleaver Design for a High Throughput HSPA/LTE Multi-Standard Turbo Decoder. *IEEE Transactions on Circuits and Systems. I, Regular Papers*, *61*(5).

Yoo. (2015, December). Reverse Rate Matching for Low-Power LTE-Advanced Turbo Decoders. *IEEE Transactions on Circuits and Systems. I, Regular Papers*, *62*(12).

Zhang, L., & Li, Y. (2011). Implementing and Optimizing a Turbo Decoder on a TI TMS320C64x Device. *ICCP2011 IEEE Proceedings*, 401-04. 10.1109/ICCPS.2011.6092297

Compilation of References

Light, J. (2020). Green Networking: A Simulation of Energy Efficient Methods. *Procedia Computer Science*, *171*, 1489–1497. doi:10.1016/j.procs.2020.04.159

5G and blockchain: The building blocks of the shared economy. (n.d.). Available: https://www.ericsson.com/en/blog/2019/10/5G-blockchain-shared-economy

Abhishek, S. K., & Chakrabarti, S. (2011). Performance Evaluation of Asymmetric Turbo codes using Log-MAP decoding technique. *Proceedings of the IEEE International Conference on Devices and Communications*, 1-5. 10.1109/ICDECOM.2011.5738457

Abidi, M. H., Alkhalefah, H., Moiduddin, K., Alazab, M., Mohammed, M. K., Ameen, W., & Gadekallu, T. R. (2021). Optimal 5G network slicing using machine learning and deep learning concepts. *Computer Standards & Interfaces*, *76*, 103518. Advance online publication. doi:10.1016/j.csi.2021.103518

Adat, V., & Gupta, B. (2018). Security in internet of things: Issues, challenges, taxonomy, and architecture. *Telecommunication Systems*, *67*(3), 423–441. doi:10.100711235-017-0345-9

Aguilar, S., Vidal, R., & Gomez, C. (2017). Opportunistic Sensor Data Collection with Bluetooth Low Energy. *MDPI Sensors*, *17*(12), 159. doi:10.339017010159 PMID:28124987

Agyapong, P. K., Iwamura, M., Staehle, D., Kiess, W., & Benjebbour, A. (2014). Design Considerations for a 5G Network Architecture. *IEEE Communications Magazine*, *52*(11), 65–65. doi:10.1109/MCOM.2014.6957145

Ahad, A., Tahir, M., & Yau, K. L. A. (2019). 5G-based smart healthcare network: Architecture, taxonomy, challenges and future research directions. *IEEE Access: Practical Innovations, Open Solutions*, *7*, 100747–100762. doi:10.1109/ACCESS.2019.2930628

Ahlawat, & Dave, M. (2017, June). A Hybrid Approach for Path Vulnerability Matrix on Random Key Predistribution for Wireless Sensor Networks. *Wireless Personal Communications*, *94*(4), 3327–3353. doi:10.100711277-016-3779-6

Ahmad, Z., Khan, A. S., Shiang, C. W., Abdullah, J., & Ahmad, F. (2020). *Network intrusion detection system: A systematic study of machine learning and deep learning approaches*. https://onlinelibrary.wiley.com/doi/full/10.1002/ett.4150

Ahmed, M., Mahmood, A., & Hu, J. (2016). *A survey of network anomaly detection techniques*. https://www.sciencedirect.com/science/article/abs/pii/S1084804515002891

Ai-Mohandes, I., & Elmasry, M. (2004). A low-power 5mb/s turbo decoder for third generation wireless terminals. *Canadian Conference on Electrical and Computer Engineering*, 2387-2390.

Ai, Y., Peng, M., & Zhang, K. (2017). *Edge cloud computing technologies for Internet of Things: A primer*. Digital Communications and Networks.

Akinyele, J. A., Pagano, M. W., Green, M. D., Lehmann, C. U., Peterson, Z. N. J., & Rubin, A. D. (2011). Securing electronic medical records using attribute-based encryption on mobile devices. *Proceedings of the 1st ACM Workshop on Security and Privacy in Smartphones and Mobile Devices*, 75–86. 10.1145/2046614.2046628

Akpakwu, G. A., Silva, B. J., Hancke, G. P., & Abu-Mahfouz, A. M. (2017). A Survey on 5G Networks for the Internet of Things: Communication Technologies and Challenges. *IEEE Access: Practical Innovations, Open Solutions*, *6*, 3619–3647. doi:10.1109/ACCESS.2017.2779844

Alcaraz-Calero, J., Belikaidis, I. P., Cano, C. J. B., Bisson, P., Bourse, D., Bredel, M., ... Wang, Q. (2018). Leading innovations towards 5G: Europe's perspective in 5G Infrastructure Public-Private Partnership (5G-PPP). *IEEE International Symposium on Personal, Indoor and Mobile Radio Communications, PIMRC,* 1–5. 10.1109/PIMRC.2017.8292654

Al-Dunainawi. (2018, October). Green Network Costs of 5G and Beyond, Expectations Vs Reality. *IEEE Access: Practical Innovations, Open Solutions*.

Alfian, G., Syafrudin, M., Ijaz, M. F., Syaekhoni, M. A., Fitriyani, N. L., & Rhee, J. (2018). A personalized healthcare monitoring system for diabetic patients by utilizing BLE-based sensors and real-time data processing. *Sensors (Switzerland)*, *18*(7), 2183. Advance online publication. doi:10.339018072183 PMID:29986473

Alharbi, A., Alhaidari, S., & Zohdy, M. (2018). *Denial-of-Service, Probing, User to Root (U2R) & Remote to User (R2L) Attack Detection using Hidden Markov Models*. https://www.ijcit.com/archives/volume7/issue5/IJCIT070501.pdf

Ali, Wang, Bhuiyan, & Jiang. (2018). Secure data provenance in cloud-centric internet of things via blockchain smart contracts. *2018 IEEE SmartWorld, Ubiquitous Intelligence & Computing, Advanced & Trusted Computing, Scalable Computing & Communications, Cloud & Big Data Computing, Internet of People and Smart City Innovation (SmartWorld/SCALCOM/UIC/ATC/CBDCom/IOP/SCI)*, 991–998.

Allan, G., & Simmons, S. (2001). A VLSI implementation of an adaptive-effort low-power Viterbi decoder for wireless communications. *Canadian Conference on Electrical and Computer Engineering (CCECE)*, 1183-1188. 10.1109/CCECE.2001.933609

Alli, A. A., & Alam, M. M. (2020). The fog cloud of things: A survey on concepts, architecture, standards, tools, and applications. *Internet of Things*, *9*, 100177. doi:10.1016/j.iot.2020.100177

Amato, F., Mazzocca, N., Moscato, F., & Vivenzio. (2017). *Multilayer Perceptron: An Intelligent Model for Classification and Intrusion Detection*. https://ieeexplore.ieee.org/document/7929765

Amin, R., & Biswas, G. (2016). A secure Light weight scheme for user authentication & key agreement in Multi gateway based wireless sensor networks. *Ad Hoc Networks*, *36*, 58–80. doi:10.1016/j.adhoc.2015.05.020

Ammar, M., Russello, G., & Crispo, B. (2018). Internet of things: A survey on the security of iot frameworks. *Journal of Information Security and Applications*, *38*, 8–27. doi:10.1016/j.jisa.2017.11.002

Andrew, J. (2002). A 690-mW 1-Gb/s 1024-b, Rate-1/2 Low-Density Parity-Check Code Decoder. *IEEE Journal of Solid-State Circuits*, *37*(3), 404–412. doi:10.1109/4.987093

Andrew, J. V. (1998). An Intuitive Justification and a Simplified Implementation of the MAP Decoder for Convolutional Codes. *IEEE Journal on Selected Areas in Communications*, *16*(2), 260–264. doi:10.1109/49.661114

Andrews, J. G., Buzzi, S., Choi, W., Hanly, S. V., Lozano, A., Soong, A. C., & Zhang, J. C. (2014). What Will 5G Be? *IEEE Journal on Selected Areas in Communications*, *32*(6), 1065–1082. doi:10.1109/JSAC.2014.2328098

Anju, M., & Gawas, U. (2015). An Overview on Evolution of Mobile Wireless Communication Networks : 1G-6G. *Ijritcc. Org*, *3*(5), 3130–3133.

Arasteh, H., Hosseinnezhad, V., Loia, V., Tommasetti, A., Troisi, O., Shafie-Khah, M., & Siano, P. (2016). Iot-based smart cities: A survey. *EEEIC 2016 - International Conference on Environment and Electrical Engineering*. 10.1109/EEEIC.2016.7555867

Ardakani & Shabany. (2015b). *An Efficient Max-Log MAP Algorithm for VLSI Implementation of Turbo Decoders*. IEEE.

Ardakani, & Shabany. (2015a, June). A Novel Area-Efficient VLSI Architecture for Recursion Computation in LTE Turbo Decoders. *IEEE Transactions on Circuits and Wystems. II, Express Briefs*, *62*(6).

Arora, P., Singh, A., & Tiyagi, H. (2012). Evaluation and comparison of security issues on cloud computing environment. *World of Computing and Knowledge Technology Journal*, *2*(5), 179–183.

Arun, C., & Rajamani, V. (2007). Minimized memory architecture for low latency Viterbi decoder using Zig-Zag algorithm. *International Journal on Wireless and Optical Communications*, *4*(3), 313–323. doi:10.1142/S0219799507000667

Aziz, Abdel-Kader, & Youssef. (2011). *Implementation of a Smart and Power Efficient Turbo Decoder Using SDR algorithm*. IEEE.

Baek, J., Safavi-Naini, R., & Susilo, W. (2008a). Public key encryption with keyword search revisited. *Computational Science and Its …*, 1–15. Retrieved from https://link.springer.com/chapter/10.1007/978-3-540-69839-5_96

Baek, J., Safavi-Naini, R., & Susilo, W. (2008b). Public key encryption with keyword search revisited. In *Computational Science and Its Applications—ICCSA 2008* (pp. 1249–1259). Springer. doi:10.1007/978-3-540-69839-5_96

Baker, J., & Stanley, A. (2018). Telemedicine Technology: A Review of Services, Equipment, and Other Aspects. *Current Allergy and Asthma Reports*, *18*(11), 60. Advance online publication. doi:10.100711882-018-0814-6 PMID:30259201

Ballard, L., Kamara, S., & Monrose, F. (2005). Achieving efficient conjunctive keyword searches over encrypted data. In *Information and Communications Security* (pp. 414–426). Springer. doi:10.1007/11602897_35

Bao, F., Deng, R. H., Ding, X., & Yang, Y. (2008). Private query on encrypted data in multi-user settings. *International Conference on Information Security Practice and Experience*, 71–85. 10.1007/978-3-540-79104-1_6

Baoid, Light, & Mahanti. (2021). Blockchain Technology and its Applications Across Multiple Domains: A Survey. *The Journal of International Technology and Information Management*.

Baratè, A., Haus, G., Ludovico, L. A., Pagani, E., & Scarabottolo, N. (2019). 5G Technology for Augmented and Virtual Reality in Education. *Education and New Developments*, *2019*(1), 512–516. doi:10.36315/2019v1end116

Belov, & Mosin. (2017). FPGA Implementation of LTE Turbo Decoder Using MAX-log MAP Algorithm. *6th Mediterranean Conference on Embedded Computing (MECO)*.

Bera, B., Saha, S., Das, A. K., Kumar, N., Lorenz, P., & Alazab, M. (2020). Blockchain-Envisioned Secure Data Delivery and Collection Scheme for 5G-Based IoT-Enabled Internet of Drones Environment. *IEEE Transactions on Vehicular Technology*, *69*(8), 9097–9111. doi:10.1109/TVT.2020.3000576

Beverage Industry Environmental Roundtable. (2011, December). *A Practical Perspective on Water Accounting in the Beverage Sector*. Retrieved from Water footprint: https://www.waterfootprint.org/media/downloads/BIER-2011-WaterAccountingSectorPerspective.pdf

Bhalla, M. R., & Bhalla, A. V. (2010). Generations of Mobile Wireless Technology: A Survey. *International Journal of Computers and Applications*, *5*(4), 26–32. doi:10.5120/905-1282

Bianzino, Chaudet, Rossi, & Rougier. (n.d.). *A Survey of Green Networking Research*. Institut TELECOM, TELECOM ParisTech, CNRS LTCI UMR 5141.

Biatek, Hamidouche, Travers, & Deforges. (2016). Optimal bit rate allocation in the scalable HEVC extension for the deployment of UHD services. *IEEE Trans. Broadcast., 62*(4), 826–841. . doi:10.1109/TBC.2016.2599266

Bilal, Khan, & Zomaya. (2013). Green Data Center Networks: Challenges and Opportunities. *11th International Conference on Frontiers of Information Technology.*

Bi, S., & Zhang, Y. J. (2018). Computation rate maximization for wireless powered mobile-edge computing with binary computation offloading. *IEEE Transactions on Wireless Communications, 17*(6), 4177–4190.

Blockchain: A key enabler for 5G. (n.d.). Available: https://www.standardsuniversity.org/e-magazine/may-2019-volume-9-issue-1-blockchain-standards/blockchain-a-key-enabler-for-5G /

Boccardi, F., & Heath, R. W. J., Marzetta, T. L., Popovski, P., & Lozano Solsona, A. (2013). Five Disruptive Technology Directions for 5G. *IEEE Communications Magazine*, (February), 74–80.

Bogale, T. E., & Le, L. B. (2016). Massive MIMO and mmWave for 5G Wireless HetNet: Potential Benefits and Challenges. *IEEE Vehicular Technology Magazine, 11*(1), 64–75. doi:10.1109/MVT.2015.2496240

Böhle, M., Eitel, F., Weygandt, M., & Ritter, K. (2019). Layer-wise relevance propagation for explaining deep neural network decisions in MRI-based Alzheimer's disease classification. *Frontiers in Aging Neuroscience, 10*(JUL), 194. Advance online publication. doi:10.3389/fnagi.2019.00194 PMID:31417397

Bojic, D., Sasaki, E., Cvijetic, N., Ting Wang, T., Kuno, J., Lessmann, J., Schmid, S., Ishii, H., & Nakamura, S. (2013). Advanced wireless and optical technologies for small-cell mobile backhaul with dynamic software-defined management. *IEEE Communications Magazine, 51*(9), 86–93. doi:10.1109/MCOM.2013.6588655

Boneh, D., Di Crescenzo, G., Ostrovsky, R., & Persiano, G. (2004). Public key encryption with keyword search. *Advances in Cryptology-Eurocrypt, 2004*, 506–522.

Boneh, D., & Waters, B. (2007). Conjunctive, subset, and range queries on encrypted data. In *Theory of cryptography* (pp. 535–554). Springer. doi:10.1007/978-3-540-70936-7_29

Bostrom, N. (2013). Existential risk prevention as global priority. *Global Policy, 4*(1), 15–31. doi:10.1111/1758-5899.12002

Britel, M. (2018). *Big Data Analytic for Intrusion Detection System.* https://ieeexplore.ieee.org/document/8610578

Broich, M. (2014). *Optimal Data Path Widths for Energy- and Area efficient Max-Log-MAP Based LTE Turbo Decoders.* IEEE.

Bromley-Challenor, K., Kowalski, M., Barnard, R., & Lynn, S. (2013). *Water use in the UK food and drink industry - A review of water use in the food and drink industry in 2007 and 2010, by sub-sector and UK nations.* Banbury: WRAP.

Byun, J. W., Lee, D. H., & Lim, J. (2006). Efficient conjunctive keyword search on encrypted data storage system. *European Public Key Infrastructure Workshop*, 184–196. 10.1007/11774716_15

Calabuig, Monserrat, & Gómez-Barquero. (2015). 5th generation mobile networks: A new opportunity for the convergence of mobile broadband and broadcast services. *IEEE Commun. Mag., 53*(2), 198–205. . doi:10.1109/MCOM.2015.7045409

Canêdo, D., & Romariz, A. (2019). *Data Analysis of Wireless Networks Using Classification Techniques.* https://arxiv.org/abs/1908.07329 doi:10.5121/csit.2019.90905

Capozzi, F., Piro, G., Grieco, L. A., Boggia, G., & Camarda, P. (2013). Downlink packet scheduling in LTE cellular networks: Key design issues and a survey. *IEEE Communications Surveys and Tutorials, 15*(2), 678–700. doi:10.1109/SURV.2012.060912.00100

Casani, S., Rouhany, M., & Knøchel, S. (2005). A discussion paper on challenges and limitations to water reuse and hygiene in the food industry. *Water Research, 39*(6), 1134–1146. doi:10.1016/j.watres.2004.12.015 PMID:15766968

Chaer, A., Salah, K., Lima, C., Ray, P. P., & Sheltami, T. (2019). Blockchain for 5G: Opportunities and challenges. *Proc. IEEE Globecom Workshops (GC Wkshps),* 1–6. 10.1109/GCWkshps45667.2019.9024627

Chamola, V., Hassija, V., Gupta, V., & Guizani, M. (2020). A Comprehensive Review of the COVID-19 Pandemic and the Role of IoT, Drones, AI, Blockchain, and 5G in Managing its Impact. *IEEE Access: Practical Innovations, Open Solutions, 8*, 90225–90265. doi:10.1109/ACCESS.2020.2992341

Chao-yang, Z. (2011, August 1). *DOS Attack Analysis and Study of New Measures to Prevent.* https://ieeexplore.ieee.org/document/5997473

Chaudhry, M. A. R., & Soptimizer, Z. A. (2019). Blockchain: A key enabler for 5G. *IEEE Standards Univ., 10*(1). Available: https://www.standardsuniversity.org/e-magazine/may-2019-volume-9- issue-1-blockchain-standards/blockchain-a-key-enabler-for-5g/

Chen, Z., Wu, C., Wang, D., & Li, S. (2012). Conjunctive keywords searchable encryption with efficient pairing, constant ciphertext and short trapdoor. *Pacific-Asia Workshop on Intelligence and Security Informatics*, 176–189.

Cheng, J., Chen, W., Tao, F., & Lin, C. L. (2018). Industrial IoT in 5G environment towards smart manufacturing. *Journal of Industrial Information Integration, 10*, 10–19. doi:10.1016/j.jii.2018.04.001

Cheng, X., Chen, C., Zhang, W., & Yang, Y. (2017). 5G-enabled cooperative intelligent vehicular (5GenCIV) Framework: When Benz Meets Marconi. *IEEE Intelligent Systems, 32*(3), 53–59. doi:10.1109/MIS.2017.53

Chen, M., Yang, J., Zhou, J., Hao, Y., Zhang, J., & Youn, C. H. (2018). 5G-Smart Diabetes: Toward Personalized Diabetes Diagnosis with Healthcare Big Data Clouds. *IEEE Communications Magazine, 56*(4), 16–23. doi:10.1109/MCOM.2018.1700788

Chen, T.-L., Chung, Y.-F., & Lin, F. Y. S. (2012). A study on agent-based secure scheme for electronic medical record system. *Journal of Medical Systems, 36*(3), 1345–1357. doi:10.100710916-010-9595-8 PMID:20857325

Chen, W., Ma, M., Ye, Y., Zheng, Z., & Zhou, Y. (2018). IoT service based on jointcloud blockchain: The case study of smart traveling. *2018 IEEE Symposium on Service-Oriented System Engineering (SOSE)*, 216–221. 10.1109/SOSE.2018.00036

Chen, X., Jiao, L., Li, W., & Fu, X. (2016). Efficient multi-user computation offloading for mobile-edge cloud computing. *IEEE/ACM Transactions on Networking, 24*(5), 2795–2808. doi:10.1109/TNET.2015.2487344

Chiang, M., & Zhang, T. (2016). Fog and IoT: An overview of research opportunities. *IEEE Internet of Things Journal, 3*(6), 854–864. doi:10.1109/JIOT.2016.2584538

China Telecom's 5G Helped Beijing Jishuitan Hospital Perform the World's First Remote Robotic Surgery. (2019). Retrieved May 17, 2021, from https://carrier.huawei.com/en/success-stories/Industries-5G/Medical/beijing

Chi, Q., Yan, H., Zhang, C., Pang, Z., & Da Xu, L. (2014). A reconfigurable smart sensor interface for industrial wsn in iot environment. *IEEE Transactions on Industrial Informatics, 10*(2), 1417–1425. doi:10.1109/TII.2014.2306798

Chiussi, F. M., Khotimsky, D. A., & Krishnan, S. (2002). Mobility management in third-generation all-IP networks. *IEEE Communications Magazine, 40*(9), 124–135. doi:10.1109/MCOM.2002.1031839

Choi, P. J., Oskouian, R. J., & Tubbs, R. S. (2018). Telesurgery: Past, Present, and Future. *Cureus, 10*(5). https://www.ncbi.nlm.nih.gov/pmc/articles/PMC6067812/ PMID:30079282

Chowdhury, Shahjalal, Ahmed, & Jang. (2020). 6G Wireless Communication Systems: Applications, Requirements, Technologies, Challenges, and Research Directions. *IEEE Open Journal of the Communication Society*.

Cognizant Digital Business. (2019). *The Five Essential IoT Requirements and How to Achieve Them*. Whitepaper.

Cohen, W. W. (2009). *Enron email dataset*. Academic Press.

Corminardi, L. (2018). Opportunities and Challenges of Joint Edge and Fog Orchestration. *IEEE Wireless Communication and Networking Conf.*

Cox, C. (2012). An Introduction to LTE: LTE, LTE-Advanced, SAE and 4G Mobile Communications. In An Introduction to LTE: LTE, LTE-Advanced, SAE and 4G Mobile Communications. doi:10.1002/9781119942825

Curwen, P., & Whalley, J. (2014). Mobile Telecommunications Networks: Restructuring as a Response to a Challenging Environment. Academic Press.

Dai, H., Wong, R., Wang, H., Zheng, Z., & Vasilakos, A. (2019). *Big Data Analytics for Large-scale Wireless Networks: Challenges and Opportunities*. https://dl.acm.org/doi/fullHtml/10.1145/3337065

Dai, Wang, & Yang. (2012). Next-generation digital television terrestrial broadcasting systems: Key technologies and research trends. *IEEE Commun. Mag., 50*(6), 150–158. . doi:10.1109/MCOM.2012.6211500

Dai, Y., Xu, D., Maharjan, S., Chen, Z., He, Q., & Zhang, Y. (2019, May). Blockchain and deep reinforcement learning empowered intelligent 5G beyond. *IEEE Network*, *33*(3), 10–17. doi:10.1109/MNET.2019.1800376

Dananjayan, S., & Raj, G. M. (2020). 5G in healthcare: How fast will be the transformation? *Irish Journal of Medical Science*. Advance online publication. doi:10.100711845-020-02329-w PMID:32737688

Daneels, G., Municio, E., Spaey, K., Vandewiele, G., Dejonghe, A., Ongenae, F., ... Famaey, J. (2017). Real-Time data dissemination and analytics platform for challenging IoT environments. *2017 Global Information Infrastructure and Networking Symposium, GIIS 2017,* 23–30. 10.1109/GIIS.2017.8169799

Darrell, M. W. (2009). How 5G technology enables the health Internet of Things. *Cyber Resilience of Systems and Networks*, (July), 1–150. Retrieved from https://link.springer.com/10.1007/978-3-319-77492-3_16

Das. (2012). A random key establishment scheme for multi-phase deployment in large-scale distributed sensor networks. *International Journal of Information Security, 11*(3), 189-211.

Das, A. K. (2012). A random key establishment scheme for multi-phase deployment in large-scale distributed sensor networks. *International Journal of Information Security*, *11*(3), 189–211. doi:10.100710207-012-0162-9

Das, A. K., Zeadally, S., & He, D. (2018). Taxonomy and analysis of security protocols for internet of things. *Future Generation Computer Systems*, *89*, 110–125. doi:10.1016/j.future.2018.06.027

Das, M. L. (2009). Two-factor user authentication in Wireless sensor network. *IEEE Transactions on Wireless Communications*, *8*(3), 1086–1090. doi:10.1109/TWC.2008.080128

Davidson, R., Townsend, K., Cufí, C., & Akiba. (2014). *Getting started with Bluetooth Low Energy - Tool and techniques for low-power networking*. Academic Press.

De Caro, A., & Iovino, V. (2011). jPBC: Java pairing based cryptography. *Proceedings of the 16th IEEE Symposium on Computers and Communications, ISCC 2011*, 850–855. 10.1109/ISCC.2011.5983948

Dias, D., & Cunha, J. P. S. (2018). Wearable health devices—Vital sign monitoring, systems and technologies. *Sensors (Switzerland)*, *18*(8), 2414. Advance online publication. doi:10.339018082414 PMID:30044415

Ding, M., Gao, F., Jin, Z., & Zhang, H. (2012). An efficient public key encryption with conjunctive keyword search scheme based on pairings. *2012 3rd IEEE International Conference on Network Infrastructure and Digital Content*, 526–530.

Dutta, R., Barua, R., & Sarkar, P. (2004). Pairing-based cryptography: A survey. *Cryptology Research Group, Stat-Math and Applied Statistics Unit, 203*.

Einfalt, T., Arnbjerg-Nielsen, K., Golz, C., Jensen, N.-E., Quirmbach, M., Vaes, G., & Vieux, B. (2004). Towards a roadmap for use of radar rainfall data in urban drainage. *Journal of Hydrology (Amsterdam)*, *299*(3–4), 186–202. doi:10.1016/S0022-1694(04)00365-8

Elminaam, D. S. A., Abdual-Kader, H. M., & Hadhoud, M. M. (2010). Evaluating the Performance of Symmetric Encryption Algorithms. *International Journal of Network Security*, *10*, 216–222.

Eom, J., Lee, D. H., & Lee, K. (2016). Patient-controlled attribute-based encryption for secure electronic health records system. *Journal of Medical Systems*, *40*(12), 253. doi:10.100710916-016-0621-3 PMID:27714562

Epstein, R. H., Dexter, F., & Patel, N. (2015). Influencing Anesthesia Provider Behavior Using Anesthesia Information Management System Data for Near Real-Time Alerts and Post Hoc Reports. *Anesthesia and Analgesia*, *121*(5), 1404. doi:10.1213/ANE.0000000000001038 PMID:26262500

Estrin, D., Govindan R., Heidemann, J., & Kumar, S. (1999). Next Century Challenges. *MobiCom '99 Seattle Washington USA*, 263-270.

European Telecommunications Standards Institute. (2016). *Network Functions Virtualisation (NFV); Acceleration Technologies; VNF Interfaces Specification*. ETSI GS NFV-IFA 002.

Eze, Sadiku, & Musa. (2018). 5G Wireless Technology: A Primer. International *Journal of Science, Engineering and Technology*, *7*(July), 62–64.

Fan, Y., & Zhang, R. (2014). *Research on Network Security and Identity Authentication*. https://www.scientific.net/AMR.926-930.2046

Farash, M. S., Turkanovic, M., Kumari, S., & Holbl, M. (2016). An efficient user authentication and key agreement scheme for heterogeneous wireless sensor network tailored for the internet of things environment. *Ad Hoc Networks*, *36*, 152–176. doi:10.1016/j.adhoc.2015.05.014

Farnaaz, N., & Jabbar, M. (2016). *Random Forest Modeling for Network Intrusion Detection System*. doi:10.1016/j.procs.2016.06.047

Forouzan. (n.d.). *Data Communications and Networking*. McGraw-Hill.

Freitag, F. (2018). On the collaborative governance of decentralized edge microclouds with blockchain-based distributed ledgers. *2018 IEEE/WIC/ACM International Conference on Web Intelligence (WI)*, 709–712. 10.1109/WI.2018.000-7

García Moro, F. (2020). The Death and Life of Hong Kong's Illegal Façades. *ARENA Journal of Architectural Research*, *5*(1), 2. Advance online publication. doi:10.5334/ajar.231

Ghetas, M., Yong, C. H., & Sumari, P. (2015). Harmony-based monarch butterfly optimization algorithm. *International Conference on Control System, Computing and Engineering (ICCSCE), 2015 IEEE International Conference on. IEEE*, 156–161.

Ghosh, A., Maeder, A., Baker, M., & Chandramouli, D. (2019). 5G Evolution: A View on 5G Cellular Technology beyond 3GPP Release 15. *IEEE Access: Practical Innovations, Open Solutions*, *7*, 127639–127651. doi:10.1109/ACCESS.2019.2939938

Golle, P., Staddon, J., & Waters, B. (2004). Secure conjunctive keyword search over encrypted data. *Applied Cryptography and Network Security*, 31–45.

Goosen. (2014). *Design and Implementation of a Bluetooth 4.0 LE Infrastructure for Mobile Devices* (Thesis). ULM University, California.

Goyal, V., Pandey, O., Sahai, A., & Waters, B. (2006). Attribute-based encryption for fine-grained access control of encrypted data. *Proceedings of the 13th ACM Conference on Computer and Communications Security - CCS '06*, 89. 10.1145/1180405.1180418

Gupta, N., Juneja, P. K., Sharma, S., & Garg, U. (2021). Future Aspect of 5G-IoT Architecture in Smart Healthcare System. *2021 5th International Conference on Intelligent Computing and Control Systems (ICICCS)*, 406–411.

Habibzadeh, H., Dinesh, K., Rajabi Shishvan, O., Boggio-Dandry, A., Sharma, G., & Soyata, T. (2020). A Survey of Healthcare Internet of Things (HIoT): A Clinical Perspective. *IEEE Internet of Things Journal*, *7*(1), 53–71. doi:10.1109/JIOT.2019.2946359 PMID:33748312

Hamid, Y., Sugumaran, M., & Journaux, L. (2016). *A Comparative Analysis. Machine Learning Techniques for Intrusion Detection*. https://dl.acm.org/doi/10.1145/2980258.2980378

Han, J. (2012). *Concepts and Techniques*. Morgan Kaufmann.

Han, R., Ghanem, M. M., Guo, L., Guo, Y., & Osmond, M. (2014). Enabling cost-aware and adaptive elasticity of multi-tier cloud applications. *Future Generation Computer Systems*, *32*, 82–98. doi:10.1016/j.future.2012.05.018

Haskell, H. (2020). Cumberlege review exposes stubborn and dangerous flaws in healthcare. *BMJ (Clinical Research Ed.)*, *370*, m3099. Advance online publication. doi:10.1136/bmj.m3099 PMID:32763955

Hossain, S. (2013). 5G wireless communication systems. *American Journal of Engineering Research*, *2*(10), 344–353.

Höyhtyä, M., Corici, M., Covaci, S., & Guta, M. (2021). 5G and beyond for new space: vision and research challenges. *Advances in Communications Satellite Systems: Proceedings of the 37th International Communications Satellite Systems Conference (ICSSC-2019)*, 387–402. 10.1049/PBTE095E_ch30

Huang, H., Du, J., Wang, H., & Wang, R. (2016). A Multi-keyword Multi-user Searchable Encryption Scheme Based on Cloud Storage. *Trustcom/BigDataSE/I SPA, 2016 IEEE*, 1937–1943.

Huang, X., Xiang, Y., Bertino, E., Zhou, J., & Xu, L. (2014). Robust Multi-factor authentication for fragile communications. *IEEE Transactions on Dependable and Secure Computing*, *11*(6), 568–581. doi:10.1109/TDSC.2013.2297110

Hwang, M.-S., Hsu, S.-T., & Lee, C.-C. (2014). A new public key encryption with conjunctive field keyword search scheme. *Information Technology and Control*, *43*(3), 277–288. doi:10.5755/j01.itc.43.3.6429

Hwang, Y. H., & Lee, P. J. (2007). Public key encryption with conjunctive keyword search and its extension to a multi-user system. In *Pairing-Based Cryptography—Pairing 2007* (pp. 2–22). Springer. doi:10.1007/978-3-540-73489-5_2

Irving, K. E. (2006). The impact of technology on the 21st century. *Teaching Science in the 21st Century*, (March), 3–19. Retrieved from http://books.google.com/books?id=g5NflcuxkJcC&pgis=1

Jagadeeswari, V., Subramaniyaswamy, V., Logesh, R., & Vijayakumar, V. (2018). A study on medical Internet of Things and Big Data in personalized healthcare system. *Health Information Science and Systems*, *6*(1), 14. Advance online publication. doi:10.100713755-018-0049-x PMID:30279984

Jagtap, S. (2019). *Utilising the internet of things concepts to improve the resource efficiency of food manufacturing* (Doctoral dissertation). Loughborough University.

Jagtap, S., & Rahimifard, S. (2018). Real-time data collection to improve energy efficiency in food manufacturing. In *International Congress on Organizational Management, Energy Efficiency and Occupational Health and Safety in Agrifood Industry (+ AGRO 2018), Castelo Branco, Portugal* (pp. 3-4). Academic Press.

Jagtap, S., Bader, F., Garcia-Garcia, G., Trollman, H., Fadiji, T., & Salonitis, K. (2021a). Food logistics 4.0: Opportunities and challenges. *Logistics*, *5*(1), 2. doi:10.3390/logistics5010002

Jagtap, S., Garcia-Garcia, G., & Rahimifard, S. (2021b). Optimisation of the resource efficiency of food manufacturing via the Internet of Things. *Computers in Industry*, *127*, 103397. doi:10.1016/j.compind.2021.103397

Jain, R. (2016). Introduction to 5G. *Washington University in St. Louis*. Retrieved from https://www.cse.wustl.edu/~jain/cse574-16/

Jha, J., & Ragha, L. (2013). *Intrusion Detection System using Support Vector Machine.* https://research.ijais.org/icwac/number3/icwac1342.pdf

Jia, M., Gu, X., Guo, Q., Xiang, W., & Zhang, N. (2016). Broadband hybrid satellite-terrestrial communication systems based on cognitive radio toward 5G. *IEEE Wireless Communications*, *23*(6), 96–106. doi:10.1109/MWC.2016.1500108WC

Jiang, J., Das, R., Ananthanarayanan, G., Chou, P. A., Padmanabhan, V. N., Sekar, V., . . . Zhang, H. (2016). VIA: Improving internet telephony call quality using predictive relay selection. *SIGCOMM 2016 - Proceedings of the 2016 ACM Conference on Special Interest Group on Data Communication*, 286–299. 10.1145/2934872.2934907

Jiang, Q., Ma, J., Lu, X., & Tian, Y. (2015). An efficient two-factor user authentication scheme with unlinkability for wireless sensor networks. *Peer-to-Peer Networking and Applications*, *8*(6), 1070–1081. doi:10.100712083-014-0285-z

Jingzhang, S., Chunjie, C., & Hui, L. (2018). Searchable Encryption Scheme Based on CPABE with Attribute Update in a Cloud Medical Environment. *International Conference on Cloud Computing and Security*, 265–276. 10.1007/978-3-030-00012-7_25

Johnsson, L. (n.d.). *Overview of data centers energy efficiency evolution*. Available: https://pdfs.semanticscholar.org/559f/5b4bb297999ed00d4a787cf9317ec515afa1.pdf

Karim & Chakrabarti. (2010). An Improved Low-Power High-Throughput Log-MAP Turbo Decoder. *IEEE Transactions on Consumer Electronics, 56*(2).

Karim & Chakrabarti. (2011). Design of Efficient High Throughput Pipelined Parallel Turbo Decoder Using QPP Interleaver. *International Conference on Multimeadia, Signal Processing and Communication Technologies, IEEE 2011*, 248-51. 10.1109/MSPCT.2011.6150486

Kar, U. N., & Sanyal, D. K. (2018). An overview of device-to-device communication in cellular networks. *ICT Express*, *4*(4), 203–208. doi:10.1016/j.icte.2017.08.002

Khader, R., & Eleyan, D. (2021). Survey of DoS/DDoS attacks in IoT. *Sustainable Engineering and Innovation*, *3*(1), 23–28. doi:10.37868ei.v3i1.124

Khan, M. K., & Alghathbar, K. (2010). Cryptanalysis and security improvements of two-factor user authentication in wireless sensor networks. *Sensors (Basel)*, *10*(3), 2450–2459. doi:10.3390100302450 PMID:22294935

Khan, R., Kumar, P., Jayakody, D. N. K., & Liyanage, M. (2020). A Survey on Security and Privacy of 5G Technologies: Potential Solutions, Recent Advancements, and Future Directions. *IEEE Communications Surveys and Tutorials*, *22*(1), 196–248. doi:10.1109/COMST.2019.2933899

Kiayias, A., Oksuz, O., Russell, A., Tang, Q., & Wang, B. (2016). Efficient encrypted keyword search for multi-user data sharing. *European Symposium on Research in Computer Security*, 173–195. 10.1007/978-3-319-45744-4_9

Kibria, M., Nguyen, K., Villardi, G., Zhao, O., Ishizu, K., & Kojima, F. (2018). *Big Data Analytics, Machine Learning, and Artificial Intelligence in Next-Generation Wireless Networks*. https://ieeexplore.ieee.org/document/8360430

Kim & Kim. (2013). Design of Early Stopping Unit in Parallel Turbo Decoder based on Galois Field Operation. *IEEE Proceedings ISOCC*.

Kim, H., Lee, Y., & Kim, J.-H. (2015, October 8). Low-complexity CRC-aided early stopping unit for parallel turbo decoder. *Electronics Letters*, *51*(21), 1660–1662. doi:10.1049/el.2015.2262

Kitanov, S., Monteiro, E., & Janevski, T. (2016). 5G and the Fog – Survey of Related Technologies and Research Directions. *Proceedings of the 18th Mediterranean IEEE Electrotechnical Conference MELECON*, 1-6.

Korhonen, J. (2003). Introduction to 3G Mobile Communications. Artech House.

Krishna, P., Yenduri, S., & Ariwa, E. (2020). *Data analytics in wireless systems and IoT issues and challenges*. https://onlinelibrary.wiley.com/doi/full/10.1002/dac.4522

Kumar, S., Viinikainen, A., & Hamalainen, T. (2016). *Machine learning classification model for Network based Intrusion Detection System*. https://ieeexplore.ieee.org/document/7856705

Kumar, A., Jakhar, S., & Makkar, S. (2012, July). Comparative analysis between des and RSA algorithms. *International Journal of Advanced Research in Computing and Software Engineer in G*, *2*(7), 386–391.

Kumari, S., & Om, H. (2016). Authentication protocol for wireless sensor networks applications like safety monitoring in coal mines. *Computer Networks*, *104*, 137–154. doi:10.1016/j.comnet.2016.05.007

Lahre, M. K., Diwan, M. T., Kashyap, S., & Agrawal, P. (2013). *Analyze Different approaches for IDS using KDD 99 Data Set*. https://www.academia.edu/4823609/Analyze_Different_approaches_for_IDS_using_KDD_99_Data_Set

Lai, J., Zhou, X., Deng, R. H., Li, Y., & Chen, K. (2013). Expressive search on encrypted data. *Proceedings of the 8th ACM SIGSAC Symposium on Information, Computer and Communications Security*, 243–252.

Lama, P., & Zhou, X. (2013). Autonomic provisioning with self-adaptive neural fuzzy control for percentile-based delay guarantee. *ACM Transaction. Autonomous. Adapt. Syst. (TAAS)*, *8*(2). doi:10.1109/ICNISC.2015.91

Latif, S., Qadir, J., Farooq, S., & Imran, M. A. (2017). How 5G wireless (and Concomitant Technologies) will revolutionize healthcare? *Future Internet*, *9*(4), 93. Advance online publication. doi:10.3390/fi9040093

Lee, C.-C., Hsu, S.-T., Hwang, M.-S., & ... (2013). A Study of Conjunctive Keyword Searchable Schemes. *International Journal of Network Security*, *15*(5), 321–330.

Lee, I., & Lee, K. (2015). The Internet of Things (IoT): Applications, investments, and challenges for enterprises. *Business Horizons*, *58*(4), 431–440. doi:10.1016/j.bushor.2015.03.008

Lee, J., Tejedor, E., Ranta-Aho, K., Wang, H., Lee, K. T., Semaan, E., Mohyeldin, E., Song, J., Bergljung, C., & Jung, S. (2018). Spectrum for 5G: Global Status, Challenges, and Enabling Technologies. *IEEE Communications Magazine*, *56*(3), 12–18. doi:10.1109/MCOM.2018.1700818

Lei, L., Zhong, Z., Lin, C., & Shen, X. (2012). Operator controlled device-to-device communications in LTE-advanced networks. *IEEE Wireless Communications*, *19*(3), 96–104. doi:10.1109/MWC.2012.6231164

Li, Maunder, Al-Hashimi, & Hanzo. (2013). A Low-Complexity Turbo Decoder Architecture for Energy-Efficient Wireless Sensor Networks. *IEEE Transactions on Very Large Scale Integration (VLSI) Systems, 21*(1).

Light, Selvi, Li, & Malali. (2013). Fall Pattern Classification from Brain Signals using Machine Learning Models. *Journal of Selected Areas in Health Informatics (JSHI), in the Cyber Journals: Multidisciplinary Journals in Science and Technology, 3*(12).

Li, H., Yang, Y., Dai, Y., Bai, J., Yu, S., & Xiang, Y. (2017). *Achieving secure and efficient dynamic searchable symmetric encryption over medical cloud data. IEEE Transactions on Cloud Computing.*

Li, J., & Chen, X. (2013). Efficient multi-user keyword search over encrypted data in cloud computing. *Computer Information, 32*(4), 723–738.

Lin & Der-shieh. (2015). *Efficient highly parallel turbo decoder for 3GPP LTE-advanced.* IEEE.

Lin. (2015). An Area Efficient Radix-4 Reciprocal Dual Trellis Architecture for a High-Code-Rate Turbo Decoder. *IEEE Transactions on Circuits and Systems—II: Express Briefs, 62*(1). . doi:10.1109/TCSII.2014.2362733

Lincoln, P., Blate, A., Singh, M., Whitted, T., State, A., Lastra, A., & Fuchs, H. (2016). From Motion to Photons in 80 Microseconds: Towards Minimal Latency for Virtual and Augmented Reality. *IEEE Transactions on Visualization and Computer Graphics, 22*(4), 1367–1376. doi:10.1109/TVCG.2016.2518038 PMID:26780797

Li, S., Da Xu, L., & Zhao, S. (2018). 5G Internet of Things: A survey. *Journal of Industrial Information Integration, 10*, 1–9. doi:10.1016/j.jii.2018.01.005

Liu, T., Li, J., Kim, B., Lin, C.-W., Shiraishi, S., Xie, J., & Han, Z. (2018). Distributed file allocation using matching game in mobile fog-caching service network. *IEEE INFOCOM 2018-IEEE Conference on Computer Communications Workshops (INFOCOM WKSHPS)*, 499–504. 10.1109/INFCOMW.2018.8406854

Liu, Y., Yu, F. R., Li, X., Ji, H., & Leung, V. C. (2018). Hybrid computation offloading in fog and cloud networks with non-orthogonal multiple access. *IEEE INFOCOM 2018-IEEE Conference on Computer Communications Workshops (INFOCOM WKSHPS)*, 154–159. 10.1109/INFCOMW.2018.8406940

Liu, Yang, Yang, Wang, & Mao. (2018). DATS: Dispersive stable task scheduling in heterogeneous fog networks. *IEEE Internet of Things Journal.*

Liu, E., Effiok, E., & Hitchcock, J. (2020). Survey on health care applications in 5G networks. *IET Communications, 14*(7), 1073–1080. doi:10.1049/iet-com.2019.0813

Liu, J., Huang, X., & Liu, J. K. (2015). Secure sharing of personal health records in cloud computing: Ciphertext-policy attribute-based signcryption. *Future Generation Computer Systems, 52*, 67–76. doi:10.1016/j.future.2014.10.014

Liu, J., Zhang, L.-M., & Zhong, Z.-G. (2014). Research on Low Latency Decoding Scheme of Turbo Codes. *2014 International Conference on Wireless Communication and Sensor Network.* 10.1109/WCSN.2014.10

Li, X., Ma, J., Wang, W., Xiong, Y., & Zhang, J. (2013). A novel smart card and dynamic ID based remote user authentication scheme for multi-server environments. *Mathematical and Computer Modelling, 58*(1), 85–95. doi:10.1016/j.mcm.2012.06.033

Löhr, H., Sadeghi, A.-R., & Winandy, M. (2010). Securing the e-health cloud. *Proceedings of the 1st Acm International Health Informatics Symposium*, 220–229. 10.1145/1882992.1883024

Lounis, K., & Zulkernine, M. (2020). Attacks and defenses in short-range wireless technologies for IoT. *IEEE Access: Practical Innovations, Open Solutions, 8*, 88892–88932. doi:10.1109/ACCESS.2020.2993553

Loven, Peltonen, Leppänen, & Partala. (2019). Edge AI: A vision for distributed, edge-native artificial intelligence in future 6G networks. *Proc. 6G Wireless Summit*, 1-2.

Luckmann, J., Grethe, H., McDonald, S., Orlov, A., & Siddig, K. (2014). An integrated economic model of multiple types and uses of water. *Water Resources Research*, *50*(5), 3875–3892. doi:10.1002/2013WR014750

Luo, H., & Zhang, Y. (2017). Low Latency Parallel Turbo Decoding Implementation for Future Terrestrial Broadcasting Systems. *IEEE Transactions on Broadcasting*, (Jan), 1–8.

Lv, Z., Zhang, M., & Feng, D. (2014). Multi-user searchable encryption with efficient access control for cloud storage. *2014 IEEE 6th International Conference on Cloud Computing Technology and Science*, 366–373.

M, F., G, G., K, B., N.d, P., & L, H. (2013). P4 medicine: How systems medicine will transform the healthcare sector and society. *Personalized Medicine*, 565–576.

Macdonald, A. J. R. (1986). An introduction to medical manipulation. *Pain*, *24*(1), 124. doi:10.1016/0304-3959(86)90035-7

Magsi, H., Sodhro, A. H., Chachar, F. A., Abro, S. A. K., Sodhro, G. H., & Pirbhulal, S. (2018). Evolution of 5G in Internet of medical things. *2018 International Conference on Computing, Mathematics and Engineering Technologies: Invent, Innovate and Integrate for Socioeconomic Development, ICoMET 2018 - Proceedings*, 1–7. 10.1109/ICOMET.2018.8346428

Mahasty, Thompson, & Soleimani. (2018). A Concise Temporal Data Representation Model for Prediction in Biomedical Wearable Devices. *IEEE Internet of Things Journal, 6*(2), 1438 - 1445.

Manashty, A., & Light, J. (2019, March). Life Model: A novel representation of life-long temporal sequences in health predictive analytics. *Future Generation Computer Systems*, *92*, 141–156. doi:10.1016/j.future.2018.09.033

Mandal, A. K., Parakash, C., & Tiwari, A. (2012). Performance evaluation of cryptographic algorithms: DES and AES. In *IEEE Students' Conference on Electrical, Electronics and computing (SCEECS), 2012* (pp. 1–5). IEEE.

Manjunatha, K. N., & Kiran, B. (2012). Design and ASIC Implementation of a 3GPP LTE Advance Turbo Encoder and Turbo Decoder. *International Journal of Engineering Research and Applications, 2*(4), 6-10.

Manjunatha, K. N., & Lohith Kumar, H. G. (2011). Design and Performance analysis of a 3GPP LTE/LTE-Advance turbo decoder using software reference models. *International Journal of Scientific & Engineering Research, 2*(7).

Manjunatha, K. N., & Vaibhav, A. (2017). Design and FPGA Implementation of Power Efficient Turbo Decoder for 4G LTE Standards. *International Journal of Applied Engineering Research, 12*(21), 10921-10925.

Marchant, G. E., & Lindor, R. A. (2012). The Coming Collision Between Autonomous Vehicles and the Liability System. *Santa Clara Law Review*, *52*(4), 1321.

Marković, G. Z. (2017). Routing and spectrum allocation in elastic optical networks using bee colony optimization. *Photonic Network Communications*, *34*(3), 356–374. doi:10.100711107-017-0706-z

Martin, T. (n.d.). *How blockchain will disrupt your industry*. https://www.slalom.com/in-sights/how-blockchain-will-disrupt-your-industry

Mathew, A., & Terence, J. S. (2017, April). A survey on various detection techniques of sinkhole attacks in WSN. In *2017 International Conference on Communication and Signal Processing (ICCSP)* (pp. 1115-1119). IEEE. 10.1109/ICCSP.2017.8286550

Memon, I., Fazal, H., Ahmed Shaikh, R., Muhammad, G., Arain, Q. A., & Khatri, T. K. (2019). *Big Data, Cloud, 5G Networks Create Smart and Intelligent World: A Survey*. Academic Press.

Mercola, J. (2020). *EMFD 5G, Wi-Fi Cell Phones Hidden Harms and How to Protect Yourself.* Hay House.

Mi, H., Wang, H., Yin, G., Zhou, Y., Shi, D., & Yuan, L. (2010). Online self reconfiguration with performance guarantee for energy-efficient large-scale cloud computing data centers. In Services *Computing (SCC), 2010 IEEE International Conference on 2010*, (pp. 514–521). IEEE. 10.1109/SCC.2010.69

Minoli, D., Sohraby, K., & Occhiogrosso, B. (2017). IoT Considerations, Requirements, and Architectures for Smart Buildings-Energy Optimization and Next-Generation Building Management Systems. *IEEE Internet of Things Journal*, *4*(1), 269–283. doi:10.1109/JIOT.2017.2647881

Miraz, M. H., Ali, M., Excell, P. S., & Picking, R. (2018). *Internet of nano-things, things and everything: Future growth trends*. ArXiv.

Muhammad, Algehyne, Usman, Ahmad, Chakraborty, & Mohammed. (2020). *Supervised Machine Learning Models for Prediction of COVID-19 Infection using Epidemiology Dataset.* Springer Nature Singapore Pte Ltd.

Mullen, K. (2012). *Information on Earth's water.* Retrieved September 11, 2017, from https://www.ngwa.org/Fundamentals/teachers/Pages/information-on-earth-water.aspx

Nadeem, A., & Javed, M. Y. (2005). A performance comparison of knowledge encryption algorithms. In *Information and communication technologies. ICICT. (2005). First international conference on, 2005* (pp. 84–89). Academic Press.

Nakamura, M., Nakamura, J., Lopez, G., Shuzo, M., & Yamada, I. (2011). Collaborative processing of wearable and ambient sensor system for blood pressure monitoring. *Sensors (Basel)*, *11*(7), 6760–6770. doi:10.3390110706760 PMID:22163984

Narayanan, A., Ramadan, E., Carpenter, J., Liu, Q., Liu, Y., Qian, F., & Zhang, Z. L. (2020). A First Look at Commercial 5G Performance on Smartphones. *The Web Conference 2020 - Proceedings of the World Wide Web Conference, WWW 2020*, 894–905. 10.1145/3366423.3380169

Naveen, S., & Kounte, M. R. (2020). In Search of the Future Technologies: Fusion of Machine Learning, Fog and Edge Computing in the Internet of Things. Lecture Notes on Data Engineering and Communications Technologies, 31, 278–285. doi:10.1007/978-3-030-24643-3_33

Nayak, S., & Patgiri, R. (2020). A Vision on Intelligent Medical Service for Emergency on 5G and 6G Communication Era. *EAI Endorsed Transactions on Internet of Things*, *6*(22), 166293. doi:10.4108/eai.17-8-2020.166293

Ng, B., Peng, X., Faegh, E., & Mustain, W. E. (2020). Using nanoconfinement to inhibit the degradation pathways of conversion-metal oxide anodes for highly stable fast-charging Li-ion batteries. *Journal of Materials Chemistry. A, Materials for Energy and Sustainability*, *8*(5), 2712–2727. doi:10.1039/C9TA11708C

Nguyen, D. C., Pathirana, P. N., Ding, M., & Seneviratne, A. (2019). *Blockchain for 5G and beyond networks: A state of the art survey*. https://arxiv.org/abs/1912.05062

Nguyen, Pathirana, & Seneviratne. (2020). Blockchain for 5G and Beyond Networks: A State of the Art Survey. *Computer Science, Engineering, Mathematics, J. Netw. Comput. Appl.*

Nightingale, J., Salva-Garcia, P., Calero, J. M. A., & Wang, Q. (2018). 5G-QoE: QoE modelling for ultra-HD video streaming in 5G networks. *IEEE Transactions on Broadcasting*, *64*(2), 621–634. doi:10.1109/TBC.2018.2816786

Numan, M., Subhan, F., Khan, W. Z., Hakak, S., Haider, S., Reddy, G. T., Jolfaei, A., & Alazab, M. (2020). A Systematic Review on Clone Node Detection in Static Wireless Sensor Networks. *IEEE Access: Practical Innovations, Open Solutions*, *8*, 65450–65461. doi:10.1109/ACCESS.2020.2983091

Oleshchuk, V., & Fensli, R. (2011). Remote patient monitoring within a future 5G infrastructure. *Wireless Personal Communications*, *57*(3), 431–439. doi:10.100711277-010-0078-5

Orlosky, J., Kiyokawa, K., & Takemura, H. (2017). Virtual and augmented reality on the 5G highway. *Journal of Information Processing*, *25*(0), 133–141. doi:10.2197/ipsjjip.25.133

Othman, S. M., Ba-Alwi, F. M., Alsohybe, N. T., & Al-Hashida, A. Y. (2018). *Intrusion detection model using machine learning algorithm on Big Data environment.* doi:10.1186/s40537-018-0145-4

Oua & Phan. (2008). Traceable privacy of recent provably-secure rfid protocols. In *International conference on applied cryptography and network security*. Springer.

Oughton, E. J., Lehr, W., Katsaros, K., Selinis, I., Bubley, D., & Kusuma, J. (2021). Revisiting Wireless Internet Connectivity: 5G vs Wi-Fi 6. *Telecommunications Policy*, *45*(5), 102127. Advance online publication. doi:10.1016/j.telpol.2021.102127

Padiya, S., & Gulhane, V., (2020). Analysis of Data Aggregation Methods to avoid Data Redundancy in Wireless Sensor Network. *12th IEEE CICN 2020,* 25-26.

Palattella, M. R., Dohler, M., Grieco, A., Rizzo, G., Torsner, J., Engel, T., & Ladid, L. (2016). Internet of Things in the 5G Era: Enablers, Architecture, and Business Models. *IEEE Journal on Selected Areas in Communications*, *34*(3), 510–527. doi:10.1109/JSAC.2016.2525418

Paramita, S., Das Bebartta, H. N., & Pattanayak, P. (2021). IoT Based Healthcare Monitoring System Using 5G Communication and Machine Learning Models. *Studies in Computational Intelligence*, *932*, 159–182. doi:10.1007/978-981-15-9735-0_9

Peersman, G., Cvetkovic, S., Griffiths, P., & Spear, H. (2000). Global system for mobile communications short message service. *IEEE Personal Communications*, *7*(3), 15–23. doi:10.1109/98.847919

Peng, M., Li, Y., Jiang, J., Li, J., & Wang, C. (2014). Heterogeneous cloud radio access networks: A new perspective for enhancing spectral and energy efficiencies. *IEEE Wireless Communications*, *21*(6), 126–135.

Pereira, V., & Sousa, T. (2004). *Evolution of Mobile Communications: from 1G to 4G*. Academic Press.

Pham, Q. V., Mirjalili, S., Kumar, N., Alazab, M., & Hwang, W. J. (2020). Whale Optimization Algorithm with Applications to Resource Allocation in Wireless Networks. *IEEE Transactions on Vehicular Technology*, *69*(4), 4285–4297. doi:10.1109/TVT.2020.2973294

Phan, R. C.-W. (2009). Cryptanalysis of a new ultra lightweight rfid authentication protocols as IEEE Transactions on Dependable and secure. *Computing*, *6*(4), 316–320.

Pohlig, S., & Hellman, M. (1978). An improved algorithm for computing logarithms over GF (p) and its cryptographic significance (Corresp.). *IEEE Transactions on Information Theory*, *24*(1), 106–110. doi:10.1109/TIT.1978.1055817

Polese, M., Giordani, M., Zugno, T., Roy, A., Goyal, S., Castor, D., & Zorzi, M. (2020). Integrated Access and Backhaul in 5G mmWave Networks: Potential and Challenges. *IEEE Communications Magazine*, *58*(3), 62–68. doi:10.1109/MCOM.001.1900346

Poretti, M. (1990). Quality control of water as a raw material in the food industry. *Food Control*, *1*(2), 79–83. doi:10.1016/0956-7135(90)90089-U

Prabadevi, Deepa, Pham, Nguyen, Reddy, Reddy, Pathirana, & Dobre. (2021). Towards Blockchain for Edge-of-Things: A New Paradigm, Opportunities, and Future Directions. *IEEE Internet of Things Magazine*.

Qadri, Y. A., Nauman, A., Zikria, Y., Vasilakos, A. V., & Kim, S. W. (2020). The Future of Healthcare Internet of Things: A Survey of Emerging Technologies. *IEEE Communications Surveys and Tutorials, 22*(2), 1121–1167. doi:10.1109/COMST.2020.2973314

Qiao, J., Shen, X., Mark, J., Shen, Q., He, Y., & Lei, L. (2015). Enabling device-to-device communications in millimeter-wave 5G cellular networks. *IEEE Communications Magazine, 53*(1), 209–215. doi:10.1109/MCOM.2015.7010536

Qi, Y., Hunukumbure, M., Nekovee, M., Lorca, J., & Sgardoni, V. (2016). Quantifying data rate and bandwidth requirements for immersive 5G experience. *2016 IEEE International Conference on Communications Workshops, ICC 2016*, 455–461. 10.1109/ICCW.2016.7503829

Raddo, T. R., Rommel, S., Cimoli, B., Vagionas, C., Perez-Galacho, D., Pikasis, E., … Tafur Monroy, I. (2021). Transition technologies towards 6G networks. *Eurasip Journal on Wireless Communications and Networking, 2021*(1). doi:10.1186/s13638-021-01973-9

Rai, M., & Mandoria, H. (2019). *Network Intrusion Detection: A comparative study using state-of-the-art machine learning methods*. https://ieeexplore.ieee.org/document/8977679

Rajput, D. S., Basha, S. M., Xin, Q., Gadekallu, T. R., Kaluri, R., Lakshmanna, K., & Maddikunta, P. K. R. (2021). Providing diagnosis on diabetes using cloud computing environment to the people living in rural areas of India. *Journal of Ambient Intelligence and Humanized Computing*. Advance online publication. doi:10.100712652-021-03154-4

Ramaraju, A. (2020). Unlocking the Potential of 5G for Content Production. *Psychology and Education Journal, 57*(9), 5912-5917. Retrieved from https://www.bbc.co.uk/rd/blog/2019-03-5g-production-media-broadcasting

Rappaport, T. S., Xing, Y., Kanhere, O., Ju, S., Madanayake, A., Mandal, S., Alkhateeb, A., & Trichopoulos, G. C. (2019). Wireless communications and applications above 100 GHz: Opportunities and challenges for 6G and beyond. *IEEE Access: Practical Innovations, Open Solutions, 7*, 78729–78757. doi:10.1109/ACCESS.2019.2921522

Rau, H.-H., Hsu, C.-Y., Lee, Y.-L., Chen, W., & Jian, W.-S. (2010). Developing electronic health records in Taiwan. *IT Professional, 12*(2), 17–25. doi:10.1109/MITP.2010.53

Reddy, Clermidy, Al Khayat, Bhagdadi, & Jezquel. (2010). *Power Consumption Analysis and Energy Efficient Optimization for Turbo Decoder Implementation.* IEEE.

Ren, H., Shen, J., Tang, X., & Feng, T. (2020). 5G Healthcare Applications in COVID-19 Prevention and Control. *2020 ITU Kaleidoscope. Industry-Driven Digital Transformation, ITU K, 2020*, 1–4. Advance online publication. doi:10.23919/ITUK50268.2020.9303191

Research, A. B. I. (2016). *Bluetooth Smart Evolution Helps the Technology Break into Key IoT Market Verticals. PRNewswire.*

Riaz, M. N., Buriro, A., & Mahboob, A. (2018). Classification of attacks on wireless sensor networks: A survey. *International Journal of Wireless and Microwave Technologies, 8*(6), 15–39. doi:10.5815/ijwmt.2018.06.02

Romana, R., Lopeza, J., & Mambob, M. (2016). Mobile Edge Computing, Fog et al: A Survey and Analysis of Security threats and challenges. In Future Generation Computer Systems. Elsevier.

Rommer, S., Hedman, P., Olsson, M., Frid, L., Sultana, S., & Mulligan, C. (2019). *5G core networks: Powering digitalization. In 5G Core Networks*. Powering Digitalization. doi:10.1016/C2018-0-01335-3

Roth, C., Benkesery, C., & Huang, Q. (2014). Power-Efficient Turbo-Decoder Design based on Algorithm-Specific Power Domain Partitioning. *24th International Workshop on Power and Timing Modeling, Optimization and Simulation (PATMOS)*. 10.1109/PATMOS.2014.6951907

Rubertis, A. D., Mainetti, L., Mighali, V., Patrono, L., Sergi, I., Stefanizzi, M. L., & Pascali, S. (2013). Performance evaluation of end-to-end security protocols in an internet of things. *International Conference Software, Telecommunications and Computer Networks*, 1–6.

Sachs, J., Andersson, L. A. A., Araujo, J., Curescu, C., Lundsjo, J., Rune, G., Steinbach, E., & Wikstrom, G. (2019). Adaptive 5G low-latency communication for tactile internet services. *Proceedings of the IEEE*, *107*(2), 325–349. doi:10.1109/JPROC.2018.2864587

Sagner, M., McNeil, A., Puska, P., Auffray, C., Price, N. D., Hood, L., Lavie, C. J., Han, Z.-G., Chen, Z., Brahmachari, S. K., McEwen, B. S., Soares, M. B., Balling, R., Epel, E., & Arena, R. (2017). The P4 health spectrum–a predictive, preventive, personalized and participatory continuum for promoting healthspan. *Progress in Cardiovascular Diseases*, *59*(5), 506–521. doi:10.1016/j.pcad.2016.08.002 PMID:27546358

Saito. (2016). 8K terrestrial transmission field tests using dual polarized MIMO and higher-order modulation OFDM. *IEEE Trans. Broadcast.*, *62*(1), 306–315. . doi:10.1109/TBC.2015.2494853

Sam Lucero IHS Technology. (2016). *IoT Platforms: Enabling the Internet of Things*. Whitepaper.

Sarumi, O., Adetunmbi, A., & Adetoye, F. (2020). *Discovering computer networks intrusion using data analytics and machine intelligence*. https://www.sciencedirect.com/science/article/pii/S2468227620302386

Saxena, N., Roy, A., Sahu, B. J. R., & Kim, H. (2017). Efficient IoT Gateway over 5G Wireless: A New Design with Prototype and Implementation Results. *IEEE Communications Magazine*, *55*(2), 97–105. doi:10.1109/MCOM.2017.1600437CM

Schabetsberger, T., Ammenwerth, E., Andreatta, S., Gratl, G., Haux, R., Lechleitner, G., Schindelwig, K., Stark, C., Vogl, R., Wilhelmy, I., & Wozak, F. (2006). From a paper-based transmission of discharge summaries to electronic communication in health care regions. *International Journal of Medical Informatics*, *75*(3-4), 209–215. doi:10.1016/j.ijmedinf.2005.07.018 PMID:16112892

Schulz, P., Matthe, M., Klessig, H., Simsek, M., Fettweis, G., Ansari, J., Ashraf, S. A., Almeroth, B., Voigt, J., Riedel, I., Puschmann, A., Mitschele-Thiel, A., Muller, M., Elste, T., & Windisch, M. (2017). Latency Critical IoT Applications in 5G: Perspective on the Design of Radio Interface and Network Architecture. *IEEE Communications Magazine*, *55*(2), 70–78. doi:10.1109/MCOM.2017.1600435CM

Senders, J., & Moray, N. (1991). *Human error: Cause, prediction and reduction*. Academic Press.

Sengupta, J., Ruj, S., & Bit, S. D. (2020). A comprehensive survey on attacks, security issues and blockchain solutions for IoT and IIoT. *Journal of Network and Computer Applications*, *149*, 102481. doi:10.1016/j.jnca.2019.102481

Seth, S. M., & Mishra, R. (2011). *Comparative Analysis of Encryption Algorithms for digital communication*. Academic Press.

Seth, S. M., & Ishra, R. (2011, June). Comparative Analysis of encryption algorithms for data communication. *International Journal of Computers and Technology*, *2*(2), 292–294.

Shafi, M., Fellow, L., Molisch, A. F., Smith, P. J., Haustein, T., Zhu, P., ... Member, S. (2017). 5G: A Tutorial Overview of Standards, Trials, Challenges, Deployment, and Practice. *IEEE Journal on Selected Areas in Communications*, *35*(6), 1201–1221. doi:10.1109/JSAC.2017.2692307

Shafiq, M., Ashraf, H., Ullah, A., & Tahira, S. (2020). Systematic Literature Review on Energy Efficient Routing Schemes in WSN–A Survey. *Mobile Networks and Applications*, *25*(3), 1–14. doi:10.100711036-020-01523-5

Sharma, M., Tandon, A., Narayan, S., & Bhushan, B. (2017, September). Classification and analysis of security attacks in WSNs and IEEE 802.15. 4 standards: A survey. In *2017 3rd International Conference on Advances in Computing, Communication & Automation (ICACCA)* (pp. 1-5). IEEE.

Sharma, D., & Jinwala, D. C. (2017). Multi-User Searchable Encryption with Token Freshness Verification (MUSE-TFV). *Security and Communication Networks*, *2017*, 16. doi:10.1155/2017/6435138

Sharma, D., & Jinwala, D. C. (n.d.). Multi-Writer Multi-Reader Conjunctive Keyword Searchable Encryption. *International Journal of Information and Computer Security.*

Sharma, S. K., Bogale, T. E., Le, L. B., Chatzinotas, S., Wang, X., & Ottersten, B. (2018). Dynamic Spectrum Sharing in 5G Wireless Networks with Full-Duplex Technology: Recent Advances and Research Challenges. *IEEE Communications Surveys and Tutorials*, *20*(1), 674–707. doi:10.1109/COMST.2017.2773628

Sharmila, B., & Nagapadma, R. (2019). *Intrusion Detection System using Naive Bayes algorithm.* https://ieeexplore.ieee.org/document/9019921

She, J., Soonsawad, P., & Ng, P. (2018). BLE Beacons for the Internet of Things Applications: Survey, Challenges, and Opportunities. *IEEE IoT Journal.*

Shikhare, G., & Shaikh, A. (2014). 4G LTE Technology. *International Journal of Networking and Parallel Computing*, *2*(03), 110–117.

Shrestha & Paily. (2014). High-Throughput Turbo Decoder with Parallel Architecture for LTE Wireless Communication Standards. *IEEE Transactions on Circuits and Systems—I: Regular Papers, 61*(9). . doi:10.1109/TCSI.2014.2332266

Sigwele, T., Hu, Y. F., Ali, M., Hou, J., Susanto, M., & Fitriawan, H. (2018). Intelligent and Energy Efficient Mobile Smartphone Gateway for Healthcare Smart Devices Based on 5G. *2018 IEEE Global Communications Conference, GLOBECOM 2018 - Proceedings*. 10.1109/GLOCOM.2018.8648031

Sinha, P., Jha, V. K., Rai, A. K., & Bhushan, B. (2017, July). Security vulnerabilities, attacks and countermeasures in wireless sensor networks at various layers of OSI reference model: A survey. In *2017 International Conference on Signal Processing and Communication (ICSPC)* (pp. 288-293). IEEE. 10.1109/CSPC.2017.8305855

SJ. C., J., K., & RD, G. (2015). Teledermatology: From historical perspective to emerging techniques of the modern era: Part I: History, rationale, and current practice. *Journal of the American Academy of Dermatology, 72*(4), 563–574. Retrieved from http://www.embase.com/search/results?subaction=viewrecord&from=export&id=L604736181%255Cnhttp://dx.doi.org/10.1016/j.jaad.2014.07.061%255Cnhttp://elvis.ubvu.vu.nl:9003/vulink?sid=EMBASE&issn=10976787&id=doi:10.1016%252Fj.jaad.2014.07.061&atitle=Teledermatology%25

Sklar, B. (2001). *Digital Communications: Fundamentals and Applications with Fundamentals of Turbo Codes* (2nd ed.). Prentice-Hall.

Skouteris, G., Webb, D. P., Shin, K. L. F., & Rahimifard, S. (2018). Assessment of the capability of an optical sensor for in-line real-time wastewater quality analysis in food manufacturing. *Water Resources and Industry*, *20*, 75–81. doi:10.1016/j.wri.2018.10.002

Slinger, C., Cameron, C., & Stanley, M. (2005). Computer-generated holography as a generic display technology. *Computer*, *38*(8), 46–53. doi:10.1109/MC.2005.260

Soldani, D., Fadini, F., Rasanen, H., Duran, J., Niemela, T., Chandramouli, D., ... Nanavaty, N. (2017). 5G Mobile Systems for Healthcare. *IEEE Vehicular Technology Conference*. 10.1109/VTCSpring.2017.8108602

Spasov, Gushev, & Ristov. (2015). *Max-Log-MAP Decoding with Reduced Memory Complexity.* IEEE.

Srivastava, K., Awasthi, A. K., Kaul, S. D., & Mittal, R. (2015). A hash based mutual rfid tag authentication protocol in telecare medicine information system. *Journal of Medical Systems*, *39*(1), 153. doi:10.100710916-014-0153-7 PMID:25491577

Statista. (2018). *Production volume of aerated and soft drinks across India from FY 2015 to FY 2018 (in million liters)*. Retrieved from Statista: https://www.statista.com/statistics/762413/india-aerated-and-soft-drinks-production-volume/

Steinmetz, R. (2012). *Multimedia: Computing communications & applications*. Academic Press.

Stoica, R. A., & Abreu, G. T. F. (2019). *6G: The wireless communications network for collaborative and AI applications*. Available: arXiv:1904.03413.

Strinati, E. C., Barbarossa, S., Gonzalez-Jimenez, J. L., Ktenas, D., Cassiau, N., Maret, L., & Dehos, C. (2019). sixth generation (6G) has already started taking place, and it is expected to be implemented between 2027 and 2030. *IEEE Vehicular Technology Magazine*, *14*(3), 42–50.

Studer, Benskeser, Belfanti, & Huang. (2011). Design and Implementation of a Parallel Turbo-Decoder ASIC for 3GPP-LTE. *IEEE Journal of Solid-State Circuits, 46*(1). . doi:10.1109/JSSC.2010.2075390

Sunyaev, A., Kaletsch, A., Mauro, C., & Krcmar, H. (2009). Security Analysis of the German Electronic Health Card's Peripheral Parts. *ICEIS*, (3), 19–26. doi:10.5220/0001854000190026

Tahir, O., Habebi, M. H., Dabbagh, M., Mugheesi, A., Ahad, A., & Ahmed, K. I. (2020, July). A Review on Application of Blockchain in 5G and Beyond Networks: Taxonomy, Field-Trials, Challenges and Opportunities. *IEEE Access: Practical Innovations, Open Solutions*, *8*, 115876–115904. doi:10.1109/ACCESS.2020.3003020

Talebpour, A., & Mahmassani, H. S. (2016). Influence of connected and autonomous vehicles on traffic flow stability and throughput. *Transportation Research Part C, Emerging Technologies*, *71*, 143–163. doi:10.1016/j.trc.2016.07.007

Taleb, T., Samdanis, K., Mada, B., Flinck, H., Dutta, S., & Sabella, D. (2017). On multi-access edge computing: A survey of the emerging 5G network edge cloud architecture and orchestration. *IEEE Communications Surveys and Tutorials*, *19*(3), 1657–1681. doi:10.1109/COMST.2017.2705720

Tamimi, A. (2008). *Performance analysis of knowledge encryption algorithms*. Academic Press.

Tavallaee, M., Bagheri, E., Lu, W., & Ghorbani, A. (2009). *A detailed analysis of the KDD CUP 99 data set*. https://ieeexplore.ieee.org/document/5356528

Tikhvinskiy, V., & Bochechka, G. (2015). Prospects and QoS Requirements in 5G Networks. *Journal of Telecommunications and Information Technologies*, *1*(1), 23–26.

ToolP. (n.d.). https://prosecco.gforge.inria.fr/personal/bblanche/proverif/

Trends, G. (2021). Retrieved May 17, 2021, from https://trends.google.com/trends/explore?date=now7-d&q=5G in healthcare

Trends, G. (2021a). *5G in Internet of Things*. Retrieved May 15, 2021, from 2021 website: https://trends.google.com/trends/explore?date=today5-y&q=5GinInternetofThings

Trends, G. (2021b). *High Speed Connectivity*. Retrieved May 15, 2021, from 2021 website: https://trends.google.com/trends/explore?date=today5-y&q=HighSpeedConnectivity

Turkanovic, M., Brumen, B., & Holbl, M. (2014). A novel user authentication and key agreement scheme for heterogeneous ad hoc wireless sensor networks, based on the internet of things notion. *Ad Hoc Networks*, *20*, 96–112. doi:10.1016/j.adhoc.2014.03.009

Ullah, H., Gopalakrishnan Nair, N., Moore, A., Nugent, C., Muschamp, P., & Cuevas, M. (2019). 5G communication: An overview of vehicle-to-everything, drones, and healthcare use-cases. *IEEE Access: Practical Innovations, Open Solutions*, *7*, 37251–37268. doi:10.1109/ACCESS.2019.2905347

United Nations. (2017). *World Population Prospects 2017*. Retrieved June 29, 2017, from https://esa.un.org/unpd/wpp/DataQuery/

Vargas, D., Kim, Y. J. D., Bajcsy, J., Gomez-Barquero, D., & Cardona, N. (2015, September). A MIMO-channel-precoding scheme for next generation terrestrial broadcast TV systems. *IEEE Transactions on Broadcasting*, *61*(3), 445–456. doi:10.1109/TBC.2015.2450431

Vatandsoost, M., & Litkouhi, S. (2019). The Future of Healthcare Facilities: How Technology and Medical Advances May Shape Hospitals of the Future. *Hospital Practices and Research*, *4*(1), 1–11. doi:10.15171/hpr.2019.01

Verma, O., Agarwal, R., Dafouti, D., & Tyagi, S. (2011). Performance analysis of knowledge encryption algorithms. In *Electronics Technology (ICECT) 3rd International Conference on, 2011* (pp. 399–403). Academic Press.

Vucinic, M., Tourancheau, B., Watteyne, T., Rousseau, F., Duda, A., Guizzetti, R., & Damon, L. (2015). DTLS performance in duty-cycled networks. *International Symposium on Personal, Indoor, and Mobile Radio Communications,* 1333–1338.

Wang, D., & Wang, P. (2014). On the anonymity of two-factor authentication schemes for wireless sensor networks: Attacks, principle and solutions. *Computer Networks*, *73*, 41–57. doi:10.1016/j.comnet.2014.07.010

Wang, F., Xu, J., Wang, X., & Cui, S. (2018). Joint offloading and computing optimization in wireless powered mobile-edge computing systems. *IEEE Transactions on Wireless Communications*, *17*(3), 1784–1797.

Wang, G. (2014, May). Parallel Interleaver Design for a High Throughput HSPA/LTE Multi-Standard Turbo Decoder. *IEEE Transactions on Circuits and Systems. I, Regular Papers*, *61*(5).

Wang, S., Zhang, X., & Zhang, Y. (2016). Efficiently Multi-User Searchable Encryption Scheme with Attribute Revocation and Grant for Cloud Storage. *PLoS One*, *11*(11), e0167157. doi:10.1371/journal.pone.0167157 PMID:27898703

Wang, Y., Li, J., Huang, L., Jing, Y., Georgakopoulos, A., & Demestichas, P. (2014). 5G mobile: Spectrum broadening to higher-frequency bands to support high data rates. *IEEE Vehicular Technology Magazine*, *9*(3), 39–46. doi:10.1109/MVT.2014.2333694

Watro, R., & Kong, D. (n.d.). Securing sensor networks with public key technology. In *Proceedings of the 2nd ACM Workshop On Security of Ad hoc and Sensor Networks*. ACM.

Webb, D. P., Skouteris, G., & Rahimifard, S. (2018). In-plant real-time manufacturing water content characterisation. *Water Resources and Industry*, *20*, 37–45. doi:10.1016/j.wri.2018.08.003

Wey, J. S., & Zhang, J. (2019). Passive Optical Networks for 5G Transport: Technology and Standards. *Journal of Lightwave Technology*, *37*(12), 2830–2837. doi:10.1109/JLT.2018.2856828

Wigren, T., Colombi, D., Thors, B., & Berg, J. E. (2016). Implication of RF-EMF exposure limitations on 5g data rates above 6 GHz. *2015 IEEE 82nd Vehicular Technology Conference, VTC Fall 2015 - Proceedings*. 10.1109/VTC-Fall.2015.7390974

William, S., & Stallings, W. (2006). *Cryptography and network security*. Pearson Education.

Wong, K. H. M., Zheng, Y., Cao, J., & Wang, S. (2006). A dynamic user authentication scheme for wireless sensor networks. *IEEE International Conference on Sensor-Networks, Ubiquitous & Trust-worthy Computing*.

Woolley. (2019). *Bluetooth SIG Bluetooth Core Specification v5.1*. Bluetooth SIG.

Woolley. (2020). *Bluetooth SIG Bluetooth Core Specification v5.2.* Bluetooth SIG.

Woolley, M. (2016). *Bluetooth 5 /Go Faster. Go Further*. Bluetooth SIG.

Wu, F. (2017). An Efficient Authentication & Key agreement protocol for Multi-Gateway wireless sensor Networks. *Journal of Network and Computer Applications*, *89*, 72–85. doi:10.1016/j.jnca.2016.12.008

Wu, F., Xue, L., Kumari, S., & Li, X. (2015). A novel and provably secure biometrics-based three-factor remote authentication scheme for mobile client-server networks. *Computers & Electrical Engineering*, *45*, 274–285. doi:10.1016/j.compeleceng.2015.02.015

Wu, Y., Lu, X., Su, J., & Chen, P. (2016). An efficient searchable encryption against keyword guessing attacks for sharable electronic medical records in cloud-based system. *Journal of Medical Systems*, *40*(12), 258. doi:10.100710916-016-0609-z PMID:27722976

Xiong, K., Chen, C., Qu, G., Fan, P., & Letaief, K. B. (2017). Group cooperation with optimal resource allocation in wireless powered communication networks. *IEEE Transactions on Wireless Communications*, *16*(6), 3840–3853.

Xiong, Z., Feng, S., Wang, W., Niyato, D., Wang, P., & Han, Z. (2019, June). Cloud/fog computing resource management and pricing for blockchain networks. *IEEE Internet of Things Journal*, *6*(3), 4585–4600. doi:10.1109/JIOT.2018.2871706

Xue, K., Ma, C., Hong, P., & Ding, R. (2013). A temporal-credential-based mutual authentication and key agreement scheme for wireless sensor networks. *Journal of Network and Computer Applications*, *36*(1), 316–323. doi:10.1016/j.jnca.2012.05.010

Xu, L., Xu, C., Liu, J. K., Zuo, C., & Zhang, P. (2019). Building a dynamic searchable encrypted medical database for multi-client. *Information Sciences*.

Yang, H., Zheng, H., Zhang, J., Wu, Y., Lee, Y., & Ji, Y. (2017). Blockchainbased trusted authentication in cloud radio over fiber network for 5G. *2017 16th International Conference on Optical Communications and Networks (ICOCN)*, 1–3.

Yang, H., Wu, Y., Zhang, J., Zheng, H., Ji, Y., & Lee, Y. (2018). Blockonet: blockchain-based trusted cloud radio over optical fiber network for 5G fronthaul. In *Optical Fiber Communication Conference*. Optical Society of America. 10.1364/OFC.2018.W2A.25

Yang, J. J., Li, J., Mulder, J., Wang, Y., Chen, S., Wu, H., Wang, Q., & Pan, H. (2015). Emerging information technologies for enhanced healthcare. *Computers in Industry*, *69*, 3–11. doi:10.1016/j.compind.2015.01.012

Yang, M., Margheri, A., Hu, R., & Sassone, V. (2018). Differentially private data sharing in a cloud federation with blockchain. *IEEE Cloud Computing*, *5*(6), 69–79. doi:10.1109/MCC.2018.064181122

Yang, X., Liu, Z., & Yang, Y. (2018). Minimization of weighted bandwidth and computation resources of fog servers under per-task delay constraint. *2018 IEEE International Conference on Communications (ICC)*, 1–6. 10.1109/ICC.2018.8422318

Yaqoob, I., Khan, L. U., Kazmi, S. M. A., Imran, M., Guizani, N., & Hong, C. S. (2020). Autonomous Driving Cars in Smart Cities: Recent Advances, Requirements, and Challenges. *IEEE Network*, *34*(1), 174–181. doi:10.1109/MNET.2019.1900120

Ye, X. (2015). A Survey on Scheduling Workflows in Cloud Environment. In *Network and Information Systems for Computers (ICNISC), 2015 International Conference on*. IEEE.

Ye, J., Wang, J., Zhao, J., Shen, J., & Li, K.-C. (2016). Fine-grained searchable encryption in multi-user setting. *Soft Computing*, 1–12.

Yellowlees, P., Nakagawa, K., Pakyurek, M., Hanson, A., Elder, J., & Kales, H. C. (2020). Rapid conversion of an outpatient psychiatric clinic to a 100% virtual telepsychiatry clinic in response to COVID-19. *Psychiatric Services (Washington, D.C.)*, *71*(7), 749–752. doi:10.1176/appi.ps.202000230 PMID:32460683

Yin, C., Zhu, Y., Fei, J., & He, X. (2017). *A Deep Learning Approach for Intrusion Detection Using Recurrent Neural Networks*. https://ieeexplore.ieee.org/document/8066291

Yin, Z., & Gunnarsson, G. (2017). Received-Signal-Strength threshold optimization using Gaussian processes. *IEEE Transactions on Signal Processing*, *65*(8), 2164–2177. doi:10.1109/TSP.2017.2655480

Yoo. (2015, December). Reverse Rate Matching for Low-Power LTE-Advanced Turbo Decoders. *IEEE Transactions on Circuits and Systems. I, Regular Papers*, *62*(12).

Younis, O., & Fahmy, S. (2004). HEED: A Hybrid, Energy-Efficient, Distributed Clustering Approach for Ad Hoc Sensor Networks. *IEEE Transactions on Mobile Computing*, *3*(4), 366–379. doi:10.1109/TMC.2004.41

Yu, J., Wang, G., Mu, Y., & Gao, W. (2014). An efficient generic framework for 3-factor authentication with provably secure instantiation. *IEEE Transactions on Information Forensics and Security*, *9*(12), 2302–2313. doi:10.1109/TIFS.2014.2362979

Yu, K., Lin, L., Alazab, M., Tan, L., & Gu, B. (2020). Deep Learning-Based Traffic Safety Solution for a Mixture of Autonomous and Manual Vehicles in a 5G-Enabled Intelligent Transportation System. *IEEE Transactions on Intelligent Transportation Systems*. Advance online publication. doi:10.1109/TITS.2020.3042504

Yu, Y.-C., Huang, T.-Y., & Hou, T.-W. (2012). Forward secure digital signature for electronic medical records. *Journal of Medical Systems*, *36*(2), 399–406. doi:10.100710916-010-9484-1 PMID:20703711

Zahoor, M. I., Dou, Z., Shah, S. B. H., Khan, I. U., Ayub, S., & Gadekallu, T. R. (2020). Pilot decontamination using asynchronous fractional pilot scheduling in massive MIMO systems. *Sensors (Switzerland)*, *20*(21), 1–21. doi:10.339020216213 PMID:33143363

Zaidi, A. A., Baldemair, R., Tullberg, H., Bjorkegren, H., Sundstrom, L., Medbo, J., Kilinc, C., & Da Silva, I. (2016). Waveform and Numerology to Support 5G Services and Requirements. *IEEE Communications Magazine*, *54*(11), 90–98. doi:10.1109/MCOM.2016.1600336CM

Zhang, L., & Li, Y. (2011). Implementing and Optimizing a Turbo Decoder on a TI TMS320C64x Device. *ICCP2011 IEEE Proceedings*, 401-04. 10.1109/ICCPS.2011.6092297

Zhang, Y., Liu, L., & Wang, S. (2016). Multi-User and Keyword-Based Searchable Encryption Scheme. *Computational Intelligence and Security (CIS), 2016 12th International Conference on*, 223–227.

Zhang, B., & Zhang, F. (2011). An efficient public key encryption with conjunctive-subset keywords search. *Journal of Network and Computer Applications*, *34*(1), 262–267. doi:10.1016/j.jnca.2010.07.007

Zhang, H., Dong, Y., Cheng, J., Hossain, M. J., & Leung, V. C. M. (2016). Fronthauling for 5G LTE-U ultra-dense cloud small cell networks. *IEEE Wireless Communications*, *23*(6), 48–53. doi:10.1109/MWC.2016.1600066WC

Zhang, R., Xue, R., & Liu, L. (2017). Searchable encryption for healthcare clouds: A survey. *IEEE Transactions on Services Computing*, *11*(6), 978–996. doi:10.1109/TSC.2017.2762296

Zhang, Y., Xu, C., Lin, X., & Shen, X. S. (2019). *Blockchain-based public integrity verification for cloud storage against procrastinating auditors*. IEEE Transactions on Cloud Computing. doi:10.1109/TCC.2019.2908400

Zheng, Xiong, Fan, Zhong, & Letaief. (2019). Fog-assisted multi-user SWIPT networks: local computing or offloading. *IEEE Internet of Things Journal*.

Zhou, J., Cao, Z., Dong, X., & Vasilakos, A. V. (2017). Security and Privacy for Cloud-Based IoT: Challenges. *IEEE Communications Magazine*, *55*(1), 26–33. doi:10.1109/MCOM.2017.1600363CM

About the Contributors

Chintan Bhatt (Ph.D.) is currently working as an Assistant Professor in Computer Engineering department, Chandubhai S. Patel Institute of Technology, CHARUSAT. He is a member of IEEE, EAI, ACM, CSI, AIRCC and IAENG (International Association of Engineers). His areas of interest include Internet of Things, Data Mining, Web Mining, Networking, Security Mobile Computing, Big Data and Software Engineering. He has more than 5 years of teaching experience and research experience, having good teaching and research interests. He has chaired a track in CSNT 2015 and ICTCS 2014. He has been working as Reviewer in Wireless Communications, IEEE (Impact Factor-6.524) and Internet of Things Journal, IEEE, Knowledge-Based Systems, Elsevier (Impact Factor-2.9) Applied Computing and Informatics, Elsevier and Mobile Networks and Applications, Springer. He has delivered an expert talk on Internet of Things at Broadcast Engineering Society Doordarshan, Ahmedabad on 30/09/2015. He has been awarded Faculty with Maximum Publication in CSIC Award and Paper Presenter Award at International Conference in CSI-2015, held at New Delhi.

Neeraj Kumar received his Ph.D. in CSE from SMVD University, Katra (J & K), India, and was a postdoctoral research fellow in Coventry University, Coventry, UK. He is working as an Associate Professor in the Department of Computer Science and Engineering, Thapar Institute of Engineering & Technology, Patiala (Pb.), India since 2014. Dr. Neeraj is an internationally renowned researcher in the areas of VANET & CPS Smart Grid & IoT Mobile Cloud computing & Big Data and Cryptography. He has published more than 150 technical research papers in leading journals and conferences from IEEE, Elsevier, Springer, John Wiley, and Taylor and Francis. His paper has been published in some of the high impact factors journals such as-IEEE Transactions on Industrial Informatics, IEEE Transactions on Industrial Electronics, IEEE Transactions on Information Forensics and Security, IEEE Transactions on Dependable and Secure Computing, IEEE Transactions on Power Systems, IEEE Transactions on Vehicular Technology, IEEE Transactions on Smart Grid, IEEE Journal of Biomedical and Health Informatics, IEEE Access, IEEE Transactions on Consumer Electronics, IEEE Systems Journal, IEEE IoT Journal, IEEE Wireless Communication Magazine, IEEE Vehicular Technology Magazine, IEEE Communication Magazine, IEEE Networks Magazine etc. Apart from the journals conferences, he has also published papers in some of the core conferences of his area of specialization such as-IEEE Globecom, IEEE ICC, IEEE Greencom, IEEE CSCWD.

Ali Kashif Bashir is a Senior Lecturer/Associate Professor at the Department of Computing and Mathematics, Manchester Metropolitan University, UK. He is a senior member of IEEE, invited member of IEEE Industrial Electronic Society, member of ACM, and Distinguished Speaker of ACM. His past assignments include Associate Professor of ICT, University of the Faroe Islands, Denmark; Osaka University, Japan; Nara National College of Technology, Japan; the National Fusion Research Institute, South Korea; Southern Power Company Ltd., South Korea, and the Seoul Metropolitan Government, South Korea. He received his Ph.D. in computer science and engineering from Korea University South Korea. He has authored over 140 research articles. His research interests include internet of things, wireless networks, distributed systems, network/cyber security, cloud/network function virtualization, machine learning, etc. He is serving as the Editor-in-chief of the IEEE Future Directions Newsletter. He is leading many conferences as a chair (program, publicity, and track) and had organized workshops in flagship conferences.

Mamoun Alazab is an Associate Professor at the College of Engineering, IT and Environment at Charles Darwin University, Australia. He received his PhD degree in Computer Science from the Federation University of Australia. He is a cyber security researcher and practitioner with industry and academic experience. Alazab's research is multidisciplinary that focuses on cyber security with a focus on cybercrime detection and prevention. His research interest includes the intersection use of Artificial Intelligence (AI) and Machine Learning as essential tools for cybersecurity. He has more than 300 research papers (>90% in Q1, and more than 100 in IEEE/ACM Transactions) and 9 authored/edited books. He is a Senior Member of the IEEE, and the founding chair of the IEEE Northern Territory (NT) Subsection. Much of his work has been done through establishing partnership linkages with various Australian and overseas educational institutions and industry partners such as NTG, ACSC, Australian Crime Commission, AFP, Westpac Bank, IBM, Microsoft Digital Crimes Unit, and TrendMicro. He was a consultant for the United Nations Office of Crime and Drugs (UNODC) on cybercrime, and more recently on Cyber Terrorism. Many of his publications are referenced and cited in highly regarded outlets such as: UNODC, ITU, and the World Bank. Nineteen articles were listed in the top ten downloads and acknowledged as featured articles. He received several awards including IEEE Outstanding Leadership Award (2020), the CDU College of Engineering, IT and Environment Exceptional Researcher Award (2020) (2021), and 4 Best Research Paper Awards. He is ranked in top 2% of world's scientists in the subfield discipline of Artificial Intelligence (AI) and Networking & Telecommunications, in the study published in PLOS Biology. He was ranked in the top 10% of 30k cyber security authors. He delivered more than 120 keynote speeches, chaired 56 national events and more than 90 international events; on program committees for 200 conferences. He is an Editor of 10 academic journals [many of which are Q1 & Q2].

* * *

Jack Azzopardi is a final year student at the University of Malta currently reading a degree in BSc Business and IT. Through this, he discovered a profound interest in emerging technologies and ERP systems in which he has recently started a career in.

Jaishankar B. completed his B.E. (ECE) from University of Madras. He obtained M.E and Ph.D from Anna University, Chennai. He is having 2 years of industrial experience and 11 years of teaching as well as research experience. His areas of interest include signal processing, Image processing and Networks. He has published 10 Papers in International Journals and 4 papers in National and International Conferences. He is a Reviewer for International Journal of Advances in Engineering and Technology and an Editorial Board Member for 3 International Journals. He is a member of IEEE, ISTE, and BES.

Kiran B. received B.E. in Instrumentation Technology and M.Tech in Digital Electronics and Communication from Visvesvaraya Technological University, Belgaum, Karnataka. Currently Pursuing PhD in Electronics Engineering JAIN (Deemed-to-be University). Bengaluru. He has 9 years of teaching experience in UG and PG and Industry experience of 2 years. His current research interests include design of CMOS Data Converters and Application Specific ICs. He has published his research findings in various Scopus indexed Journals, National and International conferences. He is a member of various professional bodies.

Darren Camilleri is a Final year B.Sc. student at Computer Information System, Faculty of Information and Communication Technology at University of Malta, Msida, Malta.

Luke Camilleri is a Final year B.Sc. student at Computer Information System, Faculty of Information and Communication Technology at University of Malta, Msida, Malta.

Thomas Camilleri is a Final year B.Sc. student at Computer Information System, Faculty of Information and Communication Technology at University of Malta, Msida, Malta.

Vilendra Choudhari is currently working as a General manager for Jubiliant FoodWorks Limited, India. He has over 15 years of experience in the Food industry and in the past he worked for PepsiCo, Godfrey Philips, GSK and Nestle.

Joseph Curmi is an undergraduate student currently in the final year of my degree course, BSc in Business and IT at the University of Malta. Machine Learning and its applications being areas of interest together with other emerging technologies.

Lalit Garg is a Senior Lecturer in Computer Information Systems at the University of Malta, Malta, and an honorary lecturer at the University of Liverpool, UK. He has been a researcher at the Nanyang Technological University, Singapore, and Ulster University, UK. He has supervised 200+ Masters' dissertations, 2 DBA and 2 PhD thesis and authored 120+ high impact publications in refereed journals/conferences/books, five edited books and 20 patents. He has delivered several keynotes and organized/chaired international conferences, and consulted numerous public and private organizations for their IS implementation and management. His research interests are business intelligence, machine learning, data science, deep learning, cloud computing, mobile computing, Internet of Things (IoT), information systems, management science and their applications mainly in healthcare and medical domains. He participates in many EU, and local funded projects, including a one million euros Erasmus+ Capacity-Building project in Higher Education (CBHE) titled Training for Medical education via innovative eTechnology (MediTec). The University of Malta has awarded him the 2021-22 Research Excellence Fund.

V. S. Gulhane received the B.E. (CSE) and M.E. (CSE) from the SGBAU, Amravati, India, in 1997 and 2006 respectively. He also received a PhD (CSE) from the SGBAU, Amravati, India in 2015. He is working as a Professor at the Department of Information Technology, Sipna's College of Engineering & Technology, Amravati, India, since 2001. He has academic experience of 24 years. He is also working as the Head of the Department for the last 10 years. His current research interest includes database management systems, cloud computing, machine learning, blockchain technology, the Internet of Things, etc. He is a lifetime member of the IETE, ISTE, IEEE, ACM, IE and CSI. He has published 78 research papers in various international journals and presented 24 research proposals at international conferences. He had organized 08 workshops/ STTP/ CEP/ training programs on different research areas. He works as a member at the Board of Studies (BOS), Information Technology, SGBAU, Amravati and Chairman of CSI Amravati Chapter.

Sandeep Jagtap is a Lecturer in Smart and Green manufacturing at Sustainable Manufacturing Systems Centre, School of Aerospace, Transport and Manufacturing. He has over 15 years of combined experience within academics and industry. Dr Jagtap serves on the Editorial Advisory Board for the British Food Journal. He is a Fellow of the Institute of Food Science and Technology (FIFST) and Higher Education Academy (FHEA).

Devesh Jinwala is a professor at Sardar Vallabhbhai National Institute of Technology (SVNIT), Surat, Gujarat, India and an Adjunct Professor at IIT Jammu. He received his Ph D degree from Computer Engineering Department, SVNIT, Surat working in close association with Space Application Centre, Indian Space Research Organization, Ahmadabad. He is serving at SVNIT since 1991. He also served as a Visiting PRofessor at University of Denver (2016-17) and as a Professor-on-deputation (2018-2019) at IIT Jammu. His research interests are in Information & Communications Security, Privacy Issues, Security & Privacy Issues in Resource constrained Environments, Machine Learning for Security and Software Requirements Specifications.

Janet Light received her Bachelor degree in Electronics and Communications Engineering in 1983, Masters in Electrical and Electronics Engineering in 1990 and PhD in Computer Science in 2002. She joined UNB Saint John in 2002 in the department of Computer Science. Her research interests are in wireless & mobile computing, green computing, sensor networks, and network security. For almost a decade, her research focus has been on applied health and emergency services. Dr. Light has served as chair/vice chair locally in IEEE-NB Section (2012 to 2017) and the Chair of the IEEE Region-7 Women in Engineering Canada in 2015.

Kuldip Mori is a student of BE - Bachelor of Computer Engineering from Chandubhai S. Patel Institute of Technology join in 2018. His research areas are Cloud computing, Networking and DevOps.

Manjunatha N. has nine years of experience in Academic and Research. His research interests were in VLSI and Image Processing. He has published his research findings in various Scopus indexed Journals and International conferences. He is a member of various professional bodies.

Raghu N. received B.E in Telecommunication and M.Tech in Digital Electronics and Communication from Visvesvaraya Technological University, Belgaum, Karnataka. He Secured PhD in Electronics Engineering, JAIN (Deemed-to-be University). Bengaluru. He has 9 years of teaching experience in UG and Industry experience of 1 years. His current research interests include design of RF Communication and Image Processing. He has published his research findings in various Scopus indexed Journals, National and International conferences. He is a member of various professional bodies.

Sneha Padhiar is an Assistant professor in U & P U. Patel Department of Computer Engineering, CHARUSAT University, Gujarat, India. She received her B.E.C.E. From Gujarat Technological University in 2014 and M.E.C.E. From Gujarat Technological University in 2016. Currently, she is pursuing doctoral course in Computer Engineering at CHARUSAT. Her major area of research includes Information Security and IOT.

S. D. Padiya received the B.E. (IT) and M.Tech (IT) degrees from the SGBAU, Amravati, India, in 2009, and from the RGPV, Bhopal, India, in 2013 respectively. He is admitted for PhD in Computer Science and Engineering at SGBAU, Amravati, India in 2019. He is working as an Assistant Professor at the Department of Information Technology, Shri Sant Gajanan Maharaj College of Engineering, Shegaon, India, since 2014. His current research interest includes Discrete Structure, Wireless Sensor Network, Internet of Things, Bluetooth Low Energy, Lightweight Protocols, etc. He is a lifetime member of the IETE and ISTE. He has published 07 research papers in international journals and presented 02 research proposals at international conferences. He worked as a reviewer at The Symposium on Emerging Topics in Computing and Communications (SETCAC'20), CoCoNet'20. He also worked as Technical Program Committee (TPC) member at 2020 IEEE 12th International Conference on Computational Intelligence and Communication Networks (CICN) India, 2021 IEEE 10th International Conference on Communication Systems and Network Technologies (CSNT), India and 2021 IEEE 13th International Conference on Computational Intelligence and Communication Networks (CICN), Peru.

Shilpi Parikh is a masters candidate pursuing Masters of Science in Computer Software Engineering from Arizona State University, Tempe. She pursued my bachelor's degree in Computer Engineering major from Charotar University of Science and Technology and worked as a Technology Intern at Thomson Reuters Corporation under Tax and Accounting section. She is a strong computer science professional with an inclination towards Data Science, Data Analytics and Machine Learning Technologies. As an energetic, enthusiastic and conscientious self-starter, She would like to apply my theoretical knowledge in a firm with a professional work-driven environment where She can develop and utilize my practical, technical and interpersonal skills. She is a Motivated, Team-Oriented person, possessing Meticulous Vision with a strategic mindset.

Vijay Prakash is currently working as a Lecturer at the Department of Computer Science and Engineering at Thapar Institute of Engineering and Technology, Patiala. He has done his masters in Software engineering from Thapar Institute of Engineering and Technology, Patiala. His research area includes Cloud, Edge and Fog Computing, Internet of Things and Wireless networking. He has 6 years of teaching experience and he has published more than 10 research article in various SCI/ SCIE journals and International conferences of repute.

Shahin Rahimifard is an Industrial Engineer with over twenty years of experience in R & D projects. He has been involved in a large number of industrial projects, mainly concerned with the design and implementation of supply chain management and production planning, and environmental management systems for a variety of manufacturing companies, ranging from metalworking SMEs to the food and drink industry. His current research work is focused on sustainability issues within 'Life Cycle Engineering', including projects on sustainable product design, low carbon manufacturing, sustainable business and consumption models, product service systems, and product end-of-life management, recovery, reuse and recycling technologies.

Anushka Sandesara is a Computer Engineer and an aspiring Data Scientist who is curious to explore and work at the intersection of innovation, cutting-edge technology and creativity to make human lives better. She is inclined towards Neural Networks, Machine Learning, Deep Learning, Reinforcement Learning, Natural Language Processing and Artificial Intelligence. She is an active contributor to academic research publications in Data Science and has always been enthralled by underlying computer mechanisms and ability to make things work efficiently.

Dhruti Sharma is an assistant professor at Sarvajanik College of Engineering and Technology (SCET), Surat, Gujarat, India. She received her Ph Degree from Computer Engineering department, Sardar Vallabhbhai National Institute of Technology (SVNIT), Surat in 2019. Her current research interest includes numerous cryptographic mechanisms viz. Identity based Encryption, Attribute Based Encryption, Functional Encryption, Information Security -Privacy issues, Block-chain based protocols.

George Skouteris currently works as a Scientist at Helmholtz-Zentrum Dresden-Rossendorf in Germany. He participated in projects that mainly focused on the implementation of continuous water quality monitoring techniques in the food manufacturing industry to achieve more efficient water usage.

Jagadesh T. is research scholar at Sathyabama Institute of Science and Technology. He has done M.E VLSI Design in Kongu Engineering College and B.E ECE in Jeppiaar Engineering College. He is currently pursuing his research in Radar Signal Processing. His areas of interest are VLSI Signal Processing, ASIC Design and CAD for VLSI Circuits. He has published 16 papers in various Scopus indexed journal, 7 papers in IEEE International Conference. He has conducted 2 workshops using MATLAB. He worked as Assistant Professor in Jeppiaar Engineering College from 2014 to 2018 and joined in KPR Institute of Engineering and Technology in June 2018.

Index

www.ingramcontent.com/pod-product-compliance
Lightning Source LLC
LaVergne TN
LVHW080529300926
849162LV00007B/1571
* 9 7 8 1 7 9 9 8 6 9 8 8 7 *